T0254838

Lecture Notes in Computer Science   14493

Founding Editors

Gerhard Goos
Juris Hartmanis

The series Lecture Notes in Computer Science (LNCS), including its subseries Lecture Notes in Artificial Intelligence (LNAI) and Lecture Notes in Bioinformatics (LNBI), has established itself as a medium for the publication of new developments in computer science and information technology research, teaching, and education.

LNCS enjoys close cooperation with the computer science R & D community, the series counts many renowned academics among its volume editors and paper authors, and collaborates with prestigious societies. Its mission is to serve this international community by providing an invaluable service, mainly focused on the publication of conference and workshop proceedings and postproceedings. LNCS commenced publication in 1973.

Zahir Tari · Keqiu Li · Hongyi Wu
Editors

Algorithms and Architectures for Parallel Processing

23rd International Conference, ICA3PP 2023
Tianjin, China, October 20–22, 2023
Proceedings, Part VII

Springer

Editors
Zahir Tari
Royal Melbourne Institute of Technology
Melbourne, VIC, Australia

Keqiu Li
Tianjin University
Tianjin, China

Hongyi Wu
University of Arizona
Tucson, AZ, USA

ISSN 0302-9743   ISSN 1611-3349 (electronic)
Lecture Notes in Computer Science
ISBN 978-981-97-0861-1   ISBN 978-981-97-0862-8 (eBook)
https://doi.org/10.1007/978-981-97-0862-8

This Springer imprint is published by the registered company Springer Nature Singapore Pte Ltd.
The registered company address is: 152 Beach Road, #21-01/04 Gateway East, Singapore 189721, Singapore

Paper in this product is recyclable.

Preface

On behalf of the Conference Committee, we welcome you to the proceedings of the 2023 International Conference on Algorithms and Architectures for Parallel Processing (ICA3PP 2023), which was held in Tianjin, China from October 20–22, 2023. ICA3PP2023 was the 23rd in this series of conferences (started in 1995) that are devoted to algorithms and architectures for parallel processing. ICA3PP is now recognized as the main regular international event that covers the many dimensions of parallel algorithms and architectures, encompassing fundamental theoretical approaches, practical experimental projects, and commercial components and systems. This conference provides a forum for academics and practitioners from countries around the world to exchange ideas for improving the efficiency, performance, reliability, security, and interoperability of computing systems and applications.

A successful conference would not be possible without the high-quality contributions made by the authors. This year, ICA3PP received a total of 503 submissions from authors in 21 countries and regions. Based on rigorous peer reviews by the Program Committee members and reviewers, 193 high-quality papers were accepted to be included in the conference proceedings and submitted for EI indexing. In addition to the contributed papers, six distinguished scholars, Lixin Gao, Baochun Li, Laurence T. Yang, Kun Tan, Ahmed Louri, and Hai Jin, were invited to give keynote lectures, providing us with the recent developments in diversified areas in algorithms and architectures for parallel processing and applications.

We would like to take this opportunity to express our sincere gratitude to the Program Committee members and 165 reviewers for their dedicated and professional service. We highly appreciate the twelve track chairs, Dezun Dong, Patrick P. C. Lee, Meng Shen, Ruidong Li, Li Chen, Wei Bao, Jun Li, Hang Qiu, Ang Li, Wei Yang, Yu Yang, and Zhibin Yu, for their hard work in promoting this conference and organizing the reviews for the papers submitted to their tracks. We are so grateful to the publication chairs, Heng Qi, Yulei Wu, Deze Zeng, and the publication assistants for their tedious work in editing the conference proceedings. We must also say "thank you" to all the volunteers who helped us at various stages of this conference. Moreover, we were so honored to have many renowned scholars be part of this conference. Finally, we would like to thank all speakers, authors, and participants for their great contribution to and support for the success of ICA3PP 2023!

October 2023

Jean-Luc Gaudiot
Hong Shen
Gudula Rünger
Zahir Tari
Keqiu Li
Hongyi Wu
Tian Wang

Organization

General Chairs

Jean-Luc Gaudiot   University of California, Irvine, USA
Hong Shen   University of Adelaide, Australia
Gudula Rünger   Chemnitz University of Technology, Germany

Program Chairs

Zahir Tari   Royal Melbourne Institute of Technology,
                                     Australia
Keqiu Li   Tianjin University, China
Hongyi Wu   University of Arizona, USA

Program Vice-chair

Wenxin Li   Tianjin University, China

Publicity Chairs

Hai Wang   Northwest University, China
Milos Stojmenovic   Singidunum University, Serbia
Chaofeng Zhang   Advanced Institute of Industrial Technology,
                                     Japan
Hao Wang   Louisiana State University, USA

Publication Chairs

Heng Qi   Dalian University of Technology, China
Yulei Wu   University of Exeter, UK
Deze Zeng   China University of Geosciences (Wuhan), China

Workshop Chairs

Laiping Zhao   Tianjin University, China
Pengfei Wang   Dalian University of Technology, China

Local Organization Chairs

Xiulong Liu   Tianjin University, China
Yitao Hu   Tianjin University, China

Web Chair

Chen Chen   Shanghai Jiao Tong University, China

Registration Chairs

Xinyu Tong   Tianjin University, China
Chaokun Zhang   Tianjin University, China

Steering Committee Chairs

Yang Xiang (Chair)   Swinburne University of Technology, Australia
Weijia Jia   Beijing Normal University and UIC, China
Yi Pan   Georgia State University, USA
Laurence T. Yang   St. Francis Xavier University, Canada
Wanlei Zhou   City University of Macau, China

Program Committee

Track 1: Parallel and Distributed Architectures

Dezun Dong (Chair)   National University of Defense Technology,
                                    China
Chao Wang   University of Science and Technology of China,
                                    China
Chentao Wu   Shanghai Jiao Tong University, China

Chi Lin	Dalian University of Technology, China
Deze Zeng	China University of Geosciences, China
En Shao	Institute of Computing Technology, Chinese Academy of Sciences, China
Fei Lei	National University of Defense Technology, China
Haikun Liu	Huazhong University of Science and Technology, China
Hailong Yang	Beihang University, China
Junlong Zhou	Nanjing University of Science and Technology, China
Kejiang Ye	Shenzhen Institute of Advanced Technology, Chinese Academy of Sciences, China
Lei Wang	National University of Defense Technology, China
Massimo Cafaro	University of Salento, Italy
Massimo Torquati	University of Pisa, Italy
Mengying Zhao	Shandong University, China
Roman Wyrzykowski	Czestochowa University of Technology, Poland
Rui Wang	Beihang University, China
Sheng Ma	National University of Defense Technology, China
Songwen Pei	University of Shanghai for Science and Technology, China
Susumu Matsumae	Saga University, Japan
Weihua Zhang	Fudan University, China
Weixing Ji	Beijing Institute of Technology, China
Xiaoli Gong	Nankai University, China
Youyou Lu	Tsinghua University, China
Yu Zhang	Huazhong University of Science and Technology, China
Zichen Xu	Nanchang University, China

Track 2: Software Systems and Programming Models

Patrick P. C. Lee (Chair)	Chinese University of Hong Kong, China
Erci Xu	Ohio State University, USA
Xiaolu Li	Huazhong University of Science and Technology, China
Shujie Han	Peking University, China
Mi Zhang	Institute of Computing Technology, Chinese Academy of Sciences, China

Jing Gong	KTH Royal Institute of Technology, Sweden
Radu Prodan	University of Klagenfurt, Austria
Wei Wang	Beijing Jiaotong University, China
Himansu Das	KIIT Deemed to be University, India
Rong Gu	Nanjing University, China
Yongkun Li	University of Science and Technology of China, China
Ladjel Bellatreche	National Engineering School for Mechanics and Aerotechnics, France

Track 3: Distributed and Network-Based Computing

Meng Shen (Chair)	Beijing Institute of Technology, China
Ruidong Li (Chair)	Kanazawa University, Japan
Bin Wu	Institute of Information Engineering, China
Chao Li	Beijing Jiaotong University, China
Chaokun Zhang	Tianjin University, China
Chuan Zhang	Beijing Institute of Technology, China
Chunpeng Ge	National University of Defense Technology, China
Fuliang Li	Northeastern University, China
Fuyuan Song	Nanjing University of Information Science and Technology, China
Gaopeng Gou	Institute of Information Engineering, China
Guangwu Hu	Shenzhen Institute of Information Technology, China
Guo Chen	Hunan University, China
Guozhu Meng	Chinese Academy of Sciences, China
Han Zhao	Shanghai Jiao Tong University, China
Hai Xue	University of Shanghai for Science and Technology, China
Haiping Huang	Nanjing University of Posts and Telecommunications, China
Hongwei Zhang	Tianjin University of Technology, China
Ioanna Kantzavelou	University of West Attica, Greece
Jiawen Kang	Guangdong University of Technology, China
Jie Li	Northeastern University, China
Jingwei Li	University of Electronic Science and Technology of China, China
Jinwen Xi	Beijing Zhongguancun Laboratory, China
Jun Liu	Tsinghua University, China

Kaiping Xue   University of Science and Technology of China,
                             China
Laurent Lefevre   National Institute for Research in Digital Science
                             and Technology, France
Lanju Kong   Shandong University, China
Lei Zhang   Henan University, China
Li Duan   Beijing Jiaotong University, China
Lin He   Tsinghua University, China
Lingling Wang   Qingdao University of Science and Technology,
                             China
Lingjun Pu   Nankai University, China
Liu Yuling   Institute of Information Engineering, China
Meng Li   Hefei University of Technology, China
Minghui Xu   Shandong University, China
Minyu Feng   Southwest University, China
Ning Hu   Guangzhou University, China
Pengfei Liu   University of Electronic Science and Technology
                             of China, China
Qi Li   Beijing University of Posts and
                             Telecommunications, China
Qian Wang   Beijing University of Technology, China
Raymond Yep   University of Macau, China
Shaojing Fu   National University of Defense Technology,
                             China
Shenglin Zhang   Nankai University, China
Shu Yang   Shenzhen University, China
Shuai Gao   Beijing Jiaotong University, China
Su Yao   Tsinghua University, China
Tao Yin   Beijing Zhongguancun Laboratory, China
Tingwen Liu   Institute of Information Engineering, China
Tong Wu   Beijing Institute of Technology, China
Wei Quan   Beijing Jiaotong University, China
Weihao Cui   Shanghai Jiao Tong University, China
Xiang Zhang   Nanjing University of Information Science and
                             Technology, China
Xiangyu Kong   Dalian University of Technology, China
Xiangyun Tang   Minzu University of China, China
Xiaobo Ma   Xi'an Jiaotong University, China
Xiaofeng Hou   Shanghai Jiao Tong University, China
Xiaoyong Tang   Changsha University of Science and Technology,
                             China
Xuezhou Ye   Dalian University of Technology, China
Yaoling Ding   Beijing Institute of Technology, China

Yi Zhao	Tsinghua University, China
Yifei Zhu	Shanghai Jiao Tong University, China
Yilei Xiao	Dalian University of Technology, China
Yiran Zhang	Beijing University of Posts and Telecommunications, China
Yizhi Zhou	Dalian University of Technology, China
Yongqian Sun	Nankai University, China
Yuchao Zhang	Beijing University of Posts and Telecommunications, China
Zhaoteng Yan	Institute of Information Engineering, China
Zhaoyan Shen	Shandong University, China
Zhen Ling	Southeast University, China
Zhiquan Liu	Jinan University, China
Zijun Li	Shanghai Jiao Tong University, China

Track 4: Big Data and Its Applications

Li Chen (Chair)	University of Louisiana at Lafayette, USA
Alfredo Cuzzocrea	University of Calabria, Italy
Heng Qi	Dalian University of Technology, China
Marc Frincu	Nottingham Trent University, UK
Mingwu Zhang	Hubei University of Technology, China
Qianhong Wu	Beihang University, China
Qiong Huang	South China Agricultural University, China
Rongxing Lu	University of New Brunswick, Canada
Shuo Yu	Dalian University of Technology, China
Weizhi Meng	Technical University of Denmark, Denmark
Wenbin Pei	Dalian University of Technology, China
Xiaoyi Tao	Dalian Maritime University, China
Xin Xie	Tianjin University, China
Yong Yu	Shaanxi Normal University, China
Yuan Cao	Ocean University of China, China
Zhiyang Li	Dalian Maritime University, China

Track 5: Parallel and Distributed Algorithms

Wei Bao (Chair)	University of Sydney, Australia
Jun Li (Chair)	City University of New York, USA
Dong Yuan	University of Sydney, Australia
Francesco Palmieri	University of Salerno, Italy

George Bosilca	University of Tennessee, USA
Humayun Kabir	Microsoft, USA
Jaya Prakash Champati	IMDEA Networks Institute, Spain
Peter Kropf	University of Neuchâtel, Switzerland
Pedro Soto	CUNY Graduate Center, USA
Wenjuan Li	Hong Kong Polytechnic University, China
Xiaojie Zhang	Hunan University of Technology and Business, China
Chuang Hu	Wuhan University, China

Track 6: Applications of Parallel and Distributed Computing

Hang Qiu (Chair)	Waymo, USA
Ang Li (Chair)	Qualcomm, USA
Daniel Andresen	Kansas State University, USA
Di Wu	University of Central Florida, USA
Fawad Ahmad	Rochester Institute of Technology, USA
Haonan Lu	University at Buffalo, USA
Silvio Barra	University of Naples Federico II, Italy
Weitian Tong	Georgia Southern University, USA
Xu Zhang	University of Exeter, UK
Yitao Hu	Tianjin University, China
Zhixin Zhao	Tianjin University, China

Track 7: Service Dependability and Security in Distributed and Parallel Systems

Wei Yang (Chair)	University of Texas at Dallas, USA
Dezhi Ran	Peking University, China
Hanlin Chen	Purdue University, USA
Jun Shao	Zhejiang Gongshang University, China
Jinguang Han	Southeast University, China
Mirazul Haque	University of Texas at Dallas, USA
Simin Chen	University of Texas at Dallas, USA
Wenyu Wang	University of Illinois at Urbana-Champaign, USA
Yitao Hu	Tianjin University, China
Yueming Wu	Nanyang Technological University, Singapore
Zhengkai Wu	University of Illinois at Urbana-Champaign, USA
Zhiqiang Li	University of Nebraska, USA
Zhixin Zhao	Tianjin University, China

Ze Zhang   University of Michigan/Cruise, USA
Ravishka Rathnasuriya   University of Texas at Dallas, USA

Track 8: Internet of Things and Cyber-Physical-Social Computing

Yu Yang (Chair)   Lehigh University, USA
Qun Song   Delft University of Technology, The Netherlands
Chenhan Xu   University at Buffalo, USA
Mahbubur Rahman   City University of New York, USA
Guang Wang   Florida State University, USA
Houcine Hassan   Universitat Politècnica de València, Spain
Hua Huang   UC Merced, USA
Junlong Zhou   Nanjing University of Science and Technology,
                                        China
Letian Zhang   Middle Tennessee State University, USA
Pengfei Wang   Dalian University of Technology, China
Philip Brown   University of Colorado Colorado Springs, USA
Roshan Ayyalasomayajula   University of California San Diego, USA
Shigeng Zhang   Central South University, China
Shuo Yu   Dalian University of Technology, China
Shuxin Zhong   Rutgers University, USA
Xiaoyang Xie   Meta, USA
Yi Ding   Massachusetts Institute of Technology, USA
Yin Zhang   University of Electronic Science and Technology
                                        of China, China
Yukun Yuan   University of Tennessee at Chattanooga, USA
Zhengxiong Li   University of Colorado Denver, USA
Zhihan Fang   Meta, USA
Zhou Qin   Rutgers University, USA
Zonghua Gu   Umeå University, Sweden
Geng Sun   Jilin University, China

Track 9: Performance Modeling and Evaluation

Zhibin Yu (Chair)   Shenzhen Institute of Advanced Technology,
                                        Chinese Academy of Sciences, China
Chao Li   Shanghai Jiao Tong University, China
Chuntao Jiang   Foshan University, China
Haozhe Wang   University of Exeter, UK
Laurence Muller   University of Greenwich, UK

Lei Liu	Beihang University, China
Lei Liu	Institute of Computing Technology, Chinese Academy of Sciences, China
Jingwen Leng	Shanghai Jiao Tong University, China
Jordan Samhi	University of Luxembourg, Luxembourg
Sa Wang	Institute of Computing Technology, Chinese Academy of Sciences, China
Shoaib Akram	Australian National University, Australia
Shuang Chen	Huawei, China
Tianyi Liu	Huawei, China
Vladimir Voevodin	Lomonosov Moscow State University, Russia
Xueqin Liang	Xidian University, China

Reviewers

Dezun Dong	Xiaolu Li
Chao Wang	Shujie Han
Chentao Wu	Mi Zhang
Chi Lin	Jing Gong
Deze Zeng	Radu Prodan
En Shao	Wei Wang
Fei Lei	Himansu Das
Haikun Liu	Rong Gu
Hailong Yang	Yongkun Li
Junlong Zhou	Ladjel Bellatreche
Kejiang Ye	Meng Shen
Lei Wang	Ruidong Li
Massimo Cafaro	Bin Wu
Massimo Torquati	Chao Li
Mengying Zhao	Chaokun Zhang
Roman Wyrzykowski	Chuan Zhang
Rui Wang	Chunpeng Ge
Sheng Ma	Fuliang Li
Songwen Pei	Fuyuan Song
Susumu Matsumae	Gaopeng Gou
Weihua Zhang	Guangwu Hu
Weixing Ji	Guo Chen
Xiaoli Gong	Guozhu Meng
Youyou Lu	Han Zhao
Yu Zhang	Hai Xue
Zichen Xu	Haiping Huang
Patrick P. C. Lee	Hongwei Zhang
Erci Xu	Ioanna Kantzavelou

Jiawen Kang	Yongqian Sun
Jie Li	Yuchao Zhang
Jingwei Li	Zhaoteng Yan
Jinwen Xi	Zhaoyan Shen
Jun Liu	Zhen Ling
Kaiping Xue	Zhiquan Liu
Laurent Lefevre	Zijun Li
Lanju Kong	Li Chen
Lei Zhang	Alfredo Cuzzocrea
Li Duan	Heng Qi
Lin He	Marc Frincu
Lingling Wang	Mingwu Zhang
Lingjun Pu	Qianhong Wu
Liu Yuling	Qiong Huang
Meng Li	Rongxing Lu
Minghui Xu	Shuo Yu
Minyu Feng	Weizhi Meng
Ning Hu	Wenbin Pei
Pengfei Liu	Xiaoyi Tao
Qi Li	Xin Xie
Qian Wang	Yong Yu
Raymond Yep	Yuan Cao
Shaojing Fu	Zhiyang Li
Shenglin Zhang	Wei Bao
Shu Yang	Jun Li
Shuai Gao	Dong Yuan
Su Yao	Francesco Palmieri
Tao Yin	George Bosilca
Tingwen Liu	Humayun Kabir
Tong Wu	Jaya Prakash Champati
Wei Quan	Peter Kropf
Weihao Cui	Pedro Soto
Xiang Zhang	Wenjuan Li
Xiangyu Kong	Xiaojie Zhang
Xiangyun Tang	Chuang Hu
Xiaobo Ma	Hang Qiu
Xiaofeng Hou	Ang Li
Xiaoyong Tang	Daniel Andresen
Xuezhou Ye	Di Wu
Yaoling Ding	Fawad Ahmad
Yi Zhao	Haonan Lu
Yifei Zhu	Silvio Barra
Yilei Xiao	Weitian Tong
Yiran Zhang	Xu Zhang
Yizhi Zhou	Yitao Hu

Zhixin Zhao
Wei Yang
Dezhi Ran
Hanlin Chen
Jun Shao
Jinguang Han
Mirazul Haque
Simin Chen
Wenyu Wang
Yitao Hu
Yueming Wu
Zhengkai Wu
Zhiqiang Li
Zhixin Zhao
Ze Zhang
Ravishka Rathnasuriya
Yu Yang
Qun Song
Chenhan Xu
Mahbubur Rahman
Guang Wang
Houcine Hassan
Hua Huang
Junlong Zhou
Letian Zhang
Pengfei Wang
Philip Brown
Roshan Ayyalasomayajula

Shigeng Zhang
Shuo Yu
Shuxin Zhong
Xiaoyang Xie
Yi Ding
Yin Zhang
Yukun Yuan
Zhengxiong Li
Zhihan Fang
Zhou Qin
Zonghua Gu
Geng Sun
Zhibin Yu
Chao Li
Chuntao Jiang
Haozhe Wang
Laurence Muller
Lei Liu
Lei Liu
Jingwen Leng
Jordan Samhi
Sa Wang
Shoaib Akram
Shuang Chen
Tianyi Liu
Vladimir Voevodin
Xueqin Liang

Contents – Part VII

An Efficient Scheduling Algorithm for Multi-mode Tasks on Near-Data Processing SSDs

Guo Li, Xianzhang Chen$^{(\boxtimes)}$, Duo Liu$^{(\boxtimes)}$, Jiali Li, Yujuan Tan, and Ao Ren

Chongqing University,Chongqing400044, China
liguo@stu.cqu.edu.cn, {xzchen,liuduo,lijiali,tanyujuan,ren.ao}@cqu.edu.cn

Abstract. Near-Data Processing (NDP) architectures have been proposed to alleviate the large overhead of data movement between the host and the Computational Storage Device (CSD) by offloading tasks to the CSD. In NDP architectures, each task can run in multiple modes according to the resource it takes for computing, such as the CPU of the host, the accelerator or the processor of the CSD. However, existing task scheduling algorithms on NDP architectures are unaware of the multi-mode tasks, leading to increased completion time of tasks and low resource utilization. In this paper, we propose a Multi-Mode Task Scheduling (MMTS) algorithm to optimize the completion time of the multi-mode tasks in NDP architectures. MMTS employs a greedy strategy to fully use the computing resources in the host and the CSD and align the completion time of the tasks by picking the proper modes. Our experimental results show that MMTS achieves 20.6% performance improvement on average over the state-of-the-art task scheduling algorithm on NDP-based system.

Keywords: Near data processing · Computational storage · Multi-mode task scheduling

1 Introduction

In the context of the vast amount of data being generated, lots of data-intensive tasks are posing significant challenges to existing storage systems [10]. One promising solution called Near-Data Processing (NDP) architecture is proposed to address these challenges. NDP-based system alleviates the issue of large data movement between the host and the Computational Storage Device (CSD) by offloading computing tasks to the CSD.

In NDP architectures, a task can be executed in multiple modes that correspond to different types of computing resources in the system, such as the CPU of the host, the accelerator or the embedded processor of the CSD [9,11,13], denoted by Host CPU, CSD accelerator, and CSD CPU, respectively. However, the performance of different resources are various. For example, the performance of Host CPU is generally higher than that of the CSD CPU. Also, the internal

Z. Tari et al. (Eds.): ICA3PP 2023, LNCS 14493, pp. 1–16, 2024.
https://doi.org/10.1007/978-981-97-0862-8_1

I/O bandwidth of the CSD is higher than the bandwidth between the host and the CSD [7]. Therefore, the completion time of the same task in different modes is different. In addition, different tasks compete for I/O resources and computing resources, resulting in reduced resource utilization and increased task completion time. Therefore, it is critical to design a scheduling strategy to effectively improve the performance of NDP system.

However, existing NDP efforts do not focus on scheduling strategies from a system perspective. Some focuses on the offloading task mechanism. This part of the work can be roughly divided into two categories. One is to study offloading domain-specific tasks to CSD accelerator, such as data retrieval [9]. The other aims to explore general-purpose near-data processing architecture, which supports offloading user-defined computational logic to be executed on general-purpose processors on CSD [2,13]. However, tasks are often performed directly offloaded to the CSD in these work, ignoring that the computing resources of the host can also be used.

Several transaction or request work scheduling approaches have been developed within CSD to mitigate average request delays [8,12]. However, these approaches are not designed with a system-level perspective in mind. λ-IO [14] takes into account the utilization of the Host CPU. It models the time overhead of the task on the host and CSD, and distributes the task to the side with the minimum time. However, under this strategy, the computing resources on the other side are still underutilized, when tasks are always offloaded to one side for execution. Therefore, we can observe that some computing resources are not taken into account and are idle in existing work.

Besides, we find that not all tasks are suitable for offloading to CSD. There is a trade-off between the computation overhead associated with near SSD processing and the reduction in communication overhead to the host system in NDP architecture [6]. The internal I/O bandwidth of the CSD is higher than the bandwidth between the host and the CSD while the performance of the Host CPU is generally stronger than that of the CSD embedded processor [7]. Therefore, when we offload computationally complex tasks to disk, the increase in computation time may be higher than the decrease in data transfer time, so there is no performance improvement.

The observations can be summarized as follows:

1. Tasks are multi-mode and can be executed on the Host CPU or offloaded to CSD accelerators or CSD CPU in near-data systems.
2. Not all tasks are suitable for offloading, e.g. task with high computational complexity is processed faster on the Host CPU than offloaded to CSD embedded processor.
3. In certain greedy·scheduling strategies, the Host CPU and CSD embedded processor may remain idle for extended periods of time, as these components are not optimally suited for offloading single tasks.

Based on the above observations, this paper proposes a multi-mode task scheduling algorithm from the perspective of NDP system, which improves resource utilization and reduces the waiting time of tasks by scheduling the mode and order of tasks. First, this paper formally defines the Multi-Mode NDP-task Scheduling Problem (MNSP). In this problem, we model a task into two phases, I/O phase and computing phase. Based on the general NDP architecture, we abstract four type of resource, I/O channels, CSD processor, CSD accelerator and Host CPU. Secondly, this paper proposes a heuristic multi-mode task scheduling algorithm, called MMTS. MMTS first adopts a greedy strategy to select the mode with the minimum time for each task. On this basis, we schedule the mode of the task to make the completion time of the queue of different computing resources as equal as possible. We further optimize the average waiting time of tasks by adjusting the order of tasks. Finally, we build a simulator to test the performance of MMTS. The experimental results show that MMTS improves the performance of 20.6% and 29.34% on average compared with existing scheduling strategies.

In summary, we make the following contributions:

1. We formally define the multi-mode NDP-task scheduling problem (MNSP) and show that it is an NPC problem.
2. We propose a schedule algorithm, called MMTS, to solve this problem. Based on the idea of parallel and concurrency, MMTS schedules the mode and order of the task, which effectively reduces the average waiting time of tasks and improves the parallelism. Based on our understanding, MMTS is the first work to study multi-mode task scheduling in the near-data process architecture.
3. We construct workloads based on real NDP-based SSD implementations and built a simulator. The experiment results show that MMTS technology significantly reduces the makespan.

The rest of the article is arranged as follows: Sect. 2 describes the background of MNSP and illustrate MNSP Problem by a motivational example. Section 3 shows a motivation example and formally defines MNSP problem. Section 4 introduces the MMTS algorithm. Section 5 introduces experimental setup and evaluates the performance of MMTS. Finally, we summarize this work in Sect. 6.

2 Background and Motivation

In this section, we first introduce the general architecture of the NDP-based SSD and analyze the processing flow of task in different modes. Then, we illustrate the MNSP Problem in NDP-based SSDs by a motivational example.

2.1 Task Processing of NDP-Based SSDs

NDP-based SSDs are designed to mitigate the storage wall phenomenon by reducing the data movement between the host and the SSD. Compared with traditional SSDs, NDP-based SSDs implement accelerators to accelerate application-specific tasks, or perform tasks on the CSD embedded processor, making full use of the internal bandwidth of CSD and avoiding huge data movement between host and CSD [2,3,9].

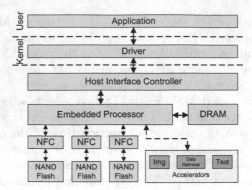

Fig. 1. The general architecture of NDP-based SSDs.

Typically, an NDP-based SSD usually consists of the following six types of components, as Fig. 1 shows.

1. Host interface controller: It provides a communication channel between a host system and Cosmos+ OpenSSD, which runs NVMe or SATA protocol.
2. Embedded processor: It is usually ARM processor, which runs firmware communicating with host interface controller and storage controller in traditional SSD. In addition, it is also used to process data in some NDP-based SSDs [2,13].
3. DRAM memory: It supports for running firmware and provides a temporary store for data transfer between the host system and flash device. The rest space could be used for data processing in NDP architecture.
4. NAND flash array: It acts as storage media, composed of multiple NAND flash devices, whose basic unit of operation is page, which is usually 16 KB [7].
5. Flash channel controller: There are multiple parallel flash channel controllers in SSDs, which is responsible for orchestrating the operation of flash devices attached to the same channel.
6. Domain-specific accelerator: It is used to accelerate domain-specific task in NDP-based SSDs, such as cognitive SSD [9].

In architecture of NDP-based SSDs, each task can be split into two phases, data transfer phase and computing phase. In the data transfer phase, the task occupies I/O resources. Each task contains many requests. Therefore, we can assume that data transfer phase monopolizes all channels. In the computing phase, the task occupies one of the computing resources, such as the CPU of host (denoted by Host-CPU), the accelerator in the CSD (denoted by CSD-Acce), and the embedded processor in the CSD for data processing (denoted by CSD-CPU).

When a task is dispatched to Host-CPU, the Host-CPU first send normal request to CSD. And the data for this task is transferred from Nand flash array to Host-DRAM. And then Host-CPU accesses Host-DRAM for data-processing.

While when a task is dispatched to CSD-CPU or CSD-Acce, the data is transferred from Nand flash array to CSD-DRAM, avoiding the data movement between host and CSD. And then the CSD-CPU or CSD-Acce accesses the CSD-DRAM for data processing.

A task could be dispatched to computing resource when I/O resource is available and the previous task of corresponding task queue has completed as shown in Fig. 2.

We assume that the NDP architecture composed of one Host-CPU, one CSD-CPU and one CSD-Acce in this work.

2.2   Motivation Example

In this section, we illustrate the MNSP Problem in NDP-based SSDs by a motivational example.

Table 1. Detailed requirements of example Tasks.

Task	h-io-time	h-exe-time	d-io-time	d-cpu-time	d-acce-time
T1	299	552	138	805	276
T2	200	759	115	1265	345
T3	345	920	184	1840	483

Table 1 shows the attribution of the three tasks. The time unit in the table is us. The h-io-time indicates the transfer time from the NAND flash array to host DRAM, and the h-exe-time indicates the computing time when task is processed on Host-CPU. The d-io-time indicates the data transfer time from Nand flash array to NAND memory in CSD. The d-cpu-time indicates the computing time when task is processed on CSD-CPU. The d-acce-time indicates computing time when task is processed on CSD-Acce.

In general NDP-based SSDs architecture, a task has three modes, corresponding to be processed on the Host-CPU, CSD-CPU, and CSD-Acce, respectively. In this paper, we assume that there is no data dependency between tasks.

Let us try to schedule the mode and order of tasks in Table 1. Each mode corresponds to a computation resource with a queue. λ-IO adopts shortest-dispatching strategy to determine the mode of task. Under this strategy, each task chooses the mode that has minimum process time, and then be appended in corresponding queue. Figure 2(a) shows the result of adopting shortest-dispatching strategy, where the total process time is 1518 us. We also give the optimal scheduling as Fig. 2(b) shown, whose total process time process time is 1097 us, reduced by 27.73%.

In summary, we can get a significant performance improvement by scheduling the mode and order of tasks. The scheduling for MNSP problem should considers the resource type and duration consumed by the task in different modes, as well as the order of tasks.

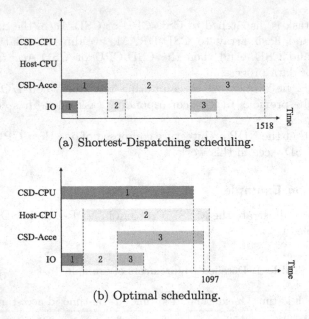

(a) Shortest-Dispatching scheduling.

(b) Optimal scheduling.

Fig. 2. Motivation example.

3 Problem Definition

The input of the MNSP problem contains three sets: resources, tasks, and modes, and the output is the execution order of the tasks and the modes of each task. The optimization goal of the MNSP problem is to reduce the makespan.

The MNSP problem can be formulated as follows. The input of MNSP is an instance $<R, J, M>$. $R = \{r_1, r_2, ..., r_{RC}\}$ is the set of resource, where RC is the count of resources. $J = \{j_1, j_2, ..., j_{JC}\}$ is the set of task, where JC is the number of tasks. $M = \{m_1, m_2, ..., m_{MC}\}$ is the set of mode, where MC is the count of execution modes. The definition of a task is $j_i = \{DS, T^{io}, T^{compute}\}$. DS is the data size that task j_i needs to transfer and compute. T^{io} and $T^{compute}$ are the sets of data transmission time and computation time, respectively, where each element T_m^{io} and $T_m^{compute}$ denotes the transmission and computation time of task j_i under mode m, respectively.

We define some symbols as follows to represent resource constraints. st_{jm} indicates the start time of task j choosing mode m. r_{jmk} represents the amount of type k resource consumed by the task j in m mode. $AR = \{aR_k \mid k \in R, aR_k \in \mathbb{R}\}$ is the set of the total amount of k type resource. T is time slot, indicating the smallest time unit. $t_n = (n-1) \times T$ is the start time of the nth time slot. We use the following formula to represent resource constraint that a resource can only process one task at a time.

$$\sum_{(j,m,n)} x_{jmkn} \times r_{jmk} \leq aR_k, \ k \in R \tag{1}$$

where x_{jmkn} is a binary, indicating that task j choosing mode m occupies k type resource in the time interval $[t_n,\ t_n + T)$ while $x_{jmkn} = 1$.

For computing resource,

$$x_{jmkn} = \begin{cases} 1, & if\ st_{jm} \leq t_c < (st_{jm} + T_m^{io} + T_m^{compute}) \\ 0, if\ t_c < st_{jm}\ or\ t_c \geq (st_{jm} + T_m^{io} + T_m^{compute}) \end{cases} \quad (2)$$

For I/O resource,

$$x_{jmkn} = \begin{cases} 1, & if\ st_{jm} \leq t_c < (st_{jm} + T_m^{io}) \\ 0, if\ t_c < st_{jm}\ or\ t_c \geq (st_{jm} + T_m^{io}) \end{cases} \quad (3)$$

where t_c indicates the current time.

We also use following formula to represent that a task can select only one mode, where $b_m^j = 1$ refers to task j choose mode m

$$\sum_{j=1}^{|J|} \sum_{m=1}^{|M_j|} b_m^j = 1,\ b_m^j \in 0, 1 \quad (4)$$

The output of the problem is $I = (j_1, j_2, \ldots\ldots, j_N)$, a queue in which each task has chosen the mode. Our goal is to minimize the makespan of all tasks. The objective function is

$$min\ max(st_{jm} + T_m^{io} + T_m^{compute}) \quad (5)$$

The MNSP problem is a hybrid of Multi-mode Resource Constraint Project Schedule Problem [1] and multi-processors problem [4], which has been proved to be NP-complete.

4 Multi-mode Task Scheduling

In this chapter, we propose an effective heuristic algorithm, called MMTS, to solve the MNSP problem. We first analyzed the MNSP problem for optimization. Then, we cover MMTS technology in detail.

4.1 Problem Analysis

In a ideal scheduling, The utilization of resources should reach 100% as much as possible and most tasks choose the shortest mode as Fig. 3 shows. First, let's analyze the approximate ideal scheduling scheme. We observe that I/O resources are critical resources, since I/O resources are always occupied no matter what mode the task select. We could think further that MNSP problem can be simplified to multi-processors problem if we assume that the I/O resources are infinite. However, the multi-processors problem was proved to be an NPC

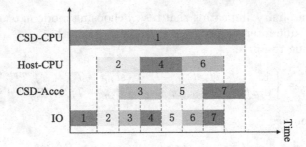

Fig. 3. Approximate ideal scheduling.

problem [4], and there is no deterministic algorithm for the optimal solution in polynomial time. We find that making the completion time of each task queue as equal as possible can be approximated as the optimal solution, assuming the number of tasks is infinite. Based on this idea, we consider dispatching tasks to queues of corresponding computing resource, and try to make the completion time of each task queue approximately equal.

Besides, in order to avoid that most tasks choose the mode that takes the longest time, although the completion time of task queue is approximately equal. We select the mode with the shortest time for each task and add it to the task queue of the corresponding computing resource before dispatching tasks to queues.

Finally, it is easy to find that the sum of the I/O time of all tasks is a lower bound for the completion time of the task set, since the I/O resource is an absolutely critical resource, occupied by each task to transfer data. Therefore, we propose a task alignment strategy to maximize the utilization of I/O resource by adjusting the order of tasks in the queue.

4.2 MMTS Algorithm

In this section, we propose a Multi-Mode Task Scheduling (MMTS) algorithm for solving the MNSP problem. The main idea of MMTS to fully use the computing resources in the host and the CSD and align the computing time and I/O time between tasks with a greedy strategy. MMTS chooses the proper mode of each task according to the completion time of task queues. MMST has three phases, i.e., Shortest-Mode-First, Try Average, and Task Alignment. We will introduce the three phases in the following subsections.

Shortest-Mode-First. We build a task queue for each computing resource corresponding to a mode. It is obviously not a good solution that most tasks choose the mode that takes the most time, although task queues has

---

Algorithm 1. Phase 1: Shortest mode first

    Input: An instance $<R, J, M>$
    Output: Task queue of compute resource:$csd\_acce_q, csd\_cpu_q, host\_cpu_q$
1:   $csd\_acce_q \leftarrow \emptyset$, $host\_cpu_q \leftarrow \emptyset$, $csd\_cpu_q \leftarrow \emptyset$
2:   **for** j in J **do**
3:       Choose mode m_j that make $(T_m^{io} + T_m^{compute})$ minimum for task j
4:       **if** $m_j == HOST$ **then**
5:          Append task j to $host\_cpu_q$
6:       **if** $m_j == CSD\_CORE$ **then**
7:          Append task j to csd_cpu_q
8:       **if** $m_j == CSD\_ACCE$ **then**
9:          Append task j to csd_acce_q

---

approximately equal completion time. Therefore, MMTS chooses the shortest mode for each task before next step, and adds it to the corresponding queue.

In Algorithm 1, we choose the mode that makes $(T_m^{io} + T_m^{compute})$ minimum for each task $j \in J$, and we append it to corresponding task queue.

Algorithm 2. Phase2–Try average

    Input: $csd\_acce_q, host\_cpu_q, csd\_cpu_q, iteration\_cnt$
    Output: $csd\_acce_q, host\_cpu_q, csd\_cpu_q$
1:   **function** $ave\_f1(list1, list2)$
2:       $flag \leftarrow 1$
3:       **while** (1) **do**
4:          **if** The completion time of list1 $<$ that of list2 **then**
5:              **if** $flag == 0$ **then**
6:                  **Break**
7:              Choose the longest task j of list2
8:              list1.append(j), list2.pop(j)
9:          **else**
10:         Choose the shortest task j of list1
11:         list2.append(j), list1.pop(j), $flag \leftarrow 0$
12: **function** $ave\_f2(list1, list2)$
13:       **if** The completion time of list1 $<$ that of list2 **then**
14:          $ave\_f1(list1, list2)$
15:       **else**
16:          $ave\_f1(list2, list1)$
17: **while** iteration_cnt > 0 **do**
18:       $ave\_f2(host\_cpu_q, csd\_acce_q)$;
19:       $ave\_f2(csd\_cpu_q, csd\_acce_q)$;
20:       $iteration\_cnt --$;

Try Average. We employ iterative procedures to minimize the variance in completion time across multiple queues.

In this phase, we schedule the mode of tasks to reduce the idle time of Host-CPU and CSD-CPU, and improve the utilization of computing resources.

In steps 1–12 of Algorithm 2, we define a function ave_f1 that makes the completion time of two input lists as equal as possible by exchange the tasks between them. We select the task with the longest time from list 1 with a longer completion time, and add it to list 2 with a shorter completion time. Then, we add the task with the minimum time taken from list 2 to list 1 until the completion time of list 1 exceeds that of list 2. The completion time of list 1 gradually approaches that of list 2 in this procedure, since we choose the task with minimum time in each iteration. In steps of 13–17 of Algorithm 2, We define a function ave_f2 that first judges the longer list, and then make them as equal as possible. In steps 18–21, we employ iterative procedures to minimize the completion time disparity between two queues.

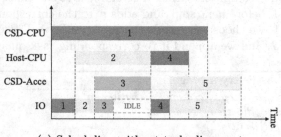

(a) Scheduling without task alignment.

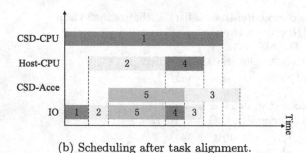

(b) Scheduling after task alignment.

Fig. 4. Example of task alignment.

Task Alignment. We found that unreasonable order of tasks would result in resources waiting. As shown in Fig. 4(a), the computational tasks are executed next to each other. This results in all computing resources being occupied and unable to execute new tasks, resulting in the waiting time of I/O resources. Figure 4(b) shows the optimized order of tasks. The computing time of task 2 is equal to the I/O time of task 5, thus filling the waiting time of I/O resources.

Therefore, MMTS considers adjusting the order of tasks to minimize the gap between the computing time of one task and the I/O time of the task following.

We only align the task queue of the CSD accelerator with the task queue of the Host-CPU without considering the order optimization of the CSD-CPU. Since the processing performance of CSD-CPU is weak, with long processing time. So that the wait time of CSD-CPU is relatively small.

Task align phase aims to improve the concurrency of computing resources and I/O resources to reduce the resource wait time. The specific details of algorithm are shown in table. The basic idea of task alignment is that for task a, we find task b, which has the minimum absolute value of difference between the I/O time of task b and the computing time of task a. And then we repeat this step for task b (called aligned task) to find task c (called patching task), until finishing traversing all the tasks.

In steps 1–3 of Algorithm 3, we define two lists $Tmp1$ and $Tmp2$ to store the aligned task and then initial $Tmp1$ by appending the first element of csd_acce_q to $Tmp1$. In steps 5–8, we find patching task b of $host\_cpu_q$ for aligned task a of csd_acce_q, and thus, task b has been the aligned task. And then in steps 9–12, we find patching task c of csd_acce_q for aligned task b of $host\_cpu_q$. We repeat this procedure until finishing traversing all the tasks.

Algorithm 3. Phase 3–Task alignment

Input: $csd\_acce_q, host\_cpu_q, csd\_cpu_q$
Output: $csd\_acce_q, host\_cpu_q, csd\_cpu_q$
1: $Tmp1 \leftarrow [], Tmp2 \leftarrow []$
2: Append the first element of csd_acce_q to $Tmp1$
3: csd_acce_q pop the first element
4: **while** (1) **do**
5:     **if** $len(csd\_acce_q)! = 0$ **then**
6:         Find an task j in $host\_cpu_q$ whose T_m^{io} has the smallest absolute value of the difference with the $T_m^{compute}$ of the last element of Tmp1
7:         Append task j to Tmp2
8:         Pop task j from $host\_cpu_q$
9:         **if** $len(host\_cpu_q)! = 0$ **then**
10:             Find an task j in csd_acce_q whose T_m^{io} has the smallest absolute value of the difference with the $T_m^{compute}$ of the last element of Tmp2
11:             Append task j to Tmp1
12:             Pop task j from csd_acce_q
13:         **else**
14:             **Break**
15:     **else**
16:         **Break**

---

5   Evaluation

In this chapter, we first introduce the experimental setup, including the platform, the workloads and the algorithms that need to be compared. We will then present the experimental results and analyze the effectiveness of the proposed MMTS.

5.1   Experimental Setups

Platform: We build a simulator to evaluate the scheduling of tasks in an NDP system with a Host and a CSD. Since We aims at a general architecture, where task could be processed on Host-CPU, CSD-Acce, and CSD-CPU. However, existing platforms either only support the offloading of application-specific tasks to CSD-Acce, or only general-purpose tasks offloaded to CSD-CPU. This simulator is built according to the architecture and parameters of Cosmos+ [7].

Workloads: As far as we know, there is currently no open-source multi-mode NDP-task workloads. Therefore, we build it based on the Cosmos platform and some real-world applications, such as embedding table operations [13], image training, and string retrieval. First, we deployed RecSSD on Cosmos+ [7] and collect execution time of the embedded table operation, as a measure of the performance gap between Host-CPU and CSD-CPU. We also implemented string retrieval accelerator on Cosmos+, and collect execution time, to measure the gap between the performance of the Host-CPU and the CSD-Acce. We construct a workload, based on the above parameters. In this paper, we define a task to be a computational task (denoted as C) if $T_m^{io} < T_m^{compute}$ when a task is processed on Host-CPU, vice versa, an I/O-type task (denoted as I/O). As shown in Fig. 5, we mix I/O-type tasks and computational tasks according to the ratio of 0.2:0.8, 0.4:0.6, 0.6:0.4 and 0.8:0.2.

Strategies: Fist-Come-First-Serve (FCFS) is widely used in existing traditional SSDs and some NDP-based SSDs [5,9,11]. FCFS strategy distributes tasks to queues based on the order in their arrival. In this experiment, each mode corresponds to a type of computing resource, Host-CPU, CSD-Acce and CSD-CPU. Each type of computing resource corresponds to a task queue. We add tasks to the corresponding queue based on the results of task mode scheduling. Strategies about the mode scheduling are as follows:

1. Shortest dispatching (SD): λ-IO [14] takes into account the utilization of the computing resources of the host. It models the time overhead of the task on the host and CSD, and dispatches the task to the side with the shortest time.
2. Dispatching when available (DWA): Task is dispatched to the computing resource instantly which becomes available, prioritized to CSD computing resource if both Host-CPU and CSD computing resource are available at the same time. Summarizer [6] proposes an approach which saturates the CSD first. We extend this strategy to DWA.

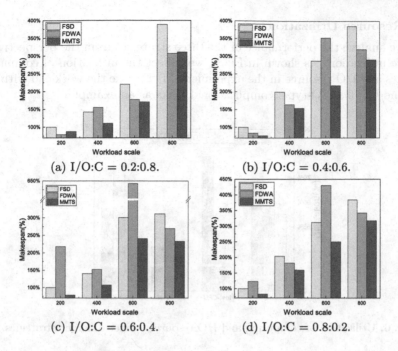

(a) I/O:C = 0.2:0.8.   (b) I/O:C = 0.4:0.6.

(c) I/O:C = 0.6:0.4.   (d) I/O:C = 0.8:0.2.

Fig. 5. Makespan of different strategies on workloads with different ratio of I/O-type tasks and computational tasks.

5.2 Effect on Performance

The proposed MMTS is compared with FCFS+SD (FSD), FCFS+DWA (FDWA). The results of the makespan of different strategy are shown in Fig. 5. The horizontal axis represents the number of tasks for the workload. The experiment results show that MMTS outperforms FSD and FDWA.

Compared with FSD, MMTS can reduce the completion time by 25.3% in best case and 20.6% on average. Since the FSD algorithm offloads all the tasks to the CSD-Acce, the Host-CPU and the CSD-CPU are both unused for the tasks. However, MMTS makes the completion time of each queue as equal as possible, improving the utilization of the Host-CPU and the CSD-CPU. Compared with FDWA, MMTS can reduce the completion time by up to 62%, and by an average of 29.34%. We find that FDWA has a wide range of completion time toggles. There are cases where a long task at the end of the queue is assigned to CSD-CPU. Furthermore, the tail delay further increases due to the weak performance of CSD-CPU. On the contrary, tasks that take longer are prioritized under MMTS, which avoids this situation. Even excluding these cases, compared with FDWA, MMTS reduces completion time by an average of 4.99%. In summary, we can observe that MMTS outperforms existing strategies on workloads with different proportions of I/O-type tasks and computational tasks.

5.3   Resource Utilization

Now, we analyze the performance of the three strategies from the perspective of resource utilization. As shown in Fig. 6, we collect the utilization of computing resources and I/O resource in the experiments. Let's take the workload with 800 tasks and 0.2: 0.8 I/O-type: computational tasks as an example.

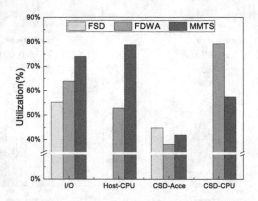

Fig. 6. Utilization of computing and I/O resources under different strategies.

We find that compared with FSD, MMTS and FDWA have higher utilization of I/O, Host-CPU and CSD-CPU. Since FSD always select the shortest mode. Therefore almost all tasks are distributed to the CSD-Acce, resulting in idle time of the Host-CPU and CSD-CPU, whose utilization is 0. Compared with FSD, FDWA and MMTS allocates some tasks to idle Host-CPU and CSD-CPU, which improves resource utilization.

Compared with FDWA, MMTS has higher utilization of I/O, Host-CPU and CSD-Acce and lower utilization of CSD-CPU. Since there is a situation where, the CSD-CPU is idle and CSD-Acce is about to become available at the current moment. Therefore, it is better to wait for a little time and then distribute the task to CSD-Acce rather than to the CSD-CPU instantly under FDWA, which results in more powerful CSD-Acce waiting for I/O resource.

As shown in Fig. 6, there is still room for improvement in the utilization of computing power by improving the utilization of CSD-Acce, although the sum of computing resource utilization of FDWA is higher than FSD and MMTS. Therefore, considering the difference in performance among computing resource, MMTS makes the completion time of task queues as equal as possible. This further enhances the utilization of computing power and reduces the overall task completion time compared with FDWA.

6   Conclusion

The NDP architecture reduces data transfer between the host and the SSD by offloading tasks to CSD. Tasks with multi modes can be executed on the Host-CPU, CSD-CPU, and CSD-Acce in a general architecture. Existing NDP related work rarely schedules tasks from a system perspective, ignoring increased computing time when offloading and idle time of Host-CPU. Therefore, it is necessary to schedule the order and mode of tasks from the perspective of the system. In this work, we propose a multi-mode task scheduling mechanism called MMTS. MMTS considers idle computing resources and task waiting time caused by task competition for I/O resources. It reduces the overall makespan by scheduling the order and mode of the task. The experimental results show that, compared with FSD and FDWA, MMTS reduces by an average of 20.6% and 29.34% of makespan, respectively.

Acknowledgement. This work is partially supported by National Natural Science Foundation of China under Grant 62072059 and 62102051, Chongqing Post-doctoral Science Foundation, China (Project No. 2021LY75), and the Funds for Chongqing Distinguished Young Scholars (No. cstc2020jcyj-jqX0012). We would like to thank the anonymous reviewers for their valuable comments and improvements to this paper.

References

1. Colak, S., Agarwal, A., Erenguc, S.: Multi-mode resource-constrained project-scheduling problem with renewable resources: new solution approaches. JBER **11**(11), 455 (2013). https://doi.org/10.19030/jber.v11i11.8193
2. Do, J., et al.: Cost-effective, energy-efficient, and scalable storage computing for large-scale AI applications. ACM Trans. Storage **16**(4), 1–37 (2020). https://doi.org/10.1145/3415580
3. HeydariGorji, A., Torabzadehkashi, M., Rezaei, S., Bobarshad, H., Alves, V., Chou, P.H.: Stannis: low-power acceleration of DNN training using computational storage devices. In: 2020 57th ACM/IEEE Design Automation Conference (DAC), San Francisco, CA, USA, pp. 1–6. IEEE, July 2020. https://doi.org/10.1109/DAC18072.2020.9218687
4. Hou, E., Ansari, N., Ren, H.: A genetic algorithm for multiprocessor scheduling. IEEE Trans. Parallel Distrib. Syst. **5**(2), 113–120 (1994). https://doi.org/10.1109/71.265940
5. Hu, Y., Jiang, H., Feng, D., Tian, L., Luo, H., Ren, C.: Exploring and exploiting the multilevel parallelism inside SSDs for improved performance and endurance. IEEE Trans. Comput. **62**(6), 1141–1155 (2013). https://doi.org/10.1109/TC.2012.60
6. Koo, G., et al.: Summarizer: trading communication with computing near storage. In: Proceedings of the 50th Annual IEEE/ACM International Symposium on Microarchitecture, Cambridge, Massachusetts, pp. 219–231. ACM, October 2017. https://doi.org/10.1145/3123939.3124553
7. Kwak, J., Lee, S., Park, K., Jeong, J., Song, Y.H.: Cosmos+ OpenSSD: rapid prototype for flash storage systems. ACM Trans. Storage **16**(3), 1–35 (2020). https://doi.org/10.1145/3385073
8. Li, J., et al.: Horae: a hybrid I/O request scheduling technique for near-data processing based SSD. IEEE Trans. Comput.-Aided Des. Integr. Circuits Syst. 1 (2022). https://doi.org/10.1109/TCAD.2022.3197518
9. Liang, S., Wang, Y., Lu, Y., Yang, Z., Li, H., Li, X.: Cognitive SSD: a deep learning engine for in-storage data retrieval, p. 17 (2019)

10. Reinsel, D., Gantz, J., Rydning, J.: The digitization of the world from edge to core (2018)
11. Ruan, Z., He, T., Cong, J.: INSIDER: designing in-storage computing system for emerging high-performance drive (2019)
12. Tavakkol, A., et al.: FLIN: enabling fairness and enhancing performance in modern NVMe solid state drives. In: 2018 ACM/IEEE 45th Annual International Symposium on Computer Architecture (ISCA), pp. 397–410, June 2018. https://doi.org/10.1109/ISCA.2018.00041
13. Wilkening, M., et al.: RecSSD: near data processing for solid state drive based recommendation inference. In: Proceedings of the 26th ACM International Conference on Architectural Support for Programming Languages and Operating Systems, Virtual USA, pp. 717–729. ACM, April 2021. https://doi.org/10.1145/3445814.3446763
14. Yang, Z., et al.: L-IO: a unified IO stack for computational storage. Performance Improvement (2023)

HR-kESP: A Heuristic Algorithm for Robustness-Oriented k Edge Server Placement

Haiquan Hu⊕, Jifu Chen⊕, and Chengying Mao$^{(\boxtimes)}$⊕

School of Software and IoT Engineering, Jiangxi University of Finance and
Economics, Nanchang 330013, China
maochy@yeah.net

Abstract. Mobile edge computing develops a new paradigm to deliver
low-latency services to mobile users by deploying edge servers at base
stations or access points close to the users. From the perspective of service providers, in this paper, we consider how to deploy k edge servers
on suitable base stations to maximize network robustness and user coverage. Accordingly, the above two metrics are focused on constructing
an optimization model for deploying a given number (i.e., k) of edge
servers in the edge computing network. In order to solve the model efficiently, a **H**euristic algorithm for **R**obustness-oriented k **E**dge **S**erver
Placement (HR-kESP) is designed. In the proposed algorithm, the initial base station is first selected in a greedy manner. Then, a heuristic
policy by considering the increment of network robustness is designed to
gradually expand the deployment solution until the server budget constraint is satisfied. The comparison experiments have been conducted on
a public dataset, showing that our HR-kESP algorithm achieves a better
optimization effect than the other four benchmark algorithms in most
cases.

Keywords: Mobile Edge Computing · Edge Server Placement ·
Robustness · Heuristic Algorithm

1 Introduction

In recent years, the massive access to mobile devices has led to a surge of data
volume [1]. As a result, traditional cloud-based mobile networks have struggled
to meet this challenge. Mobile edge computing (MEC) [2], which applies the
edge computing (EC) [3] technology to mobile network scenarios, deploys edge
servers at the base stations near mobile users or some critical network nodes.
Generally speaking, the paradigm of migrating computing and storage resources
to the edge side of a mobile network can conveniently provide computing power
and storage capacity for mobile users with low service latency [4]. This type of
cloud-edge-user hierarchical network structure can effectively deal with the storage, concurrency, and data transfer pressures caused by massive data. However,

Z. Tari et al. (Eds.): ICA3PP 2023, LNCS 14493, pp. 17–33, 2024.
https://doi.org/10.1007/978-981-97-0862-8_2

building a cost-effective and robust network architecture in the MEC environment is still challenging.

Due to the high cost of building a new network, MEC usually makes full use of existing infrastructure and data links to provide services, that is, try to make some minor adjustments and conversions to the existing network structure. As the critical step in constructing an MEC network, edge server placement (ESP) is to select several suitable base stations as the server deployment locations in a given mobile network architecture. Among the existing studies of ESP, most of them focus on metrics of network service quality and service provider profits [5], such as response latency [6,7], energy consumption [8,9] and deployment cost [10,11]. A key requirement in providing high-quality network services is the ability of the edge server network to run stably for a long period. If a network's anti-failure capability is too weak, its performance is easily affected by some sudden mishaps. Even if an MEC solution performs well on some standard metrics, it is still likely to encounter challenges when operating in real-world network environments. Frequent server network failures may bring high maintenance costs to service providers and significantly affect users' quality of service (QoS) [12,13]. At present, the research on network robustness has not received enough attention. As a representative work, Cui et al. [14] initiated the discussion on constructing a highly robust server network with a fixed edge server budget. In their solution, the robustness of the MEC network is quantified as the number of mobile users who are served simultaneously by two edge servers, i.e., it is guaranteed that in an emergency situation where one server is unable to serve a user, there are still other servers that can work for that user. Furthermore, their study proposed a means to trade off network robustness and user coverage by combining them into a unified objective, and aimed to find a solution that maximises the overall objective [15]. It is worth noting that in the experimental analysis, they only used the robustness metric to statically measure the reliability of each edge server deployment solution, and lacked the analysis of the robustness of the deployment from the perspective of actual application scenarios (i.e., the dynamic and random failures of edge servers).

In this paper, we attempt to design a cost-effective algorithm for solving the Robustness-oriented k Edge Server Placement (kESP) problem in a heuristic manner. The robustness increment is used as heuristic information to find the appropriate base stations to deploy edge servers. During the process of gradually generating the server deployment scheme, the candidate deployment locations are selected in the way of greedy from two subsets of base stations, which are differentiated and evaluated separately according to the neighbouring relationship with the deployed edge servers. In addition, for evaluating the effectiveness of solutions, most of the existing methods mainly use some static metrics to measure the user coverage and robustness of each placement solution. However, this way lacks a solid theoretical foundation, and it cannot precisely reflect the network performance in real-world scenarios. In this paper, we propose a simulation-based network robustness evaluation method, which simulates the possible emergencies (some servers crash randomly) to evaluate the real performance of a network. Through the experimental and comparative analysis, it is confirmed that our proposed algorithm shows significant superiority in both the basic static metric and the simulation evaluation of network robustness. The artifacts of this study for reproducibility and future research activities are available at https://github.com/maochy/HR-kESP.

The main contributions of this paper can be summarized below:

- A **H**euristic algorithm for **R**obustness-oriented k **E**dge **S**erver **P**lacement named **HR-kESP** is proposed to effectively solve the k edge server placement problem from the viewpoint of multi-objective optimization.
- A simulation-based evaluation method for the network robustness is proposed. Through simulating the server failures randomly, the network robustness of a given ESP solution can be indirectly measured by evaluating the impact of failures on network performance.
- Intensive experiments are conducted on the public dataset by considering various application scenarios, and the results indicate that HR-kESP outperforms other representative algorithms in most cases.

The rest of this paper is structured as follows. In Sect. 2, we describe the ESP problem and formulate the optimization model for maximizing the network robustness and user coverage. Section 3 addresses the model-solving framework and the corresponding heuristic algorithm, namely HR-kESP. The detailed comparative experiments are conducted, and the experimental results are analyzed and discussed in Sect. 4. Section 5 describes the related work. Finally, Sect. 6 concludes this study and discusses some future research directions.

2 Problem and Model Formulation

2.1 Problem Statement

In the MEC scenario shown in Fig. 1, base stations communicate with each other using wireless connections. If two base stations satisfy that one is within the coverage of the other base station, they can directly access it, but not in other cases. The service relationship between the user and the base station follows the similar rule. If the user u is within the coverage of the base station b, the user u can directly request the service from the base station b. This direct accessing relationship can be denoted as $a(b, u) \in \{0, 1\}$ where the value 1 indicates user u can directly access base station b, otherwise 0. Similarly, $a(b_i, b_j) \in \{0, 1\}$ indicates whether the base stations b_i and b_j can directly access to each other. In particular, $a(b_i, b_i) = 1$.

According to the above accessibility rules, the mobile network environment shown in Fig. 1 can be modelled as the topology network as shown in Fig. 2. The communication relationship network of base stations is represented by a weighted undirected graph $G = (B, E)$, where $B = \{b_1, b_2, \ldots, b_m\}$ is a set of m base stations, $E = \{\langle b_i, b_j \rangle | a(b_i, b_j) = 1, b_i, b_j \in B\}$ is a set composed of accessing relations between all base stations. For a given mobile user set

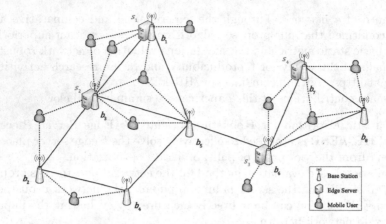

Fig. 1. A Typical Mobile Edge Computing Scenario.

$U = \{u_1,\ u_2,\ \ldots,u_n\}$ and the base station set B, the accessibility matrix is defined as

$$A = \left\{(a(b_i,b_j) - 1) + a(b_i,b_j) \times \sum_{u \in U} a(b_i,u) \times a(b_j,u)\right\}_{i \times j}, \tag{1}$$

where $1 \leq i, j \leq m$. It contains the following three types of elements.

- When $i = j$, $A_{i,j}$ (or $A_{i,i}$) represents the number of mobile users within the coverage area of the base station b_i.
- When $i \neq j$ and $a(b_i,b_j) = 1$, $A_{i,j}$ (or $A_{j,i}$) represents the number of mobile users with direct access to base stations b_i and b_j at the same time.
- When $i \neq j$ and $a(b_i,b_j) = 0$, $A_{i,j}$ (or $A_{j,i}$) is -1, indicating that the base stations b_i and b_j are not directly connected.

As illustrated by Fig. 1, the base station b_1 covers four users, then the value of $A_{1,1}$ is 4. Meanwhile, the base stations b_1 and b_3 jointly cover three users, so the weight of edge $\langle b_1, b_3 \rangle$ is 3, i.e., $A_{1,3} = A_{3,1} = 3$. However, the base stations b_1 and b_2 have not the direct communication relation, so $A_{1,2} = A_{2,1} = -1$.

Based on the communication relationship network of base stations, the Edge Server Placement (ESP) in the MEC can be expressed as an assignment of k out of m base stations (the set B) to deploy edge servers. The complete server deployment scheme is expressed as an m-dimensional binary vector $\mathbf{p} = (p_1,\ldots,p_i,\ldots,p_m)$, where $p_i \in \{0,1\}$. $p_i = 1$ means that i-th base station has been selected as the deployment location of the edge server, otherwise 0. Since the number of deployed servers is set to k, the server deployment scheme should satisfy the constraint $\sum_{i=0}^{m} p_i = k$. Taking the Fig. 1 for an example, the corresponding deployment scheme can be denoted as 0–1 vector $\mathbf{p} = (1,0,1,0,0,1,1,0)$, that is, the base stations b_1, b_3, b_6, and b_7 are selected as the deployment location of edge servers.

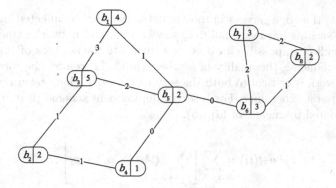

Fig. 2. The Base Station Communication Network for the Scenario in Fig. 1.

According to the above definitions, given a mobile edge network, the k Edge Server Placement (kESP) problem is to find a 0–1 vector $\mathbf{p}$ whose first-order norm (i.e. Manhattan norm) $\|\mathbf{p}\|_1$ is k, while satisfying the basic constraints in the actual application scenarios. In this paper, we investigate the optimization objectives of kESP problem from the following two aspects: network robustness and user coverage.

2.2 User Coverage

For the edge server network, providing mobile users with low-latency services is the most important requirement that needs to be considered. In general, more users being served means a wider coverage of the network, thus better meeting the access requirements of mobile users. For this reason, we consider the number of users severed by edge servers (user coverage for short) as an optimization objective. Given a base station b_i, the set of users covered by the base station is defined as $c(b_i)$, that is,

$$c(b_i) = \{u_k | a(b_i, u_k) = 1,\ u_k \in U\}. \tag{2}$$

Then, the user coverage of the base station b_i is $|c(b_i)|$. For a server deployment scheme $\mathbf{p}$, the set of mobile users served by edge servers is defined as $C(\mathbf{p})$, which can be formally represented as

$$C(\mathbf{p}) = \bigcup_{b_i \in B \wedge p_i = 1} c(b_i). \tag{3}$$

Finally, the user coverage of an edge server placement solution can be formulated as

$$coverage(\mathbf{p}) = |C(\mathbf{p})|. \tag{4}$$

2.3 Network Robustness

In actual network scenarios, if the robustness is low, the edge server network may not be able to achieve the desired effect for providing services for mobile users. In fact, the edge servers in mobile networks may be affected by different emergencies such as hardware failures, service loss, and network attacks. It can make it difficult or impossible for users to connect to the network of edge servers, significantly reducing the quality of service (QoS). Therefore, the robustness of the edge network is critical to both the service provider's long-term revenue and the mobile user's experience. For a server deployment scheme $\mathbf{p}$, its robustness can be calculated by means of Eq. (5).

$$robust(\mathbf{p}) = \sum_{i=0}^{m} \left[\sum_{j=i+1}^{m} (A_{i,j} \times p_j) \times p_i \right], \tag{5}$$

where m is the number of base stations. $A_{i,j}$ represents the number of users who have an accessing relationship with base stations b_i and b_j at the same time. p_i (or p_j) indicates whether an edge server has been deployed on the base station b_i (or b_j).

2.4 Robustness-Oriented kESP Optimization Model

According to the above description, to maximize the user coverage and network robustness at the same time, the robustness-oriented k edge server placement problem can be formulated as follows:

$$\text{maximize } CR(\mathbf{p}) = \sqrt{\omega_c \cdot (coverage(\mathbf{p}))^2 + \omega_r \cdot (robust(\mathbf{p}))^2}, \tag{6}$$

while satisfying the following constraints:

$$p_i \in \{0, 1\}, \tag{7}$$

$$\sum_{i=0}^{m} p_i = k. \tag{8}$$

The ω_c and ω_r are the adjustment weights of $coverage(\mathbf{p})$ and $robust(\mathbf{p})$ respectively, and their default values are set to $\omega_c = \omega_r = 0.5$. In this optimization model, constraint (7) describes the deployment decision of the base station b_i, and $p_i = 1$ means a server is deployed on the base station b_i. Otherwise, $p_i = 0$. Constraint (8) ensures that the total number of edge servers keeps the budget k.

3 The Heuristic Algorithm for kESP

In practical application scenarios, the distribution of base stations and mobile users is often not uniform. Therefore, from the perspective of edge server placement, priority should be given to deploying servers in high-density areas to achieve high user coverage. At the same time, in order to ensure the robustness of the deployment scheme, the deployment of servers needs to be gradually shifted from high-density regions to low-density regions. Based on the above ideas and the defined optimisation model, in this section we will discuss how to apply the heuristic strategy and heuristic information (the robustness increment) to gradually generate a server deployment scheme. In detail, the whole process can be divided into two stages: the first is to determine the initial base station for server placement, and the second is to select the candidate placement locations (i.e., base stations) for the subsequent server placement.

3.1 Selection of Initial Base Station

At first, we select an appropriate base station as the initial deployment scheme for edge servers. Then, the scheme can be gradually expanded until the number of the deployed edge servers (also known as budget) reaches k. For a given mobile network, the initial location of server placement should be as close as possible to the area where base stations and users are most densely distributed. Here, given a base station b_i $(1 \leq i \leq m)$, its area density can be measured by Eq. (9).

$$dense(b_i) = degree(b_i) \times A_{i,i}, \tag{9}$$

where $degree(b_i)$ represents the degree of base station b_i in the communication network, $A_{i,i}$ represents the number of mobile users within the coverage scope of base station b_i.

3.2 Gradual Generation of Server Deployment Scheme

Once the initial base station is picked out for deploying an edge server, the remaining $n-1$ base stations will be divided into two subsets: a set (N_{near}) of base stations adjacent to the server and a set (N_{far}) of base stations non-adjacent to the server, according to whether they have direct access to the initial base station. With the gradual generation of the server deployment scheme, the above two subsets will be updated in time. For example, if the base station b is selected as a new location for deploying an edge server, b is firstly removed from the subset N_{near} or N_{far}. Then, its neighbours $N(b)$ will be added into the subset N_{near}, and removed from the subset N_{far} simultaneously.

For the subset N_{near}, the base station with the largest $\triangle robust$ is selected as the candidate base station b_{near}. Given a base station $b_i \in N_{near}$, its $\triangle robust$ can be calculated by Eq. (10).

$$\triangle robust_{near}(b_i) = \sum_{b_j \in N(b_i)} A_{i,j} \times p_j \tag{10}$$

For the subset N_{far}, since it cannot directly calculate $\triangle robust$ for the base stations in it, we estimate the robustness increment $(\triangle robust)$ of base station $b_i \in N_{far}$ by the average number of the jointly-severed mobile users between b_i and the direct neighbours of b_i in the communication network. Specifically, the estimation is shown in Eq. (11).

$$\triangle robust_{far}(b_i) = \frac{\sum_{b_j \in N(b_i)} A_{i,j}}{2(|N(b_i)| + 1)}, \tag{11}$$

where $N(b_i)$ represents a set of direct neighbors of base station b_i, and $|N(b_i)|$ denotes the cardinality of the set $N(b_i)$. Similarly, the base station with the largest $\triangle robust$ in N_{far} is selected as the other candidate location, namely b_{far}, for server placement.

3.3   HR-kESP Algorithm

Given a base station communication network $G = (B, E)$ and the accessing relation matrix A, our heuristic algorithm for solving the robustness-oriented kESP problem is to build a deployment scheme for k edge servers. The corresponding pseudo code is presented in Algorithm 1.

Algorithm 1. HR-kESP

Inputs: (1) the base station communication network $G = (B, E)$;
  (2) the multi-objective optimization model in Section 2.4;
  (3) the server budget k.
Output: the edge server placement solution **p** with $\sum_{i=0}^{m} p_i = k$.
1. initialize the kESP solution **p** $= (0, 0, \ldots, 0)$, and set the subset of adjacent base stations $N_{\text{near}} = \emptyset$ and the subset of non-adjacent base stations $N_{\text{far}} = B$;
2. evaluate the area densities of all base stations in network G, and select the base station b_i with the largest density;
3. update **p** with $p_i = 1$ and $N_{\text{far}} = N_{\text{far}} - \{b_i\}$;
4. obtain b_i's neighboring base stations, denoted as $N(b_i)$;
5. update $N_{\text{near}} = N_{\text{near}} \cup N(b_i)$ and $N_{\text{far}} = N_{\text{far}} - N(b_i)$;
6. **while** $\sum_{i=0}^{m} p_i \neq k$ **do**
7.   evaluate the base stations in N_{near} and N_{far} respectively according to Equations (10) and (11), and select two candidate base stations from N_{near} and N_{far} respectively (i.e., b_{near} and b_{far}), whose $\triangle robust$ are the largest in the two subsets respectively;
8.   compare b_{near} with b_{far} to select the one with highest increment of robustness as $b^\star$;
9.   update **p** by $b^\star$, and update $N_{\text{near}} = N_{\text{near}} - \{b^\star\}$ if $b^\star$ is b_{near}, otherwise update $N_{\text{far}} = N_{\text{far}} - \{b^\star\}$;
10.   obtain $b^\star$'s neighboring base stations, denoted as $N(b^\star)$;
11.   update $N_{\text{near}} = N_{\text{near}} \cup N(b^\star)$ and $N_{\text{far}} = N_{\text{far}} - N(b^\star)$;
12. **end while**
13. **return** the final solution of edge server placement, that is, **p**;

The algorithm starts with an initial kESP solution **p** $= (0 \ldots 0)$, and the two base station subsets $N_{\text{near}} = \emptyset$ and $N_{\text{far}} = B$ (Line 1). Then, for each base station $b_i \in B$, its area density is calculated, and the base station with the largest value is selected as the initial location for deploying the edge server (Line 2). Next, update the kESP solution **p**, and remove the selected base station from N_{far} (Line 3). Meanwhile, move the neighbours of the initial base station from N_{far} to N_{near} (Lines 4 and 5). Subsequently, a while loop is used to expand the solution **p** for satisfying the server budget k (Lines 6–12). First, for subsets N_{near}

and N_{far}, evaluate each base station in them according to Eqs. (10) and (11), and select two candidate base stations b_{near} and b_{far}, whose $\triangle robust$ are the largest one in the two subsets respectively (Line 7). Then, by comparing b_{near} and b_{far}, the base station with the highest increment of robustness is used for deploying an edge server (Line 8). Next, the solution $\mathbf{p}$, the subsets N_{near} and N_{far} are updated accordingly (Lines 9–11). Finally, the expansion process (the while loop) is repeated until the number of deployed edge servers reaches k.

In the above HR-kESP algorithm, to select a better location of the base station for deploying the next edge server, we divide the candidate base stations into two different subsets according to their neighbouring relations to the existing edge servers and measure their robustness increments in two different ways. Simply, if the candidates are only considered from the base stations which are the direct neighbours of the deployed edge servers, a basic version of the algorithm can be obtained, i.e., a degraded version of the HR-kESP algorithm, which we denote here as HR-kESP$_0$. It should be pointed out that, the HR-kESP$_0$ algorithm also uses the information of robustness increment to guide the selection of deployment locations of edge servers. The key difference is that the HR-kESP$_0$ algorithm only considers the base stations in N_{near} as the candidate locations for deploying new edge servers, so only Eq. (10) is used in it to compute the robustness increments of candidate base stations.

4 Experimental Evaluation

4.1 Experimental Settings

To verify the performance of the proposed HR-kESP algorithm, the experiments are conducted on the widely-used EUA dataset [16] that contains the locations of 1465 real-world base stations within metropolitan Melbourne in Australia. The coverage range of each base station is randomly set from 450 to 700 m.

To comprehensively evaluate the performance of the algorithms for solving the kESP problem, the networks in different scenarios are taken into consideration in the experiments. Specifically, the following three kinds of experimental parameters are used and varied for representing different network environments: (1) the number of base stations m; (2) the server budget k; (3) the number of mobile users n. The detailed parameter settings are shown in Table 1.

For evaluating the robustness of the kESP solutions generated by different algorithms, **simulation-based network robustness evaluation** are designed in this paper. Given a network scenario, some servers are randomly shut down to simulate unexpected situations in the real-world network environment such as server hardware failure. In our current experiments, the default proportion of failed servers was set to 50%. Based on this simulation, the **User Survival Rate**, that is, the ratio of the number of the covered users after some server failures to the number of the covered users before server failures, is defined to measure the robustness of each kESP solution. Each experiment was repeated 100 times, and the results were reported in the form of average values.

Table 1. Experimental Parameter Settings.

Parameters	m	k	n
Group 1	5, 10, ..., 35	4	80
Group 2	20	1, 2, ..., 8	80
Group 3	20	4	20, 40, ..., 160

In the experiments, we selected the following four benchmark algorithms for the experimental and comparative analysis. Among them, Random and Greedy are the basic algorithms for the kESP problem. ESP-A is an approximation method proposed by Cui et al. [15]. HR-kESP$_0$ is the basic version of our HR-kESP algorithm.

– *Random*, randomly select k base stations to deploy edge servers;
– *Greedy*, select the top-k base stations to deploy edge servers according to the metric $CR(\mathbf{p})$;
– *ESP-A*, the method is proposed in [15] for finding an approximate solution for kESP problem;
– *HR-kESP$_0$*, based on the selection of the initial base station, the subsequent base stations used to deploy edge servers are selected only from the direct neighbours of the deployed servers until reaching the edge server budget k.

All the above algorithms were implemented in Java. In addition, all the experiments were performed on the PC with an Intel Core i5-10210U CPU processor, 16 GB RAM, and Windows 10 operating system.

4.2   Comparison on Different Numbers of Base Stations

By comparison with the four benchmark algorithms, Table 2 shows the algorithm's performance, which is the value of the comprehensive optimization objective $CR(\mathbf{p})$, under different numbers of base stations. It is easy to see that, our proposed HR-kESP and HR-kESP$_0$ algorithms have significant advantages over other benchmark algorithms. As the degraded version of HR-kESP, HR-kESP$_0$ also has good performance but is slightly worse than the HR-kESP. For different numbers of base stations, HR-kESP is the best one, which is 0.49% higher than the HR-kESP$_0$ on average. Compared with ESP-A and Greedy algorithm, its $CR(\mathbf{p})$ is 2.24% and 2.80% higher on average respectively. For the Random algorithm, the advantage of HR-kESP is as high as 139.95%.

To evaluate the network robustness, we used the simulate-based evaluation method, which randomly switches 50% of edge servers to the unavailable state. Then, the network robustness is represented by the proportion of mobile users who can still obtain services from the remaining edge servers, that is, the *user survival rate* defined above. As shown in Fig. 3, in terms of the user survival rate, the algorithms of HR-kESP and HR-kESP$_0$ have roughly the same effects and have significant advantages over ESP-A and Greedy algorithms. The Random

Table 2. The Performance $(CR(\mathbf{p}))$ of Algorithm with Different Numbers of Base Stations.

Algorithm\\m	5	10	15	20	25	30	35	40
Random	185.86	174.18	178.29	180.18	198.11	180.05	168.11	184.35
Greedy	239.02	350.36	393.46	429.27	472.57	466.55	495.26	514.79
ESP-A	228.72	345.92	398.10	436.62	478.43	476.33	506.70	522.41
HR-kESP$_0$	233.34	350.42	400.68	436.08	**493.53**	489.34	518.96	**531.86**
HR-kESP	**235.38**	**353.79**	**404.71**	**441.10**	491.34	**490.65**	**520.93**	530.33

algorithm performed the worst. Specifically, HR-kESP has an improvement of 0.26% compared with the HR-kESP$_0$ algorithm. For the other three benchmark algorithms (ESP-A, Greedy and Random), our algorithm improves on average by 1.85%, 2.40% and 18.81%, respectively.

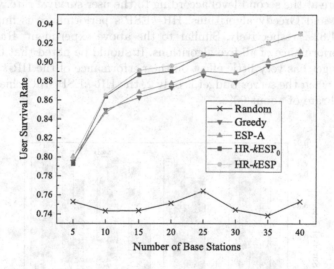

Fig. 3. User Survival Rate after Failures with Different Numbers of Base Stations.

4.3 Comparison on Different Server Budgets

Generally speaking, increasing the number of servers tends to improve user coverage and to make the network more robust. For different server budgets, both HR-kESP$_0$ and ESP-A algorithms have certain advantages when the server budget is less than 4. But as the budget keeps increasing, HR-kESP gradually shows significant improvement and performs best. As shown in Table 3, compared with HR-kESP$_0$, ESP-A and Greedy algorithms, HR-kESP has an average improvement of 0.42%, 2.00% and 3.15% respectively. The $CR(\mathbf{p})$ of the Random algorithm is still the worst. Overall, the performance of HR-kESP remains the best.

Table 3. The Performance ($CR(\mathbf{p})$) of Algorithm with Different Server Budgets.

Algorithm\k	1	2	3	4	5	6	7	8
Random	31.68	62.32	112.29	186.74	287.55	426.51	569.98	763.73
Greedy	52.02	105.01	237.88	433.37	693.00	929.68	1244.87	1597.61
ESP-A	**53.00**	**106.20**	246.50	441.28	704.48	933.31	1248.36	1570.70
HR-kESP$_0$	**53.00**	105.62	**251.01**	449.66	716.03	967.16	1272.51	1603.55
HR-kESP	51.91	105.49	250.49	**455.72**	**720.87**	**982.07**	**1286.63**	**1621.23**

In terms of network robustness, the user survival rate tends to increase gradually as the server budget increases, and tends to be stable when the value of the budget (k) is large. As the experimental results shown in Fig. 4, the performance of HR-kESP and HR-kESP$_0$ algorithms is almost the same, where the former is just 0.29% higher than the latter. In addition, the ESP-A and Greedy algorithms are at the second level according to the user survival rate. Compared with ESP-A and Greedy algorithms, HR-kESP's performance is improved by 1.32% and 1.82% respectively. Similar to the above experiment, Random has the worst performance of all five algorithms. It should be noted that the change in server budget has very little effect on the performance of the HR-kESP algorithm. Even when the server budget is only 2, the HR-kESP still achieves a high user survival rate of about 90%.

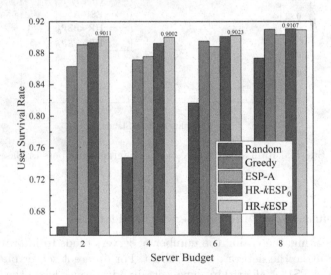

Fig. 4. User Survival Rate after Failures with Different Server Budgets.

4.4   Comparison on Different Numbers of Users

In general, the $CR(\mathbf{p})$ of each algorithm shows an upward trend as the number of users continues to increase. It is not difficult to see from Table 4 that, the HR-kESP algorithm performs optimally in almost all cases of the numbers of users. Even when the number of users is 20 and 140, HR-kESP as the second is very close to the first HR-kESP$_0$ algorithm. In other cases, it significantly outperforms the four benchmark algorithms (namely HR-kESP$_0$, ESP-A, Greedy and Random) on the $CR(\mathbf{p})$ by 0.42%, 2.39%, 4.31% and 155.01%, respectively.

Table 4. The Performance ($CR(\mathbf{p})$) of Algorithm with Different Numbers of Users.

Algorithm\\n	20	40	60	80	100	120	140	160
Random	43.65	91.16	141.12	188.31	219.96	253.56	309.23	350.02
Greedy	118.34	221.56	323.57	434.84	533.93	652.92	737.79	849.94
ESP-A	119.94	226.03	329.31	440.27	546.91	666.93	751.18	868.78
HR-kESP$_0$	**123.61**	231.40	335.98	446.82	554.25	671.64	**776.13**	880.68
HR-kESP	122.76	**234.73**	**337.69**	**447.61**	**558.88**	**675.64**	771.60	**890.16**

As the number of users increases, as shown in Fig. 5, the trend of user survival rates of all algorithms remains stable. The average user survival rates of HR-kESP and HR-kESP$_0$ algorithms are around 89.7%. Those of the Greedy and ESP-A algorithms are about 87.5%. The corresponding value of the Random algorithm is maintained at about 74.5%. Apparently, HR-kESP and HR-kESP$_0$ algorithms always keep an advantage, and HR-kESP has an improvement of

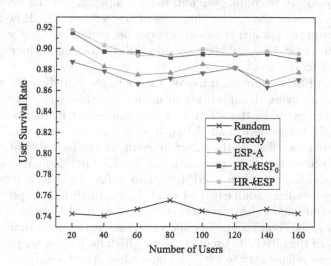

Fig. 5. User Survival Rate after Failures with Different Numbers of Users.

0.29% compared to HR-kESP$_0$ algorithm. At the same time, compared with the other three benchmark algorithms (i.e., ESP-A, Greedy, and Random), the improvements are 2.09%, 2.85%, and 20.72%, respectively.

5   Related Work

Due to the complexity of application scenarios and the diversity of network service requirements, the ESP problem in different network environments is often studied from different aspects. In recent years, many scholars have carried out research on server deployment cost [11], server provider's profits [5,9], average response latency and load balancing [17], network robustness [14,15] etc.

For the cost of deploying servers, Ren et al. [11] focused on minimizing the number of deployed edge servers while meeting user response latency constraints. However, service providers often would like an accurate profit model to further estimate the cost and revenue of servers. Therefore, some scholars incorporate the server's energy consumption (cost) and service quality (revenue) into the profit model and maximize profits by weighing cost and revenue [5,9]. Different from service providers, mobile users often pay more attention to the QoS, such as the average response latency of users. Since the response speed of network services is related to network performance, the balanced distribution of load can make full use of server resources and effectively reduce the average latency. Therefore, in the QoS-oriented research, user average response latency and load balancing indicators are usually jointly considered. Typically, Chen et al. [17] presented a preference-aware edge server placement approach that offers better workload distribution in terms of both minimizing query latency and balancing the load of edge servers. At present, there is relatively little research on network robustness. Cui et al. [14] first discussed the robustness-oriented k Edge Server Placement (kESP) problem. To maximize the anti-failure capability of the network, their strategy is to increase the overlapping area of the server range. However, server deployment requires a proper trade-off between network robustness and user coverage [15], otherwise, the total coverage area will be too small, thereby reducing the number of mobile users served.

For a given ESP optimization model, the methods of finding the exact solution by directly solving the optimization model are widely used in the research community. For example, the solver is a very convenient model-solving tool to solve a specific mathematical model such as a (mixed) integer programming model. The solution effect of the solver depends on the formal definition of the problem. According to the characteristics of the ESP problem, Wang et al. [18] proved that the ESP problem is NP-hard and expressed it as a mixed integer programming problem. Similarly, Cui et al. [14] formalized the problem as an integer programming model. However, these specific models make it difficult to accurately describe complex application scenarios. At the same time, due to the low efficiency of the solver, its time cost is very high for large-scale problems [15]. Therefore, some scholars try to carry out equivalent transformations of problems and use existing knowledge to solve new problems, such as the cooperative game problem [19] in game theory and the shortest path problem [20] in graph theory. However, these approaches are less general.

Although the exact solution method can already obtain a good solution, it is easy to be trapped in a local optimum. Therefore, by adding randomization factors, there is a chance to find the optimal solution in the global search space. The meta-heuristic algorithm represented by swarm intelligence is a typical randomization-based solution method, which realizes multi-point optimization in the global search space by simulating certain behaviours of creatures in nature. In recent years, scholars have integrated the K-means algorithm and Tabu search into the genetic algorithm (GA) [17] or ant colony optimization (ACO) [21] to solve the ESP problem. In addition, some other classic search algorithms, such as particle swarm optimization algorithm [5,8], hill climbing method, simulated annealing [22], whale optimization [23] have also been applied to this problem. The meta-heuristic algorithm has the ability of global search and simple reuse of methods. However, it is necessary to readjust the parameters, and even adjust the specific search strategy. Recently, reinforcement learning (RL) has performed well on the combinatorial optimization problem (COP) [24]. As a COP problem, edge server placement is similar to a serialized decision problem, which is to find suitable server deployment locations one by one. Therefore, Mazloomi et al. [25] proposed a new reinforcement learning framework, which modelled the edge server placement as a Markov decision process, and modelled the state space, action space and penalty function. Similarly, RL methods also have certain limitations, including changeable action spaces, difficulties in defining penalty functions for multi-objective problems, and the risk of state space dimension explosion. Most importantly, all randomization-based search methods cannot ignore the high time cost of optimizing the search.

To avoid the heavy computation overhead and the trap of local optimum, in this paper, we propose a heuristic algorithm HR-kESP to solve the robustness k edge server placement problem. In our algorithm, the increment of robustness is used as the heuristic information to guide the search process to maximize network robustness and user coverage.

6   Conclusion and Future Work

In this paper, we formally define two optimization objectives (i.e., network robustness and user coverage) and a comprehensive optimization model for the robustness-oriented k Edge Server Placement (kESP) problem. To efficiently find the server deployment scheme, we designed a robustness-oriented heuristic algorithm named HR-kESP. In the algorithm, according to the distribution of base stations and users in the network environment, the base station close to the user-intensive area is first selected as the initial location for deploying the edge server. Subsequently, the candidate with the largest robustness increase is selected from the subset of adjacent base stations and the subset of non-adjacent base stations respectively, and the optimal one of the two candidate base stations is selected to expand the server deployment scheme. The above expansion process continues until the number of deployed edge servers reaches the budget constraint. To verify the effectiveness of the proposed HR-kESP algorithm, the comparative experimental analysis is conducted on the widely-used EUA dataset. The experimental results confirm that, in most cases, the HR-kESP algorithm is significantly better than all other benchmark algorithms in both two evaluation metrics.

At present, the proposed HR-kESP algorithm in this paper is only verified and compared in the small-scale network environment, and the scenarios in the large-scale MEC environment can be further investigated. Furthermore, the current robustness evaluation method only covers simple server failure cases. In the ongoing study, the simulation-based evaluation can be improved for handling some more complex situations.

Acknowledgement. We are grateful to Guangming Cui and Qiang He for providing the EUA dataset and the implementation of their ESP-A algorithm. This work is supported in part by the National Natural Science Foundation of China (Grant No. 62172195), the Science Foundation of Jiangxi Educational Committee (Grant No. GJJ180276), and the Graduate Innovation Project of Jiangxi Province (No. YC2022-S459).

References

1. Jonsson, P., Carson, S., Davies, S., et al.: Ericsson Mobility Report. Ericsson, Stockholm (2022)
2. Hu, Y., Patel, M., Sabella, D., et al.: Mobile edge computing-a key technology towards 5G. ETSI White Paper **11**(11), 1–16 (2015)
3. Satyanarayana, M.: The emergence of edge computing. Computer **50**(1), 30–39 (2017)
4. Mao, Y., You, C., Zhang, J., et al.: A survey on mobile edge computing: the communication perspective. IEEE Commun. Surv. Tutor. **19**(4), 2322–2358 (2017)
5. Li, Y., Zhou, A., Ma, X., et al.: Profit-aware edge server placement. IEEE Internet Things J. **9**(1), 55–67 (2022)
6. Tiwary, M., Puthal, D., Sahoo, K.S., et al.: Response time optimization for Cloudlets in mobile edge computing. J. Parallel Distrib. Comput. **119**, 81–91 (2018)
7. Jin, X., Gao, F., Wang, Z., et al.: Optimal deployment of mobile Cloudlets for mobile applications in edge computing. J. Supercomput. **78**, 7888–7907 (2022)
8. Li, Y., Wang, S.: An energy-aware edge server placement algorithm in mobile edge computing. In: Proceedings of 2018 IEEE International Conference on Edge Computing (EDGE 2018), pp. 66–73 (2018)
9. Cong, P., Zhou, J., Li, L., et al.: A survey of hierarchical energy optimization for mobile edge computing: a perspective from end devices to the cloud. ACM Comput. Surv. **53**(2), 1–44 (2020)
10. Bhatta, D., Mashayekhy, L.: Cost-aware Cloudlet placement in edge computing systems. In: Proceedings of the 4th ACM/IEEE Symposium on Edge Computing (SEC 2019), pp. 292–294 (2019)
11. Ren, Y., Zeng, F., Li, W., et al.: A low-cost edge server placement strategy in wireless metropolitan area networks. In: Proceedings of the 27th International Conference on Computer Communication and Networks (ICCCN 2018), pp. 1–6 (2018)

12. Mao, C., Zhao, Z.: Predicting QoS for cloud services through prefilling-based matrix factorization. In: Proceedings of the 7th International Conference on Advanced Cloud and Big Data (CBD 2019), pp. 25–30 (2019)
13. Chen, J., Mao, C., Song, W.W.: QoS prediction for Web services in cloud environments based on swarm intelligence search. Knowl.-Based Syst. **259**, 110081:1–110081:16 (2023)
14. Cui, G., He, Q., Xia, X., et al.: Robustness-oriented k edge server placement. In: Proceedings of the 20th IEEE/ACM International Symposium on Cluster, Cloud and Internet Computing (CCGRID 2020), pp. 81–90 (2020)
15. Cui, G., He, Q., Chen, F., et al.: Trading off between user coverage and network robustness for edge server placement. IEEE Trans. Cloud Comput. **10**(3), 2178–2189 (2022)
16. Lai, P., He, Q., Abdelrazek, M., et al.: Optimal edge user allocation in edge computing with variable sized vector bin packing. In: Proceedings of the 16th International Conference on Service-Oriented Computing (ICSOC 2018), pp. 230–245 (2018)
17. Chen, Y., Lin, Y., Zheng, Z., et al.: Preference-aware edge server placement in the Internet of Things. IEEE Internet Things J. **9**(2), 1289–1299 (2022)
18. Wang, S., Zhao, Y., Xu, J., et al.: Edge server placement in mobile edge computing. J. Parallel Distrib. Comput. **127**, 160–168 (2019)
19. Cao, K., Li, L., Cui, Y., et al.: Exploring placement of heterogeneous edge servers for response time minimization in mobile edge-cloud computing. IEEE Trans. Industr. Inf. **17**(1), 494–503 (2021)
20. Xiang, H., Xu, X., Zheng, H., et al.: An adaptive Cloudlet placement method for mobile applications over GPS big data. In: Proceedings of 2016 IEEE Global Communications Conference (GLOBECOM 2016), pp. 1–6 (2016)
21. Guo, F., Tang, B., Zhang, J.: Mobile edge server placement based on meta-heuristic algorithm. J. Intell. Fuzzy Syst. **40**(5), 8883–8897 (2021)
22. Kasi, S.K., Kasi, M.K., Ali, K., et al.: Heuristic edge server placement in industrial Internet of Things and cellular networks. IEEE Internet Things J. **8**(13), 10308–10317 (2021)
23. Asghari, A., Azgomi, H., Darvishmofarahi, Z.: Multi-objective edge server placement using the whale optimization algorithm and game theory. Soft Comput. 1–15 (2023)
24. Mazyavkina, N., Sviridov, S., Ivanov, S., et al.: Reinforcement learning for combinatorial optimization: a survey. Comput. Oper. Res. **134**, 105400:1–105400:15 (2021)
25. Luo, F., Zheng, S., Ding, W., et al.: An edge server placement method based on reinforcement learning. Entropy **24**(3), 317, 1–14 (2022)

A Hybrid Kernel Pruning Approach
for Efficient and Accurate CNNs

Xiao Yi[1]([envelope]), Bo Wang[1], Shengbai Luo[1], Tiejun Li[1], Lizhou Wu[1],
Jianmin Zhang[1], Kenli Li[2], and Sheng Ma[1]([envelope])

[1] College of Computer Science and Technology, National University of Defense
Technology, Changsha, China
{yixiao,bowang,luoshengbai,tjli,lizhou.wu,jmzhang,masheng}@nudt.edu.cn
[2] Hunan University, Changsha, China
lkl@hnu.edu.cn

Abstract. To reduce the overhead of neural network training and inference, several techniques have been widely used to prune neural network models. Pruning algorithms can significantly reduce the number
of parameters in the model, which in turn reduces the amount of computation required during model training and inference. Currently, the most
popular pruning algorithm is the structured pruning algorithm, which
prunes the model at the kernel level. Researchers usually use norm-based
criteria to determine which kernels to prune. While this type of algorithms works well, there are some shortcomings. First, the effectiveness
of the norm-based pruning algorithm lacks support from mathematical
theories. Second, this pruning algorithm requires certain conditions to
work well. To address these shortcomings, we propose a novel kernel
pruning algorithm. Based on the observation that convolution kernels
act as feature extractors, we design a functional similarity-based pruning algorithm as the criteria for selecting pruned kernels. Our experimental results show that when pruning ResNet with a high pruning ratio,
this algorithm can obtain a sparse model with high accuracy. Moreover,
when combined with the norm-based pruning algorithm, our functional
similarity-based pruning algorithm can produce a more accurate model
than either algorithm alone, even at the same pruning ratio.

Keywords: Kernel Pruning Algorithm · Functional-Similarity-Based
Pruning · Hybrid Pruning

1 Introduction

Deep learning pruning algorithms can be categorized into two types: unstructured pruning algorithms [1] [2] [12] and structured pruning algorithms [10] [15]
[8]. Unstructured pruning algorithms prune models at the parameter granularity. First, they select the pruned parameters according to a particular pruning

---

X. Yi and B. Wang—Have contributed equally to this research.

criterion. Then they generate a model with sparse parameter matrices. Accelerating the computation of the pruned model in the unstructured pruning algorithms requires support from specific sparse matrix operation libraries or hardware accelerators. In contrast, structural pruning algorithms prune models at the convolution kernel or channel levels. The computation of the pruned model does not require support from special software or hardware. Meanwhile, based on whether the pruning algorithm uses training data to select the pruned convolution kernel, we can divide convolution kernel pruning algorithms into data-based and non-data-based pruning algorithms. Compared to data-based pruning algorithms, non-data-based pruning algorithms are more effective, because using training data in data-based pruning adds extra computational overhead.

Liu et al. [9] use the scaling factor of the batch normalization layer in the convolutional neural network to select which convolutional kernel to be pruned. First, it calculates the L1 norm of the scaling factor in the batch normalization layer and then prunes the kernel channels with small L1 norm of the scaling factor. Experimental results show that this method can reduce the model's computation cost by up to 95%. Luo et al. [11] selected the pruned kernel based on the channel information from the next layer. This method reduces the computational overload by 67.99% and compresses the model by 16.63 times when pruning the VGG16 model. Additionally, It can reduce the calculation amount by half when pruning the ResNet50 model. Li et al. [7] assumed that convolution kernels with small L1 norms have a low impact on the output of the model, and therefore can be pruned. When adopting this pruning algorithm, the computation of VGG and ResNet models can be reduced by 10% and 30% without compromising the model accuracy. He et al. [5] prunes the convolution kernel based on the L2 norm of the kernel. In other studies, the pruned kernel is usually not updated after pruning. However, the pruned convolution kernel in the [5] can still be updated. This pruning algorithm reduces the computational amount of the ResNet101 model by 42%. Zhuo et al. [17] utilizes spectral clustering to perform correlation analysis on convolution kernels. They grouped similar kernels into clusters and select pruned convolution kernels from each cluster. This algorithm also adopted a soft pruning strategy, where the pruned kernel can still be updated. Experimental results confirmed the effectiveness of this method.

Data-dependent kernel pruning algorithms typically require the participation of training data, which results in a lot of computational overhead. In comparison, non-data-dependent kernel pruning algorithms are more efficient. The norm-based kernel pruning algorithm is simple and effective. Many studies utilize this algorithm to prune models. However, this algorithm requires the distribution of kernel's norm to satisfy two conditions. First, the deviation of the kernel's norm should be significant to make it easy to separate kernels based on the difference among their norms. Second, the kernel's norm should be small enough, even close to 0. However, previous research [6] has shown that these conditions are not always met. Under such circumstances, a new convolution kernel pruning technology is necessary. Moreover, researchers typically use a single criterion to prune the model, which means the kernel that has little effect on the model's result can't be recognized by this single-criterion-based pruning algorithm. Therefore, it is necessary to explore a hybrid kernel pruning algorithm that can prune the model to the possible greatest extent. We study the aforementioned problems in this paper. Our main contributions are as follows.

(1) We propose a distance-based pruning algorithm, which is a pruning algorithm based on functional similarity. The proposed method can achieve high sparsity in the ResNet model while maintaining a high level of accuracy.
(2) We propose a hybrid pruning algorithm, which combines distance-based and norm-based pruning algorithms. This method can maintain the higher accuracy of the model.
(3) Experimental results show that, compared with the kernel pruning algorithm based on a single criterion, the hybrid kernel pruning algorithm can obtain higher model accuracy in most cases under the same pruning ratio.

2   Functional-Similarity-Based Kernel Pruning Algorithm

In this section, we present a pruning algorithm that leverages the kernel functional similarity. The conventional approach of determining the importance of a convolution kernel based on its norm requires certain conditions to be met, such as a large variance in norm distribution for the kernels in each layer. However, if the norm of the convolution kernel in a certain layer is nearly identical, it is difficult to select a pruned convolution kernel. In such cases, a new kernel pruning criterion must be designed and adopted.

In convolutional neural networks, the convolution kernel serves as a feature extractor for images. Different kernels extract different features from pictures. Kernels near the input layer extract low-dimensional features and kernels near the output layer extract high-dimensional features. In a certain layer, there may be numerous convolution kernels, and some of the kernels have similar functions. Consequently, functional-similar kernels can be pruned without affecting the model's accuracy. We characterize the functional similarity by measuring the distance between kernels. Specifically, kernels that are close to each other are considered to share the same functionality.

2.1   Distance-Based Kernel Pruning Algorithm

We employ Euclidean distance and Manhattan distance to calculate kernel similarity in the distance-based pruning algorithm, which is detailed in Table 1. During the training process, when kernels require pruning, we first calculate the Euclidean distance or Manhattan distance between kernels within the same layer. Next, we compute the cumulative sum of distances between a given kernel and all other kernels in the same layer. Then we sort the cumulative sum of distances for all kernels within the layer. A smaller sum of distances indicates greater functional similarity between a kernel and its peers. The kernel with the smallest distance sum is more likely to be pruned. By setting the pruning ratio for each layer in advance, the number of pruned kernels in each layer is determined. We primarily present pruning experiments conducted on the ResNet and VGG models.

Table 1. Distance-based Convolution Kernel Pruning Algorithm

Algorithm
Input: training data : X
1: Given: distance pruning ratio Pd
2: Initialize: model parameter W
3: for epoch = 1; epoch != epochmax; epoch + + do
4:   Update the model parameter W
5:   for i = 1; i != levelnumber; i++ do
6:     Find Ni * Pd filters that are satisfy distance-based pruning strategy
7:     Zeroize selected filters
8:   End for
9: End for
9: Obtain the compact model W* from W
10: Output :The compact model and its parameters W* and model accuracy

2.2 Hybrid Kernel Pruning Algorithm

The norm-based kernel pruning algorithm is widely used and can achieve high model sparsity. The distance-based kernel pruning algorithm proposed in this paper can also achieve good model sparsity. Therefore, we propose a hybrid pruning algorithm that combines these two algorithms to prune models. The hybrid pruning algorithm is shown in Fig. 1. Table 2 describes this algorithm. In this paper, we mainly focus on conducting pruning experiments on the ResNet and VGG models using our proposed hybrid pruning algorithm and comparing the results with those of the single criterion kernel pruning strategy.

3 Experimental Method

The experiments in this paper are mainly performed using Pytorch [13]. We train the ResNet [4] and VGG models [14] on the CIFAR10 dataset. When training the ResNet model, we set the training duration to 300 epochs and set the batch size to 16. When training the VGG model, we set the training duration to 160 epochs and set the batch size to 160. We mainly adopt the kernel pruning algorithm based on Euclidean distances and Manhattan distances to prune the model during the training process. However, due to the overfitting phenomenon of the VGG model, we did not analyze the relationship between the VGG model

38   X. Yi et al.

Table 2. Convolution kernel pruning algorithm based on hybrid strategy

Algorithm
Input: training data : X
1: Given: norm pruning ratio Pn and distance pruning ratio Pd
2: Initialize: model parameter W
3: for epoch = 1; epoch != epochmax; epoch + + do
4:    Update the model parameter W
5:    for i = 1; i != levelnumber; i++ do
6:        Find Ni * Pn filters that are satisfy Norm-based pruning strategy
7:        Find Ni * Pd filters that are satisfy distance-based pruning strategy
8:        Zeroize selected filters
9:    End for
10: End for
11: Obtain the compact model W* from W
12: Output: The compact model and its parameters W* and model accuracy

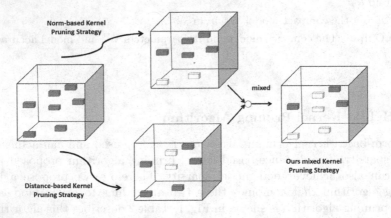

Fig. 1. Schematic diagram of hybrid convolution kernel pruning technology

training accuracy and pruning ratio. The experimental configuration is shown in Table 3. In our experiment, we use the same pruning ratio for each layer. To verify the effectiveness of the distance-based kernel pruning algorithm, experiments are conducted with a pruning ratio ranging from 0.1 to 0.9. That allows us to analyze the relationship between the model accuracy and the pruning ratio. We choose the Euclidean distance-based pruning algorithm as part of the hybrid kernel pruning algorithm. The hybrid pruning ratio settings are shown in Table 4. We adopt PFEC in [7] as the norm-based pruning algorithm.

Table 3. Distance-based pruning algorithm's pruning ratio setting

model	Distance strategy	Convolution kernel pruning ratio
ResNet	Euclidean/Manhattan	0.1–0.9
VGG	Euclidean/Manhattan	0.1–0.9

Table 4. Hybrid pruning algorithm's pruning ratio settings in ResNet model

model	Distance strategy	Distance-based pruning ratio	Norm-based pruning ratio
ResNet20	Euclidean	0.1–0.5	0.1–0.8
VGG16	Euclidean	0.1–0.5	0.1–0.8

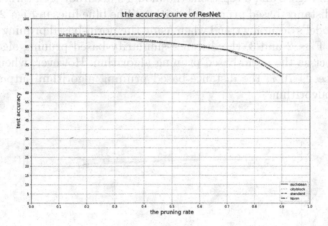

Fig. 2. The model's accuracy and sparsity during ResNet model training. (Color figure online)

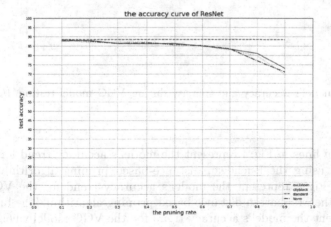

Fig. 3. The model's accuracy and sparsity during ResNet model testing. (Color figure online)

4   Experimental Results

4.1   Experimental Results for the Distance-Based Pruning Algorithm

Figures 2 and 3 show that the accuracy of the ResNet model varies with different pruning ratios when the distance-based kernel pruning algorithm is applied. The accuracy of the unpruned model serves as the baseline. Figure 2 shows the model's accuracy trend when training the model with different pruning ratios.

Figure 3 shows the model's accuracy trend when testing the model with different pruning ratios. The green lines in Figs. 2 and 3 represent the model's accuracy trend when using the Euclidean distance-based pruning algorithm. The yellow lines in Figs. 2 and 3 represent the model's accuracy trend when using the Manhattan distance-based pruning algorithm. The blue lines in Figs. 2 and 3 represent the model's accuracy trend when using the norm-based pruning algorithm. As shown in Figs. 2 and 3, the distance-similarity-based pruning algorithm has a similar effect as the norm-based pruning algorithm. However, when the pruning ratio exceeds 70%, the distance-based pruning algorithm outperforms the norm-based algorithm.

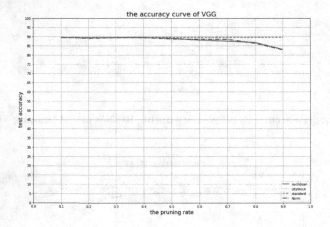

Fig. 4. The model's accuracy and sparsity during VGG model testing. (Color figure online)

The green lines in Fig. 4 represent the model's accuracy trend for the VGG model when using the Euclidean distance-based pruning algorithm. The yellow lines in Fig. 4 represent the model's accuracy trend for the VGG model when using the Manhattan distance-based pruning algorithm. The blue lines in Fig. 4 represent the model's accuracy trend for the VGG model when using the norm-based pruning algorithm. As shown in Fig. 4, the distance-based pruning algorithm performs as well as the norm-based pruning algorithm.

4.2   Experimental Results for the Hybrid Pruning Algorithm

Results on ResNet Model. Figure 5 shows the accuracy of the ResNet model when using the hybrid pruning algorithm with different hybrid pruning ratios. The x-axis of the figure is the hybrid ratio of the pruning algorithm, and the y-axis is the accuracy of the model. In the hybrid ratio, the first ratio represents the distance-based pruning ratio and the second ratio represents the norm-based pruning ratio. The total pruning ratio is the sum of the distance-based pruning ratio and the norm-based pruning ratio.

As shown in Fig. 5(a), when the total pruning ratio is 0.2, the pruning algorithm with a hybrid ratio of (0,0.2) achieves the highest accuracy. (0,0.2) means that only the norm-based pruning algorithm is used. So, in this case, the norm-based pruning algorithm obtained the highest accuracy of the model.

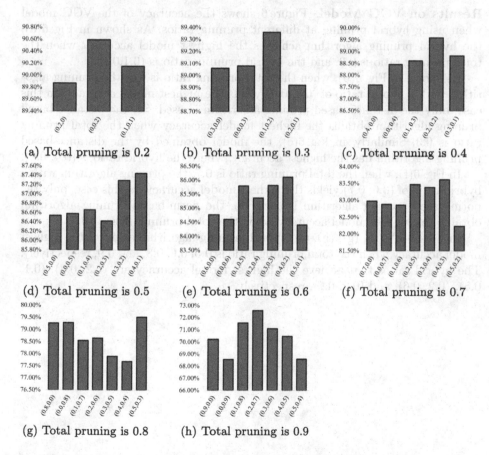

(a) Total pruning is 0.2   (b) Total pruning is 0.3   (c) Total pruning is 0.4

(d) Total pruning is 0.5   (e) Total pruning is 0.6   (f) Total pruning is 0.7

(g) Total pruning is 0.8   (h) Total pruning is 0.9

Fig. 5. ResNet model's accuracy under hybrid kernel pruning algorithm. The pair in X-axis represent hybrid pruning ratio, the first represent distance-based pruning ratio, the second represent norm-based pruning ratio. The Y-axis represent model's accuracy.

As shown in Fig. 5(b), when the total pruning ratio is 0.3, the model accuracy of the pruning algorithm is the highest when the hybrid ratio is (0.1, 0.2). In this case, the hybrid pruning algorithm is used, in which the distance-based pruning ratio is 0.1 and the norm-based pruning ratio is 0.2.

As shown in Fig. 5(c), the hybrid pruning algorithm achieves the highest model accuracy when the total pruning ratio is 0.4. In this case, the hybrid ratio is (0.2, 0.2). Similarly, when the total pruning ratio is 0.5, 0.6, 0.7, 0.8, and 0.9, the hybrid pruning algorithm achieves the highest model accuracy.

Experimental results show that a single pruning algorithm cannot achieve the highest model accuracy in most cases when different ratios of kernel pruning are performed on ResNet. The hybrid pruning algorithm can obtain a higher model accuracy than the single criterion pruning algorithm in most cases.

Results on VGG Model. Figure 6 shows the accuracy of the VGG model when using hybrid pruning at different pruning ratios. As shown in Fig. 6(a), the hybrid pruning algorithm achieves the highest model accuracy when the total pruning ratio is 0.2, and the hybrid pruning ratio is (0.1,0.1).

As shown in Fig. 6(b), when the total pruning ratio is 0.3, the pruning algorithm with a hybrid ratio of (0.3, 0.0) yields the highest model accuracy. In this case, only the distance-based pruning algorithm is used. So, the distance-based pruning algorithm obtains the highest model accuracy when the total pruning ratio is 0.3. Similarly, in Fig. 6(c), the model obtained by the distance-based pruning algorithm has the highest accuracy, where the hybrid ratio is (0.4, 0.0).

In Fig. 6(f), when the total pruning ratio is 0.7, the pruning algorithm with a hybrid ratio of (0.0, 0.7) yields the highest model accuracy. In this case, only the norm-based pruning algorithm is used. So, the norm-based pruning algorithm obtains the highest model accuracy when the total pruning ratio is 0.7.

In Figs. 6(d), (e), (g), (h), the hybrid pruning algorithm achieves the highest model accuracy when the total pruning ratio is 0.5, 0.6, 0.8, and 0.9, respectively. The hybrid ratios that achieve the highest model accuracy are (0.2, 0.3), (0.1, 0.5), (0.2, 0.6), and (0.2, 0.7), respectively.

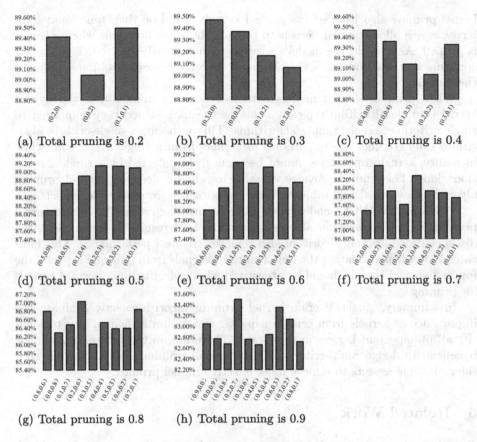

Fig. 6. VGG model's accuracy under hybrid kernel pruning algorithm. The pair in X-axis represent hybrid pruning ratio, the first represent distance-based pruning ratio, the second represent norm-based pruning ratio. The Y-axis represent model's accuracy.

Experimental results show that both the hybrid pruning algorithm and the distance-based pruning algorithm can achieve higher model accuracy when different ratios of kernel pruning are applied to VGG.

5   Analysis

In our experiment, the distance-based kernel pruning algorithm is first used to prune models, and high-sparse models were obtained. The distance-based and norm-based kernel pruning algorithms are both effective when pruning the ResNet model with a low pruning ratio. However, when the pruning ratio is high, the distance-based kernel pruning algorithm is found to be more effective than the norm-based kernel pruning algorithm. This is because the distance-based kernel pruning algorithm selects pruned kernels based on their functional similarity, which allows similar kernels to provide the same functions when a kernel is pruned. As a result, the model's accuracy is not greatly affected. This finding confirms that the norm of a kernel cannot directly represent its importance to the model.

When pruning ResNet models with increasing pruning ratios, the hybrid kernel pruning algorithm typically achieves the highest accuracy compared to single-criterion kernel pruning algorithms. This is because single-criteria algorithms only consider the importance of kernels from a single dimension, while in reality, kernel importance cannot be accurately reflected by a single dimension alone. This finding provides valuable insight for designing kernel pruning algorithms, which should evaluate kernel importance from multiple aspects.

When pruning VGG models with different pruning ratios, the hybrid kernel pruning algorithm does not always achieve the best results. In some cases, the distance-based pruning algorithm also achieves the best performance. This indicates that when evaluating the importance of kernels from multiple aspects, the functionality of kernels should be prioritized when selecting unimportant kernels for pruning.

In summary, single-criterion kernel pruning algorithms only evaluate the importance of kernels from certain aspects, which limits their ability to identify all unimportant kernels in the model. Therefore, in the future, it would be beneficial to design multi-criterion hybrid kernel pruning algorithms that consider multiple aspects to achieve more effective model pruning.

6   Related Work

To reduce the computational and storage overhead during the neural network training and inference process, many researchers have developed model sparsification algorithms to prune deep learning models. Pruning algorithms greatly reduce the number of parameters in deep learning models, thus reducing the computational overload during model training and inference process. This ultimately reduces the model's computational cost and energy consumption. Many pruning algorithms perform fine-grained pruning on deep learning models. For example, Han et al. [3] proposed to prune weights based on their importance, where weights with small values are considered to have little effect on the accuracy of models and can be pruned. After pruning, the remaining weights are fine-tuned to restore the model's accuracy. This method achieved 89% model sparsity in the AlexNet model. Zhu et al. [16] used an automatic pruning algorithm that prunes unimportant weights while minimizing retraining time. Li et al. [7] pruned trained models based on the L1 norm-based kernel pruning algorithm, and then retrained the pruned network to restore the model's accuracy. This pruning algorithm can effectively reduce the computation cost and does not require any support from sparse matrix computing libraries or dedicated accelerators. The results of this study showed that pruning algorithms can reduce the inference overhead by 34% to 38% and will not affect the model's accuracy.

7 Conclusion

In this article, we propose a convolution kernel pruning algorithm based on functional similarity. The Euclidean distance and Manhattan distance are primarily used as measures of functional similarity between kernels. Compared to the norm-based pruning algorithm, the distance-based pruning algorithm can achieve a higher accuracy pruning model when model sparsity is greater than 70%. We also introduce a hybrid pruning algorithm that combines the norm-based pruning algorithm and the distance-based pruning algorithm. The experimental results show that, under the same model sparsity, the hybrid pruning algorithm can achieve a higher accuracy than a single-criterion pruning algorithm. In the future, it is necessary to explore the optimal pruning ratio for different convolutional layers, which will lead to more efficient pruning algorithms.

Acknowledgement. This work is supported in part by the National Key R/&D Project No. 2021YFB0300300, the NSFC (62172430), the NSF of Hunan Province 2021JJ10052, the STIP of Hunan Province 2022RC3065, and the Key Laboratory of Advanced Microprocessor Chips and Systems.

References

1. Diffenderfer, J., Kailkhura, B.: Multi-prize lottery ticket hypothesis: finding accurate binary neural networks by pruning a randomly weighted network. arXiv preprint arXiv:2103.09377 (2021)
2. Frankle, J., Carbin, M.: The lottery ticket hypothesis: finding sparse, trainable neural networks. arXiv preprint arXiv:1803.03635 (2018)
3. Han, S., Pool, J., Tran, J., Dally, W.: Learning both weights and connections for efficient neural network. In: Advances in Neural Information Processing Systems 28 (2015)
4. He, K., Zhang, X., Ren, S., Sun, J.: Deep residual learning for image recognition. In: Proceedings of the IEEE Conference on Computer Vision and Pattern Recognition, pp. 770–778 (2016)
5. He, Y., Kang, G., Dong, X., Fu, Y., Yang, Y.: Soft filter pruning for accelerating deep convolutional neural networks. arXiv preprint arXiv:1808.06866 (2018)
6. He, Y., Liu, P., Wang, Z., Hu, Z., Yang, Y.: Filter pruning via geometric median for deep convolutional neural networks acceleration. In: Proceedings of the IEEE/CVF Conference on Computer Vision and Pattern Recognition, pp. 4340–4349 (2019)
7. Li, H., Kadav, A., Durdanovic, I., Samet, H., Graf, H.P.: Pruning filters for efficient convnets. arXiv preprint arXiv:1608.08710 (2016)
8. Lin, M., Ji, R., Zhang, Y., Zhang, B., Wu, Y., Tian, Y.: Channel pruning via automatic structure search. arXiv preprint arXiv:2001.08565 (2020)
9. Liu, Z., Li, J., Shen, Z., Huang, G., Yan, S., Zhang, C.: Learning efficient convolutional networks through network slimming. In: Proceedings of the IEEE International Conference on Computer Vision, pp. 2736–2744 (2017)
10. Liu, Z., Sun, M., Zhou, T., Huang, G., Darrell, T.: Rethinking the value of network pruning. arXiv preprint arXiv:1810.05270 (2018)

11. Luo, J.H., Wu, J., Lin, W.: ThiNet: a filter level pruning method for deep neural network compression. In: Proceedings of the IEEE International Conference on Computer Vision, pp. 5058–5066 (2017)

12. Malach, E., Yehudai, G., Shalev-Schwartz, S., Shamir, O.: Proving the lottery ticket hypothesis: pruning is all you need. In: International Conference on Machine Learning, pp. 6682–6691. PMLR (2020)

13. Paszke, A., et al.: Pytorch: an imperative style, high-performance deep learning library. In: Advances in Neural Information Processing Systems 32 (2019)

14. Simonyan, K., Zisserman, A.: Very deep convolutional networks for large-scale image recognition. arXiv preprint arXiv:1409.1556 (2014)

15. Wang, Z., Li, C., Wang, X.: Convolutional neural network pruning with structural redundancy reduction. In: Proceedings of the IEEE/CVF Conference on Computer Vision and Pattern Recognition, pp. 14913–14922 (2021)

16. Zhu, M., Gupta, S.: To prune, or not to prune: exploring the efficacy of pruning for model compression. arXiv preprint arXiv:1710.01878 (2017)

17. Zhuo, H., Qian, X., Fu, Y., Yang, H., Xue, X.: SCSP: spectral clustering filter pruning with soft self-adaption manners. arXiv preprint arXiv:1806.05320 (2018)

A Collaborative Migration Algorithm for Edge Services Based on Evolutionary Reinforcement Learning

Yanan Zuo, Xiuguo Zhang[✉], Bo Zhang, and Zhiying Cao[✉]

School of Information Science and Technology, Dalian Maritime University, Dalian 116026, China
{zuoyanan,zhangxg,zhang18831725009,czysophy}@dlmu.edu.cn

Abstract. Multi-access edge computing (MEC) enables users' smart devices to execute computing-intensive and delay-sensitive applications by sinking computing power to edge servers, thereby meeting users' quality of service requirements. However, due to the limited computing and storage resources on the edge server, it is impossible to migrate all user service requests to the edge server. At the same time, due to the heterogeneity of resources among edge servers and the uneven distribution of user service requirements, it can easily lead to unbalanced loads among edge servers in the edge system. Consequently, this results in low resource utilization and a decreased success rate of requests. Therefore, this paper builds a collaborative edge service migration model based on software-defined networking (SDN), which supports cloud computing centers and edge servers to collaboratively process user requests and service request migration between edge servers. Taking minimizing the response delay of user requests and the weighted sum of device energy consumption as the optimization goal and transforming the optimization problem into a Markov process. A collaborative edge service migration algorithm based on evolutionary reinforcement learning (ERL) is proposed to solve the service Migration strategies and resource allocation decisions. Experimental results show that the proposed algorithm (DEDRL) performs better than other algorithms in response delay, energy consumption and request success rate.

Keywords: Edge computing · Service migration · Reinforcement learning

1 Introduction

With the development of wireless communication technology, mobile devices are increasingly popular and integrated into people's daily life. A large amount of data is transmitted in the network, which puts a serious burden on the mobile core network and backhaul links. In addition, with the advent of the Internet of Everything era, a large number of new applications have emerged. Such as virtual reality (VR), augmented reality (AR), unmanned driving, face recognition, etc. These new applications put forward higher requirements on the computing

Z. Tari et al. (Eds.): ICA3PP 2023, LNCS 14493, pp. 47–66, 2024.
https://doi.org/10.1007/978-981-97-0862-8_4

power and storage capacity of mobile devices [1–3]. However, due to the limitations of the mobile device itself (computing power, storage capacity, battery life, etc.) cannot meet the performance requirements of these new applications, these applications cannot be executed in real time and efficiently.

In order to meet the performance requirements of new applications, the industry has introduced the Mobile Cloud Computing (MCC) service model. However, due to the long distance between the remote cloud server and the mobile device, a large transmission delay will be caused. In addition, when users centrally access the remote cloud server, the huge number of service connections will cause the core network to be blocked and cause the remote cloud server to be overloaded. These problems will affect the service request quality. In order to solve these problems of cloud computing processing service requests, mobile edge computing is proposed.

Mobile Edge Computing (MEC) is a new computing model that responds to user requests at the edge of the network. In March 2017, the European Telecommunications Standards Institute (ETSI) officially changed the name of the Mobile Edge Computing Industry Specification Working Group to Multi-Access Edge Computing (Multi-Access Edge Computing MEC), to better meet the application requirements of edge computing [4]. Compared with cloud computing, in MEC, the user's service request can be migrated to the edge server close to the mobile device, thereby reducing the long data transmission delay, improving the user's service experience quality [5,6].

The edge server has certain computing power and storage space and is closer to the mobile device. Users can migrate computing-intensive or delay-sensitive service requests to the edge server for execution. Therefore, service migration technology has been an important topic of research in mobile edge computing. The traditional edge computing service migration strategy is to migrate all computing-intensive tasks of users to edge servers for processing [5,7,8]. Literature [9,10] studied the computing offloading problem based on a single edge server, which cannot solve the migration problem of a large number of service requests. Therefore, it is necessary to coordinate the management of the remote cloud server and the edge server, and reasonably migrate service requests. Due to the uneven distribution of users [11], the load of edge servers in different regions is uneven. In previous studies, when the edge server is overloaded, requests can only be migrated to the cloud computing center. This method may cause a large transmission delay. The collaborative service migration allows high-load edge servers to migrate service requests to low-load edge servers to make full use of resources and reduce transmission delays. The heterogeneity of resources between servers and the dynamic nature of the network make it challenging to design an effective migration strategy Therefore, under the constraint of limited resources, how to make decision on the destination of user service migration and reduce service delay and device energy consumption is one of the most challenging issues.

In order to solve the above problems, this paper establishes a collaborative edge service migration model based SDN. Considering factors such as service request response time and server energy consumption, an optimal solution is proposed by solving the service migration model in a collaborative edge computing environment. The algorithm to improve the overall performance of the system, the contributions of this paper are as follows:

(1) Aiming at the resource heterogeneity of edge servers and cloud servers, this paper constructs a collaborative edge service migration model based on software-defined networks, and optimizes the response delay of service requests, energy consumption of servers, and request success rate.

(2) This paper proposes an evolutionary reinforcement algorithm based on the combination of reinforcement learning and deep neural network to solve high-dimensional nonlinear problems. Reinforcement learning improves the sampling efficiency of samples through the combination of deep neural network, but because it uses gradient information to update the strategy and its exploration efficiency for the environment is low, it may cause the agent to fall into local optimum. The evolutionary algorithm is an optimization algorithm without gradient information, which is suitable for static optimization problems, but its exploration efficiency for the environment is high, and it can better jump out of the local optimal solution. Therefore, an improved enhanced algorithm is proposed to solve the dynamic service migration strategy in the edge system.

(3) In this paper, the effectiveness of the proposed DEDRL algorithm is verified by simulation experiments. The experimental results show that the DEDRL algorithm and the baseline algorithm can effectively reduce the total cost of service migration and ensure a high request success rate in a dynamic network environment.

2 Related Work

Different service requests have different computing and data transmission requirements. In order to ensure that the performance of user requests can be improved through service migration, it is necessary to find the optimal solution in the migration strategy solution set for service migration. In addition, the service migration process is also affected by, different factors such as user preferences, software and, hardware environment, radio interference and backhaul link quality. Different systems and applications have different requirements, so it is necessary to formulate corresponding optimization goals for migration strategies according to specific environments and performance requirements. Currently, there are two main methods for solving migration strategies. One is a migration method based on swarm intelligence heuristics, and the other is an intelligent migration method based on online learning. The following summarizes the optimization objectives of the migration strategy and the solution method of the migration strategy.

(1) Traditional heuristic-based migration methods: Since the solution of the migration strategy needs to consider many factors, the solution of the migration strategy is usually expressed as a mixed integer non-linear programming (MINP) problem. Since this problem is proved to be NP-hard, many researchers use heuristic-based algorithms to carry out Quick solution. Literature [12–14] are based on heuristic algorithm to solve the migration strategy. Literature [12] considers the computing offload of mobile devices with energy harvesting function to solve the solution using the improved genetic algorithm, which has a better effect in minimizing time delay. Cao et al. [13] proposed a service migration strategy based on the particle swarm optimization algorithm, which jointly optimizes the allocation of computing and communication resources, and minimizes the total energy consumption while satisfying the user's computing delay constraints. Zhang et al. [14] proposed an algorithm based on artificial fish swarms to reduce the total energy consumption of edge systems. Yinl et al. [15] proposed a new Load Balancing and Cost Genetic Algorithm (LCGA), which can effectively reduce the cost of completing tasks while ensuring load balancing. Liu et al. [16] proposed an improved discrete particle swarm optimization algorithm, reducing the task completion time. Masadeh et al. [17] used the sea-lion optimization algorithm for cloud task scheduling, which can effectively reduce cost, energy consumption. Literature [18] and others proposed an adaptive task offloading framework to flexibly select the optimal offloading strategy.

(2) Intelligent approach based on online migration methods: Traditional computing offloading solutions based on heuristic algorithms cannot adjust strategies according to the changing environment, and cannot achieve long-term effects. Intelligent migration methods based on online learning have emerged, such as deep learning, Q-learning, and deep reinforcement learning (Deep Reinforcement Learning DRL), Federal Learning (Federated Learning FL), etc. Tang [19] modeled the multi-user computing offloading problem in an uncertain wireless environment as a PT-based non-cooperative game, and then proposed a distributed computing offloading algorithm to obtain a Nash equilibrium, thereby minimizing user overhead. Yi [20] et al. consider that tasks are randomly generated by mobile users, and propose a mechanism based on a queuing model to maximize social welfare and achieve the equilibrium of a non-cooperative game among mobile users. Wang et al. [21] introduced a new dynamic edge computing model to improve the generalization ability of offloading algorithms, and designed an online primal-dual algorithm to offload arriving tasks. Qiu et al. [22] proposed a new DRL-based online computing offloading scheme, which considered both blockchain data mining tasks and data processing tasks, and expressed the online offloading problem as a Markov decision process. Literature [23–25] all use deep reinforcement learning to solve computing offloading and task scheduling problems.

3 Service Migration Model

3.1 Scene Description

This paper introduces SDN technology into the edge system, obtains the global state of the network in the edge system with the help of the SDN controller, and makes better service migration decisions based on the dynamic information of the network state in the system, so that users can obtain better service quality experience. As shown in Fig. 1, a cloud-edge collaboration based on software-defined network (SDN) and an edge system scenario diagram of edge-edge collaboration are constructed. The system is divided into user layer, edge service layer and cloud service layer. Among them, the user layer includes all users in the system. The edge service layer includes the SDN controller equipped near the macro base station (MBS) and all edge servers equipped near the wireless access base station (gNB) within the coverage of the macro base station. Users can communicate with the edge server through the wireless channel, and migrate the service to the edge server for execution. Edge servers communicate through wired channels, enabling collaboration in service migration. The cloud service layer includes cloud servers. Users can also migrate service requests to cloud servers through edge servers for execution.

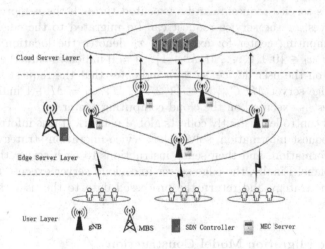

Fig. 1. Service Migration Scenario Diagram

The service migration scenario in Fig. 1 is described in detail as follows:

(1) This article assumes that the current number of users in the system is N. U_n indicates the nth user, $n \in \{1, 2, \ldots, N\}$. A set of users is denoted as $U = \{U_1, U_2, \ldots U_N\}$. Use triplets to represent users $U_n = (f_n, p_n, \gamma_n)$. f_n is the computing capability of the user devices, p_n is the processing power of the user devices, and γ_n is the static power of the user devices.

(2) A total of M edge servers and one cloud server are deployed in the system. Each edge server has a certain range, and each other's coverage does not overlap. Therefore, each user is covered by and communicates with a unique edge server. MEC_m indicates the mth edge server, $m \in \{1, 2, \ldots, M\}$. A set of users is denoted as $MEC_m = \{MEC_1, MEC_2, \ldots MEC_M\}$. A two-tuple is used to represent the edge server $MEC_m = (c_m^{MEC}, f_m^{MEC})$, c_m^{MEC} is the storage capacity of the edge server, and f_m^{MEC} is the computing capability of the edge server. Use a tuple to represent the cloud computing center $Cloud = (f_{cloud})$, and f_{cloud} represent the computing power of the cloud computing center.

(3) Assume that the number of service types requested by the user is K, and is the kth service of type, $k \in \{1, 2, \ldots, K\}$. A set of services is denoted as $S = \{S_1, S_2, \ldots S_k\}$. Use a tuple to represent each service $S_k = (c_k^s)$, and c_k^s represent the storage capacity required to deploy the service.

(4) Assume that the number of user service requests is L, and R_l is the lth service request. A set of user service requests is denoted as $l \in \{1, 2, \ldots, L\}$. Each request is represented by a quintuple $R_l = (u_l, d_l, s_l, c_l, t_l)$. u_l is the user who generated the service request, d_l is the size of the input data of the request, s_l is the type of service requested by the request, c_l is the amount of calculation (CPU cycles) required to process the request, t_l is the user's tolerance delay for processing this request.

(5) The user's service requests can be executed through the resources of their own devices, or the service request can be migrated to the edge server or cloud computing center for execution. x_l denotes the location of request execution, $x_l \in \{0, 1, 2, \ldots, M, M+1\}$. $x_l = 0$ indicates that the request is executed on the user device, $x_l = m$ indicates that the request is executed on the edge server MEC_m, $m \in \{1, 2, \ldots, M\}$. $x_l = M+1$ indicates that the request is executed on the cloud computing center.

(6) The SDN controller instantly collects global network status information and service request information, solves the service migration strategy based on above information, and then sends instructions to each server through the control plane. Finally, the edge server and cloud server execute the service migration strategy and return the processed data to the user through the data plane.

3.2 Service Migration Model Construction

In this paper, from the perspective of quality experience of user service requests. The two most important factors affecting the user's service quality experience are the service request response delay and the energy consumption of the user devices. Secondly, the request success rate is also an important factor affecting the quality of service experience.

Communication Cost. Service request data may be transmitted through wireless and wired channels in the system. In the wireless channel transmission, the system uses Time Division Multiple Access (TDMA) technology [23–25]. Users within the coverage of the same edge server use orthogonal uplink transmission sub-channels for data transmission, so signal interference between user transmissions in the same area can be ignored.

Wireless Channel Transmission. B_n is the wireless channel bandwidth between the user U_n and override its edge server, that is the local server MEC_m, Gaussian white noise is σ^2, $h_{n,m}$ is the channel gain between the user and the edge server, and p_n is the upload transmission power between the user and the edge server, then the user passes the data transfer rate at which the wireless channel transmits data to MEC_m is shown in Eq. (1).

$$R_{n,m} = B_n \log_2 \left(1 + \frac{h_{n,m}p_n}{\sigma^2}\right) \tag{1}$$

When the request is migrated to the cloud computing center for execution, the data needs to be transmitted to the cloud computing center by covering its edge server MEC_m. Edge servers and cloud computing centers need to transmit data through the core network. The channel bandwidth between the edge server MEC_m and the cloud computing center is denoted as B_m^c, the Gaussian white noise is σ_c^2, $h_{m,c}$ is the channel gain between the edge server and the cloud computing center, and p_n is the upload transmission power between the edge server and the cloud computing center. Considering the congestion of the core network caused by the competition for channel band-width, which w is the coefficient reflecting the marginal effect of communication congestion and x_c is the number of requests uploaded to the remote cloud server, and the data transmission rate uploaded from the edge server to the cloud computing center is shown in Eq. (2).

$$R_{m,c} = \frac{B_m^c}{x_c(1 + wx_c)} \log_2\left(1 + \frac{h_{m,c}p_m}{\sigma_c^2}\right) \tag{2}$$

Wired Channel Transmission. In the system, edge servers are connected by limited links, such as optical fibers. Compared with the wireless channel transmission signal interference is relatively small, so it is assumed that the signal-to-noise ratio in the wired channel is a constant value, denoted as ζ. $B_{i,j}$ is the bandwidth between MEC_i and MEC_j. Therefore, the data transmission power between edge servers is shown in Eq. (3).

$$R_{i,j} = B_{i,j} \log_2(1 + \zeta) \tag{3}$$

Response Cost. Request response latency: The response latency of a service request includes the request transmission latency and execution latency at the user device, edge server, or cloud computing center. Since user requests can be responded on their own devices, edge servers, and cloud computing centers, the following gives the service request response delays in different situations, and finally gives the calculation formula for the total response delay of system requests.

Response Latency Executed by User Devices. When the service request is executed on the user devices, the computing resource of the devices itself is used for processing without data transmission. Therefore, for the request $R_l = (u_l, d_l, s_l, c_l, t_l)$, its response delay is shown in Eq. (4).

$$T_{loc}^l = \frac{c_l}{f_n} \tag{4}$$

f_n is the computing capability of the user's devices, that is the CPU clock frequency.

Response Latency Executed by the Edge Server. When the service request is executed on the edge server, the response delay executed by the edge server includes two parts, data transmission delay and request execution delay. First, the requested data needs to be uploaded to the edge server, and the transmission delay of data uploaded to the local server is shown in Eq. (5).

$$T_{mec}^{up} = \frac{d_l}{R_{n,m}} \tag{5}$$

If the local server MEC_m and the execution server MEC_i are not the same server, that is $m \neq i$, then the data needs to be transmitted between the edge servers, and the data transmission delay between the local server and the execution server is shown in Eq. (6).

$$T_{mec}^{comm} = \frac{d_l}{R_{m,i}} \tag{6}$$

The execution delay of the request at the edge server is shown in Eq. (7).

$$T_{mec}^{exe} = \frac{c_l}{\tau_{m,l} f_m^{MEC}} \tag{7}$$

$\tau_{m,l}$ is the computing resource scaling factor for each request allocated on MEC_m. Therefore, the response delay of the request executed by the edge server is shown in Eq. (8).

$$T_{mec}^l = T_{mec}^{up} + T_{mec}^{comm} + T_{mec}^{exe} \tag{8}$$

Response Latency Executed by the Cloud Computing Center. When the service request is executed in the cloud computing center, the execution response delay includes two parts, the data transmission delay and the request execution delay, as in the edge server. The time for requesting data to be uploaded to the edge server is obtained by Eq. (5). The transmission delay of uploading the requested data from the edge server to the cloud computing center is shown in Eq. (9).

$$T_{cloud}^{up} = \frac{d_l}{R_{m,c}} \tag{9}$$

The execution delay of the request at the cloud computing center is shown in Eq. (10).

$$T_{cloud}^{exe} = \frac{c_l}{\tau_{c,l} f_m^{MEC}} \qquad (10)$$

$\tau_{c,l}$ is the computing resource scaling factor for each request allocated on the cloud computing center.

Therefore, the response delay of the request executed by the edge server is shown in Eq. (11).

$$T_{cloud}^l = T_{mec}^{up} + T_{cloud}^{up} + T_{cloud}^{exe} \qquad (11)$$

For a request R_l, the response delay T_l is shown in Eq. (12).

$$T_l = \begin{cases} T_{loc}^l & x_l = 0 \\ T_{mec}^l & x_l = m \\ T_{cloud}^l & x_l = M+1 \end{cases} \qquad (12)$$

Request Response Energy Consumption: The energy consumption of the user devices during the service request response process consists of three parts, including the energy consumption of the request executed on the device, the energy consumption of the device for data transmission, and the idle energy consumption of the device waiting for the request to be executed on the edge server or cloud computing center.

Energy Consumption Executed by User Devices. When the service request is executed on the user devices, the energy consumption when using the devices own computing resources for processing is equal to the coefficient κ multiplied by the square of the CPU clock frequency of the user devices and the number of CPU cycles required by the service is the energy consumed per CPU cycle, which depends on the chip architecture. The usual value for this is set to $\kappa = 10^{-26}$. For the request $R_l = (u_l, d_l, s_l, c_l, t_l)$, the energy consumption of its execution on the device is shown in Eq. (13).

$$E_{loc}^l = \kappa f_n^2 c_l \qquad (13)$$

Energy Consumption Executed by the Edge Server. When the user requests to migrate to the edge server, the energy consumption of the user devices is mainly composed of two parts, the energy consumption for uploading data to the edge server for transmission and the idle energy consumption for waiting for the task to be completed. Among them, the power consumption of data transmission is equal to the delay of data uploading to the edge server multiplied by the power of the user devices, and the idle energy consumption of the device is equal to the static power of the device multiplied by the delay of waiting for the edge server to execute. Therefore, when executing on the edge server, the energy consumed by the user devices is shown in Eq. (14).

$$E_{mec}^l = T_{mec}^{up} p_n + \gamma_n (T_{mec}^{comm} + T_{mec}^{exe}) \qquad (14)$$

Energy Consumption Executed by the Cloud Computing Center. When the user
requests to migrate to the cloud computing center for execution, the energy
consumption of the user devices is similar to the energy consumption of the
edge server, which is composed of data transmission energy consumption and
idle energy consumption waiting for the task to be completed. Among them, the
power consumption of data transmission is equal to the delay of data uploading
to the edge server multiplied by the power of the user devices, and the idle energy
consumption of the devices is equal to the static power of the devices multiplied
by the delay waiting for the execution of the cloud computing center. Therefore,
the energy consumption of the request when executed in the cloud computing
center is shown in Eq. (15).

$$E_{cloud}^l = T_{mec}^{up} p_n + \gamma_n \left(T_{cloud}^{up} + T_{cloud}^{exe} \right) \tag{15}$$

For a request R_l, the response energy consumption E_l is shown in Eq. (16).

$$E_l = \begin{cases} E_{loc}^l & x_l = 0 \\ E_{mec}^l & x_l = m \\ E_{cloud}^l & x_l = M+1 \end{cases} \tag{16}$$

Request Success Rate. Due to user's request has a certain delay constraint,
the successful response of the request within the user's tolerance delay is also a
factor that affects the user's service quality. If the delay requested by the user
exceeds the user's tolerance delay, a corresponding penalty will be given. For a
request $R_l = (u_l, d_l, s_l, c_l, t_l)$, if the response delay is greater than the tolerance
delay or the remaining resources of the device or server executing the request do
not meet the required resources, the penalty cost indicator function is shown in
Eq. (17).

$$\begin{cases} \delta_l{}^{loc} = \delta\{\{T_l > t_l\} \vee \{f_n < c_l\}\} \\ \delta_l{}^{mec} = \delta\{\{T_l > t_l\} \vee \{tau_{m,l} f_m^{MEC} < c_l\}\} \\ \delta_l{}^{cloud} = \delta\{\{T_l > t_l\} \vee \{tau_{c,l} f_m^{MEC} < c_l\}\} \end{cases} \tag{17}$$

If the request is not completed within the user's tolerance delay, then $\delta_l^i = 1$
which means that the request fails. If the request is completed within the user's
tolerance delay, then $\delta_l^i = 0$ which means that the request is successful.

3.3   Problem Formulation

In this paper, from the perspective of user service request quality experience, the
two most important factors affecting user service quality experience are service
request response delay and user devices energy consumption. In this paper, the
weighted sum of the response delay of all user service requests in the system
and the energy consumption of user devices is used as the service migration cost
design cost function.

Define two 0–1 variables p_l and q_l to assist in the calculation of the service
request cost. p_l indicates whether to execute on the user devices. As shown in

Eq. (18).

$$p_l = \begin{cases} 1 & \text{Executes on the user's device} \\ 0 & \text{Not executed on the user's device} \end{cases} \quad (18)$$

q_l indicates that it is executed on the edge server or cloud computing center. As shown in Eq. (19).

$$q_l = \begin{cases} 1 & \text{Executes on the edge server} \\ 0 & \text{Executes on the cloud computing center} \end{cases} \quad (19)$$

λ_t and λ_e represent the delay weight coefficient and energy consumption weight coefficient respectively. The value of the requested time weight coefficient and energy consumption weight coefficient depends on the needs in the current system. For example, when λ_e is relatively large, it means that the power of the user devices is low. When making a migration decision, it is necessary to pay more attention to the requested energy consumption size. The migration cost function of all user service requests in the system is shown in Eq. (20).

$$
\begin{aligned}
\text{Cost}(p_l, q_l, \tau_{m,l}, \tau_{c,l}) &= \sum_{i=1}^{L} \lambda_t T_l + \lambda_e E_l \\
&= \sum_{i=1}^{L} \lambda_t (p_l T_{loc}^l + (1 - p_l)(q_l T_{mec}^l + (1 - q_l) T_{cloud}^l)) \\
&\quad + \lambda_e (p_l E_{loc}^l + (1 - q_l)(q_l E_{mec}^l + (1 - q_l) E_{cloud}^l)
\end{aligned}
\quad (20)
$$

The smaller the total request cost, the better the corresponding service migration strategy at this time. Therefore, with the goal of solving the service migration decision with the minimum total cost of all requests in the system, the problem is formulated as shown in Eq. (21).

$$
\begin{aligned}
\min \quad & \sum_{i=1}^{L} \lambda_t (p_l T_{loc}^l + (1 - p_l)(q_l T_{mec}^l + (1 - q_l) T_{cloud}^l)) \\
& + \lambda_e (p_l E_{loc}^l + (1 - q_l)(q_l E_{mec}^l + (1 - q_l) E_{cloud}^l) \\
\text{s.t.} \quad & \lambda_t + \lambda_e = 1 & (C1) \\
& p_l + q_l \leq 1 & \forall l \in \{1, 2, ..., L\} \quad (C2) \\
& \sum_{i=1}^{L} p_l \tau_{m,l} \leq f_m^{MEC} & \forall m \in \{1, 2, ..., M\} \ (C3) \\
& \sum_{i=1}^{L} p_l c_k^s \leq c_m^{MEC} & \forall m \in \{1, 2, ..., M\} \ (C4) \\
& \sum_{i=1}^{L} q_l \tau_{c,l} \leq f_{cloud} & (C5) \\
& p_l, q_l \in \{0, 1\} & \forall l \in \{1, 2, ..., L\} \quad (C6) \\
& \tau_{c,l}, \tau_{m,l} \in [0, 1] & \forall m \in \{1, 2, ..., M\} \ (C7)
\end{aligned}
\quad (21)
$$

Among them $(C1)$ indicates that the delay weight coefficient and energy consumption weight coefficient add up to 1; $(C2)$ indicates that the user's service request can only be executed in one place, that is, executed on the user's own device or migrated to the edge server, cloud computing Center execution; $(C3)$

indicates that the sum of computing resources allocated by the edge server to users cannot exceed its own computing resources; ($C4$) indicates that the sum of storage resources allocated by edge servers to users cannot exceed its own storage resources; ($C5$) indicates that the sum of the computing resources allocated to users by the computing center cannot exceed the computing resources it owns.

The formalized problem is a mixed integer nonlinear programming problem (MINP), which is an NP-Hard problem. It is difficult to find a global optimal solution within a polynomial. Heuristic methods and reinforcement learning algorithms are usually used to solve such problems. As the number of user service requests increases, the solution space of the problem will increase exponentially. At the same time, the network resources in the system and the available resources of the server are dynamically changing. Evolutionary reinforcement learning combines the advantages of evolutionary computation and reinforcement learning. Therefore, this paper solves problem (21) based on the evolutionary reinforcement learning algorithm, which is detailed in Sect. 4.

4   Service Migration Algorithm Based on Evolutionary Reinforcement Learning

This paper first expresses the collaborative service migration problem in edge computing as a Markov decision process (MDP Markov Decision Process), and then designs a service migration algorithm based on DEDRL to solve this problem.

4.1   MDP Model

Considering that the service migration decision of each time slot in the edge computing system is only determined by the current system information and has nothing to do with the system state and decision at the last moment, the migration cost minimization problem is modeled as MDP, which is usually define a quintuple $<S, A, P, R, T>$, which S is the state space; A is the action space; P is the probability of going to the next state after taking an action in the state; R is the immediate reward obtained by taking an action in the state; T is a continuous period of time; the specific meanings are as follows:

Agent: The SDN controller has global network status information, including MEC server information, user's devices information, network status information in each area, and association information between users and MEC servers, etc., so we regard the SDN controller as an agent, Service migration decisions and resource allocation decisions are made by the SDN controller.

State: The system state space is $S = \{s_1, s_2, \ldots s_T\}$, $s_t = \{R_l, N_t, M_t\}$, R_l indicates the resources required by the request. N_t indicates the current network conditions of the system, including between edge servers, edge servers and cloud computing centers, and data transmission rates. M_t indicates the remaining resources of edge servers and cloud computing centers in the system.

Action: The system action space is $A = \{a_1, a_2, \ldots a_t, a_{t+1}, \ldots s_T\}$, $a_t = \{X_t, F_t\}$. $X_t = \{x_1^t, x_2^t, \ldots x_l^t\}$ indicates the migration decision vector of the current request, x_l^t is the requested service migration strategy, $F_t = \{f_1^t, f_2^t, \ldots f_l^t\}$ represents the computing resources provided by the current request, and f_l^t is the requested resource allocation strategy.

Reward: After taking action a_t in state s_t, the environment returns to the agent a immediate reward. The cost difference between the cost of non-request migration and the request migration strategy and the request completion penalty are used as instant rewards. The request completion penalty is as follows: As shown in (17), the definition method of the reward function is shown in Eq. (22).

$$r(s, a) = \cos t \, (p_l = 1) - \cos t \, (p_l, q_l, \tau_{m,l}, \tau_{c,l}) + \delta_l^i \tag{22}$$

Probability: The transition probability function shows the transition probability from the current state to the next state as shown in Eq. (23).

$$p\,(s' \mid s, a) = P\,(s_{t+1} = s' \mid s_t = s, a_t = a) \tag{23}$$

The state transition probability function of the limbic system network is unknown. Therefore, in this paper, a model-free approach is adopted to approximate the state transition probability function with a neural network.

4.2   DEDRL-Based Service Migration Process

Considering that the network resources in the system and the available resources of the server change dynamically, and the action space of the service migration strategy involves continuous variables, the reinforcement learning module of this article uses a reinforcement learning algorithm based on DDPG (Deep Deterministic Policy Gradient). The differential evolution algorithm can converge to a better solution faster than the genetic algorithm in high-dimensional problems. Therefore, the differential evolution algorithm is used in the evolution module to evaluate the fitness of the policy network in DDPG. DDPG combines Actor-Critic architecture and deep neural network. It is a deep reinforcement learning algorithm based on deterministic policy gradient, which is used to solve the reinforcement learning problem of continuous action space. The DDPG algorithm optimizes the Actor network by using deterministic policy gradients, and optimizes the Critic network by using TD error (Temporal Difference Error). By alternately updating the parameters of the Actor and Critic networks, the DDPG algorithm gradually learns and improves the strategy to obtain better reinforcement learning performance, and its framework diagram is shown in Fig. 2. The evolutionary reinforcement learning algorithm combines the advantages of both. It not only has high search efficiency in the environment, but also has the ability to learn and adapt. It is suitable for solving service migration problems in dynamic environments. As shown in the figure above, the goal of the edge service migration system is the goal of the agent SDN controller is to find the optimal policy to maximize the long-term expected return. Using the function $\mu\,(s \mid \theta^\mu)$ to fit the Actor network, and using the function $Q\,(s, a \mid \theta^Q)$ to fit the Critic

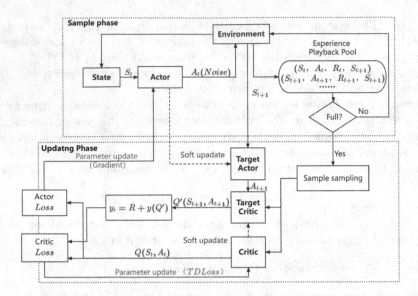

Fig. 2. DDPG algorithm frame diagram

network, where θ^μ and θ^Q are network parameters. The critic network evaluates the action by the action value function $Q(s_t, a_t)$. Its calculation method is shown in Eq. (24).

$$
\begin{aligned}
Q(s_t, a_t) &= \mathrm{E}_\pi[R(t)|s_t, a_t] \\
&= \mathrm{E}_\pi[r(t) + \gamma r(t+1) + \gamma^2 r(t+2) + ...|s_t, a_t] \\
&= \mathrm{E}_\pi[r(t) + Q(s_{t+1}, a_{t+1})|s_t, a_t]
\end{aligned}
\tag{24}
$$

$R(t)$ is the cumulative discounted reward of the agent, as shown in Eq. (25). γ is the discount factor, $\gamma \in [0, 1]$.

$$
R(t) = r(t) + \gamma r(t+1) + \gamma^2 r(t+2) + ... \gamma^{T-1} r(T-1)
\tag{25}
$$

Update the Critic network according to the $R(t)$, and input (s_t, a_t) to the Critic network to obtain the actual value $Q(s_t, a_t)$. The error of the main critic network is then calculated according to the error equation, which is updated by minimizing the error, as shown in Eq. (26).

$$
L(\theta^Q) = \frac{1}{T}\Sigma_i \left(r_i - Q^\pi\left(s_t, a_t \mid \theta^Q\right)\right)^2
\tag{26}
$$

In the update phase, the group is randomly selected from the experience reuse pool and the Actor network is updated according to the policy gradient. The calculation is shown in Eq. (27).

$$
\nabla_{\theta^\mu} J = \frac{1}{Z}\sum_t \nabla_a Q\left(s_t, a \mid \theta^Q\right) \nabla_{\theta^\mu} \mu\left(s_t \mid \theta^\mu\right)
\tag{27}
$$

In order to further increase the stability of the learning process, the parameters and of the target Actor network and the target Critic network are updated iteratively, and the update formula is shown in Eq. (28).

$$\begin{cases} \theta^{Q'} = \tau\theta^Q + (1-\tau)\theta^{Q'} \\ \theta^{\mu'} = \tau\theta^\mu + (1-\tau)\theta^{\mu'} \end{cases} \tag{28}$$

As shown in Fig. 3, the DEDRL algorithm is mainly composed of two parts: the DDPG algorithm and the Differential Evolution Algorithm (DE) algorithm. The actor network in DDPG is regarded as the individual in the population in DE algorithm. The fitness function is the cumulative reward for an individual interacting with the environment once, and the Actor network is sorted from high to low according to the fitness function value. In the part of differential evolution algorithm, there are four main steps: population initialization, selection operation, crossover operation and mutation operation. Initialize the population with a size of N_p, and use to x_p represent the individual in the population, that is, the parameters of the Actor network, $x_p = \left[\theta^{\mu'}\right]$. According to the fitness value, the individuals in the population are sorted from high to low, and the top 60% are selected each time. Set the crossover factor CR. In order to obtain better individuals, it is necessary to perform cross-over operations to increase the diversity of the population. Set the variation factor F, the search space can be increased by the variation factor. The pseudo code of the algorithm is shown in Algorithm 1.

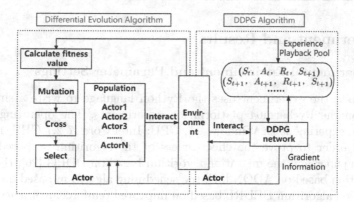

Fig. 3. DEDRL algorithm frame diagram

Algorithm 1. Service Migration Algorithm Based on DEDRL

Input: User request collection, Edge environment information, number of episode, and replay memory size
Output: Service Migration Decisions and Resource Allocation Decisions
1: Initialize the actor network θ_a and θ_c, initialize the global shared counter $t \leftarrow 1$
2: Randomly initialize the thread-specific parameters θ'_a and θ'_c
3: **repeat**
4:      Reset gradients: $d\theta_a \leftarrow 0$ and $d\theta_c \leftarrow 0$
5:      Synchronize thread-specific parameters $\theta'_a \leftarrow d\theta_a$ and $\theta'_c \leftarrow d\theta_c$
6:      $t_{start} = t$
7:      **repeat**
8:          Enter status information s_t , perform a_t according to policy $\pi(a_t|s_t; \theta'_a)$
9:          Receive reward R_t and new state s_{t+1}
10:         $t \leftarrow t + 1$
11:         $T \leftarrow T + 1$
12:     **until** t - $t_{start} == t_{max}$ or $t == t_{end}$
13:     Enter status information s_t , perform a_t according to policy $\pi(a_t|s_t; \theta'_a)$
14:     Calculate the state value of s_t under the last time series t according to $R = V(s_t; \theta'_c)$
15:     **for** $i = t - 1, ..., t_{start}$ **do**
16:         $R \leftarrow [R_i + \gamma R]$
17:         $d\theta_c \leftarrow d\theta_c + \partial[R - V(s_t; \theta'_c)]^2/\partial\theta'_c$
18:         $d\theta_a \leftarrow d\theta_a + \nabla\theta'_a log \ \pi(a_t|s_t; \theta'_a)(R - V(s_t; \theta c'))$
19:     **end for**
20:     perform asynchronous update of θ_a and θ_c using $d\theta_a$ and $d\theta_c$
21: **until** $T > T_{max}$

---

5   Experiment and Results

5.1   Experimental Environment and Parameter Settings

In this paper, the experiment uses the Python language to write a simulation program on the Pycharm64 platform for experiments. The operating system used in the experiment is: Windows 11, CPU: Intel Core i7 2.7 GHZ, memory: 16 GB. In order to prove the effectiveness of the algorithm proposed in this paper, this paper and the migration algorithm based on DDPG [26], the migration algorithm based on ADQN [27], the scheduling algorithm based on AC and the migration algorithm DEDRL based on improved reinforcement learning proposed in this paper. The average value of the comparison experiments was taken. Experimental parameter settings are shown in Table 1.

5.2   Analysis of Results

This experiment analyzes the performance of different algorithms by comparing the system's long-term rewards, system migration costs, and request success rates.

Table 1. Experimental parameter setting table.

Variable	Meaning	Value
d_l	Request input data size	Uniform [3,6] MB
f_m^{MEC}	The processing power of the edge server (CPU frequency)	Uniform [3,10] MB
C_m^{MEC}	Request input data size	5000 Megacycles
f_{cloud}	The processing power of the cloud computing center (CPU frequency)	30 GHz
B	Transmission Channel Available Bandwidth	20 MHz
σ	Noise power	−118.4 dB/Hz
p_n	Transmission power of user devices	0.6 W
λ_t	Delay weight	0.5
λ_e	Delay weight	0.5
γ	Discount factor	0.9
τ	Soft update factor	0.001
NP	Population sizer	40
F	Cross factor	0.4
CR	Variation factor	0.6

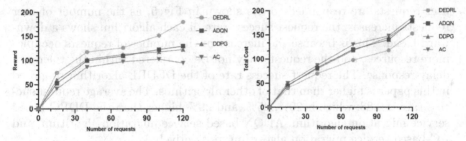

(a) Reward value under different number of (b) Total system cost for different number of
user requests   user requests

Fig. 4. Comparison results of ablation experiments

(1) Reward Value and Total System Cost: Fig. 4 shows the reward comparison
results of the four algorithms under the computing tasks with different numbers of user requests and the comparison results of the total cost of the four
algorithms in different numbers of user requests.

As shown (a) in Fig. 4, as the number of user requests increases, the total
reward value of each algorithm shows a rising trend. The total reward value
of the DEDRL algorithm proposed in this paper is higher than that of other
algorithms. This is because the method in this paper also considers the
request success rate in addition to response delay and energy consumption.
Compared with other algorithms, the penalty is lower when calculating the
reward value. The overall reward value of the AC algorithm is lower than
that of the DDPG-based service migration algorithm and the ADQN-based
service migration algorithm. Among them, the reward value of the DDPG-
based service migration algorithm and the ADQN-based service migration

algorithm are relatively close. The reward value of the DDPG-based service migration algorithm is greater than that of the ADQN-based service migration algorithm before the number of requests is 85, and the opposite is true after the number of requests is 85.

As shown (b) in Fig. 4, as the number of user requests increases, the system cost of each algorithm shows a rising trend. The system cost of the DEDRL algorithm proposed in this paper is lower than other algorithms, because the method in this paper is based on the evolutionary reinforcement algorithm and has a better ability to find the optimal solution. The system cost of the AC algorithm is higher than that of the DDPG-based service migration algorithm and the ADQN-based service migration algorithm. Among them, the reward value of the DDPG-based service migration algorithm and the ADQN-based service migration algorithm are relatively close.

(2) Request success rate: Fig. 5 shows the comparison results of the request success rate of the four algorithms under the computing tasks with different numbers of user requests. The X-axis represents different numbers of user requests, and the Y-axis represents the success rate of requests after user requests are responded to. As shown in Fig. 5, as the number of user requests increases, the request success rate of each algorithm shows a downward trend. This is because the increase in the number of requests occupies more resources, and the requests may not be completed within the tolerance delay response. The request success rate of the DEDRL algorithm proposed in this paper is higher than that of other algorithms. The average request success rate is 0.88, which is 20%, 25%, and 29% higher than the DDPG-based service migration algorithm, ADQN-based service migration algorithm, and AC-based service migration algorithm, respectively.

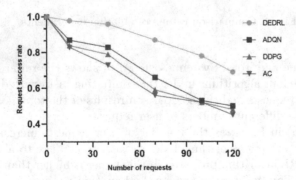

Fig. 5. Request success rate for different number of user requests

6   Conclusion

In this paper, a collaborative edge service migration model based on SDN is established, and a collaborative edge service migration algorithm based on evolutionary reinforcement learning is proposed. The simulation experiment compares the proposed DEDRL algorithm with different baseline algorithms to demonstrate its superior ability in reducing energy consumption of user devices, reducing user request delay, and achieving a high request success rate. These improvements contribute to an enhanced user service quality experience.

Acknowledgement. This work is supported by Liaoning Province Applied Basic Research Program Project (Grant No. 2023JH2/101300195).

References

1. Du, J., Yu, F.R., Lu, G., Wang, J., Jiang, J., Chu, X.: MEC-assisted immersive VR video streaming over terahertz wireless networks: a deep reinforcement learning approach. IEEE Internet Things **7**(10), 9517–9529 (2020)
2. Deng, S., Xiang, Z., Taheri, J., et al.: Optimal application deployment in resource constrained distributed edges. IEEE Trans. Mob. Comput. **20**(5), 1907–1923 (2020)
3. Dai, B., Xu, F., Cao, Y., et al.: Hybrid sensing data fusion of cooperative perception for autonomous driving with augmented vehicular reality. IEEE Syst. J. **15**(1), 1413–1422 (2020)
4. Shi, W., Cao, J., Zhang, Q., et al.: Edge computing: vision and challenges. Internet Things J. **3**(5), 637–646 (2016)
5. Zhao, J., Li, Q., Gong, Y., Zhang, K.: Computation offloading and resource allocation for cloud assisted mobile edge computing in vehicular networks. Trans. Veh. Technol. **68**(8), 7944–7956 (2019)
6. Wang, Y., Lang, P., Tian, D., et al.: A game-based computation offloading method in vehicular multiaccess edge computing networks. IEEE Internet Things J. **7**(6), 4987–4996 (2020)
7. Lyu, X., Ni, W., Tian, H., et al.: Optimal schedule of mobile edge computing for internet of things using partial information. IEEE J. Sel. Areas Commun. **35**(11), 2606–2615 (2017)
8. Basu, D., Wang, X., Hong, Y., et al.: Learn-as-you-go with Megh: efficient live migration of virtual machines. IEEE Trans. Parallel Distrib. Syst. **30**(8), 1786–1801 (2019)
9. Alameddine, H.A., Sharafeddine, S., Sebbah, S., et al.: Dynamic task offloading and scheduling for low-latency IoT services in multi-access edge computing. IEEE J. Sel. Areas Commun. **37**(3), 668–682 (2019)
10. Lim, W.Y.B., Luong, N.C., Hoang, D.T., et al.: Federated learning in mobile edge networks: a comprehensive survey. IEEE Commun. Surv. Tutor. **22**(3), 2031–2063 (2020)
11. Wang, S., Zhang, X., Yan, Z., Wang, W., et al.: Cooperative edge computing with sleep control under nonuniform traffic in mobile edge networks. IEEE Internet Things J. **6**(3), 4295–4306 (2019)
12. Zhang, G., Zhang, W., Cao, Y., et al.: Energy-delay tradeoff for dynamic offloading in mobile-edge computing system with energy harvesting devices. IEEE Trans. Industr. Inf. **14**(10), 4642–4655 (2018)

13. Cao, X., Wang, F., Xu, J., et al.: Joint computation and communication cooperation for energy-efficient mobile edge computing. IEEE Internet Things J. **6**(3), 4188–4200 (2018)

14. Zhang, H., Guo, J., Yang, L., et al.: Computation offloading considering fronthaul and backhaul in small-cell networks integrated with MEC. In: 2017 IEEE Conference on Computer Communications Workshops (INFOCOM WKSHPS), pp. 115–120. IEEE (2017)

15. Yin, S., Ke, P., Tao, L.: An improved genetic algorithm for task scheduling in cloud computing. In: 2018 13th IEEE Conference on Industrial Electronics and Applications (ICIEA), pp. 526–530. IEEE (2018)

16. Liu, S., Yin, Y.: Task scheduling in cloud computing based on improved discrete particle swarm optimization. In: IEEE 2nd International Conference on Information Systems and Computer Aided Education (ICISCAE), Dalian, China, pp. 594–597 (2019)

17. Masadeh, R., Alsharman, N., Sharieh, A., et al.: Task scheduling on cloud computing based on sea lion optimization algorithm. Int. J. Web Inf. Syst. **17**(2), 99–116 (2021)

18. Tran-Dang, H., Kim, D.S.: FRATO: fog resource based adaptive task offloading for delay-minimizing IoT service provisioning. IEEE Trans. Parallel Distrib. Syst. **32**(10), 2491–2508 (2021)

19. Tang, L., He, S.: Multi-user computation offloading in mobile edge computing: a behavioral perspective. IEEE Netw. **32**(1), 48–53 (2018)

20. Yi, C., Cai, J., Su, Z.: A multi-user mobile computation offloading and transmission scheduling mechanism for delay-sensitive applications. IEEE Trans. Mob. Comput. **19**(1), 29–43 (2019)

21. Wang, H., Xu, H., Huang, H., et al.: Robust task offloading in dynamic edge computing. IEEE Trans. Mob. Comput. **22**(1), 500–514 (2021)

22. Qiu, X., Liu, L., Chen, W., et al.: Online deep reinforcement learning for computation offloading in blockchain-empowered mobile edge computing. IEEE Trans. Veh. Technol. **68**(8), 8050–8062 (2019)

23. Ren, J., Hou, T., Wang, H., et al.: Collaborative task offloading and resource scheduling framework for heterogeneous edge computing. Wirel. Netw. 1–13 (2021)

24. Li, Z., Zhou, X., Li, T., Liu, Y.: An optimal-transport-based reinforcement learning approach for computation offloading. In: 2021 IEEE Wireless Communications and Networking Conference (WCNC), pp. 1–6. IEEE (2021)

25. Xiao, M., Shroff, N.B., Chong, E.K.P.: A utility-based power-control scheme in wireless cellular systems. IEEE/ACM Trans. Netw. **11**(2), 210–221 (2003)

26. Wang, Y., Fang, W., Ding, Y., et al.: Computation offloading optimization for UAV-assisted mobile edge computing: a deep deterministic policy gradient approach. Wirel. Netw. **27**(4), 2991–3006 (2021)

27. Chen, N., Zhang, S., Qian, Z., et al.: When learning joins edge: real-time proportional computation offloading via deep reinforcement learning. In: 2019 IEEE 25th International Conference on Parallel and Distributed Systems (ICPADS), pp. 414–421. IEEE (2019)

A Graph Generation Network with Privacy Preserving Capabilities

Yangyong Miao, Xiaoding Wang, and Hui Lin[✉]

College of Computer and Cyber Security, Fujian Normal University,
Fuzhou 350117, Fujian, China
18750768639@163.com, {wangdin1982,linhui}@fjnu.edu.cn

Abstract. The use of large datasets for data mining and analysis can stimulate progress in science and technology while also propelling economic growth. Graph-structured data is a crucial component of both data mining and analysis. However, this type of data often contains sensitive personal information, making it vulnerable to potential attacks and widespread privacy breaches. Graph data encodes sensitive information, including personal attributes (nodes) and complex interaction relationships (edges). Rényi differential privacy provides a stricter definition of privacy protection. This paper introduces the RDP-GGAN framework, which integrates Rényi differential privacy technology with generative adversarial networks to offer improved privacy protection capabilities. The framework utilizes Rényi differential privacy to establish and enforce strict privacy constraints for deep graph generative models, with a particular emphasis on preserving edge privacy in graph data to ensure connection privacy in relational data. To enhance edge differential privacy, appropriate noise is injected into the gradient of link-reconstruction-based graph generative models.

Keywords: Graph data · Rényi differential privacy · RDP-GGAN · edge differential privacy

1 Introduction

With the increasing maturity of big data mining and analysis technology, graph data that can store data attributes and link relationships at the same time is flourishing [1], but the privacy leakage of graph data has attracted increasing attention. Unlike other data types, graphs are typically used to represent data with a "many-to-many" relationship, storing information in the form of nodes and edges. In the context of social networks, for example, nodes are commonly used to represent users in the network, while edges represent the relationships between users. Therefore, protecting the privacy of these relationships is critical to safeguarding the overall privacy of graph data. The main focus of this paper is on secure graph generation, with a particular emphasis on addressing the privacy concerns surrounding the edges of the graph. Given that edges are necessary to capture the interactions between different objects in the graph data, protecting the privacy of these relationships is paramount in ensuring the overall security of sensitive graph data.

Synthetic Data Generation (SDG) is a highly promising approach for preserving privacy. This technique involves generating synthetic data that can be shared publicly without compromising individuals' privacy. SDG also offers numerous opportunities for collaborative research, including the development of prediction models and the identification of patterns. Generative Adversarial Networks (GAN) [2] have gained enormous attention in this research field due to their success in medical diagnosis [3], image processing [4], image translation [5], etc.

Differential privacy [6] is a privacy protection model that aims to protect sensitive information in datasets by adding random noise to the data. One key feature of differential privacy is its ability to provide a strong guarantee that the addition of random noise will not significantly impact the output results, even if any piece of data in the dataset is changed. This means that no matter how powerful the background knowledge is, an attacker cannot infer any user's private information by publishing the results. Compared with traditional privacy protection technologies, differential privacy provides higher privacy protection and is becoming increasingly popular in the field of graph data privacy protection. However, there are still challenges in applying differential privacy to certain types of data and achieving a balance between privacy protection and data utility.

Differential Private Stochastic Gradient Descent (DP-SGD) is a commonly used method to ensure differential privacy while maintaining model accuracy under a modest privacy budget. DP-SGD is the basis of many research works and is widely used in machine learning. DP-SGD works by limiting the sensitivity of the algorithm to individuals through the clipping of gradients, adding Gaussian noise, and performing a gradient descent optimization step. This approach enables privacy-preserving analysis of data without sacrificing the accuracy of the model.

In graph data privacy protection, perturbing the graph data to meet the statistical analysis premise can safeguard the privacy of users and their relationships. DP-SGD [7] is a promising privacy-preserving method for this purpose, as it can effectively protect sensitive information in datasets while allowing for meaningful analysis and model building.

With the increasing use of differential privacy, practical issues related to tracking the privacy budget have become a topic of discussion. Although differential privacy has many advantages, it also has some disadvantages, mainly due to the composition theorem. [8,9] This theorem states that for some composed mechanisms, the privacy cost simply adds up, leading to privacy budget restriction. This is especially problematic in deep learning, where model training is an iterative process, and each iteration adds to the privacy loss.

To address the shortcomings of differential privacy, Rényi Differential Privacy (RDP) [10] has been proposed as a natural relaxation of differential privacy.

RDP is a more robust notion of privacy that can lead to a more accurate and numerically stable computation of privacy loss.

Overall, the use of RDP can provide a more practical approach to maintaining privacy while also allowing for accurate and efficient computations. However, there are still ongoing discussions and research regarding the advantages and limitations of RDP compared to relaxed differential privacy. The main contributions of this work include:

1. The use of data augmentation techniques to improve the performance of the model. In the context of graph data processing, the dataset can be expanded by randomly removing, adding, or truncating the original graph data, thereby enhancing the training effectiveness of the model.

2. The proposed RDP-GGAN model is a novel approach to privacy-preserving graph generation. It incorporates Rényi differential privacy (RDP) into the graph generation process based on link reconstruction, and balances the need for privacy protection and data utility by adjusting the alpha parameter. By adding RDP to the GGAN framework, the model ensures that the generated graphs meet a certain level of privacy guarantee. This is achieved by introducing noise to the gradient updates during training, which makes it difficult for an attacker to infer sensitive information about the original data. Furthermore, the RDP-GGAN model provides strict privacy protection for the structured relationships within the graph, preventing various attacks while maintaining the global graph structure.

2 Related Works

As privacy, concerns in various fields are becoming increasingly important, and the availability of sensitive data sets is becoming more and more limited. Therefore, while protecting the statistical characteristics of sensitive data, it is crucial to protect the privacy of graph data to ensure that sensitive data can be reused. To achieve this goal, researchers are focusing on data privacy-preserving methods based on graph generation. These methods use techniques such as differential privacy to ensure that the resulting graph preserves the statistical properties of the original data, while also providing strong privacy guarantees. Graph generation techniques are particularly useful in situations where raw data cannot be shared due to privacy concerns. However, by using graph generation techniques, researchers can generate synthetic data that preserves the statistical properties of the original data, while also preserving individual privacy.

In 2016, Abadi et al. proposed the DPSGD algorithm [7], which aims to integrate differential privacy protection technology into the deep learning training process. The algorithm limits the sensitivity of each sample by clipping the gradient, then adds Gaussian noise to the samples in batches. This ensures that the model training process has differential privacy, which can protect the data privacy without compromising the robustness of the model, Lu et al. [12] Deploy Differential privacy in the Generative adversarial network, and re recognize the wind From a risk perspective, inference attacks are used to evaluate the difficulty of the attacker in re identifying any individual, in order to defend the model against privacy attacks.

Wang et al. [13] applied Rényi differential privacy in the deep learning model to further improve the application utility of privacy protection technology. In the deep learning training process, Wang et al. applied the Rényi differential privacy mechanism to the gradient update process of the model to protect the privacy of training data. They add noise to each gradient vector so that the sensitivity of the gradient vector does not exceed a certain threshold, thereby preserving privacy. Torkzadehmahani et al. [14] will Differential Privacy Techniques and Variations of Generative Adversarial Networks Conditional Generative Adversarial Networks (CGAN), combined with Rényi differential privacy technology, using gradient perturbation method, during the discriminator optimization process, the privacy of the real data optimization process is achieved Protect. Zhang et al. [15] proposed the dp-GAN algorithm, The algorithm applies the differential privacy technology to the deployment of the generative confrontation network, and trains the discriminant model of the generative confrontation network. During training, use the DPSGD algorithm to add noise to the gradient, and use the mutual game between the generative model and the discriminative model to improve the synthes is The quality of the dataset. Ma et al. [16] proposed the RDP-GAN method, in the discriminator update process, Using the method of target perturbation, adding the corresponding Rényi differential implicit Private, to realize the protection of the generated confrontation network model. [17] proposed PPGAN Privacy-preserving method, combined with generative adversarial network variant model WGAN, using ladder The degree perturbation method adds the corresponding Gaussian noise to the gradient to realize the hidden privacy protection. Triastcyn et al. [18] by adding Gaussian noise to the network of the discriminator layer to achieve privacy protection for the discriminator.

3   Preliminaries

In this section, we present some background knowledge of DP and GANs.

3.1   Generative Adversarial Networks

In 2014, Ian Goodfellow proposed the concept of Generative Adversarial Networks (GANs) [2], which is a type of neural network architecture consisting of two networks, a generator and a discriminator. The generator model is responsible for generating synthetic data, while the discriminator model is responsible for distinguishing between the real and generated data. The objective function of the generated confrontation network model is:

$$
\begin{aligned}
\min_{G} \max_{D} V(D,G) = {} & E_{x \sim p_{\text{data}}(x)}[\log D(x)] \\
& + E_{z \sim p_z(z)}[\log(1 - D(G(z)))]
\end{aligned}
\tag{1}
$$

where $p(z)$ is a prior distribution of latent vector z, $g(\cdot)$ is a generator function, and $d(\cdot)$ is a discriminator function whose output spans $[0, 1]$. $d(x) = 0$ (resp. $d(x) = 1$) indicates that the discriminator d classifies a sample x as generated (resp. real).

The two models are trained in a game-like manner, where the generator tries to produce synthetic data that can fool the discriminator into thinking it is real, and the discriminator tries to correctly classify the data as either real or generated. The goal of the generator is to produce data that is indistinguishable from the real data, while the goal of the discriminator is to correctly identify whether the data is real or generated. GANs have been widely used in various research fields, such as computer vision, natural language processing, and drug discovery, and have shown great research value. They have also been used for various applications, such as image and video synthesis, text-to-image generation, and style transfer. The overall structure of GAN is shown in Fig. 1.

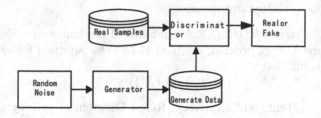

Fig. 1. The structure of GAN

3.2 Differential Privacy

Differential privacy [6] is a popular privacy protection technique used in various research fields. It is a mathematical framework that quantifies the privacy protection provided by data analysis algorithms. Differential privacy ensures that the output of an algorithm does not reveal any sensitive information about individual data points in the dataset. The relevant definitions and properties are as follows:

Definition 1 $((\epsilon, \delta) - DP)$: The randomized algorithm $\mathcal{A} : \mathcal{X} \to \mathcal{Q}$, is (ϵ, δ) - differentially private for all query outcomes $\mathcal{Q}$ and all neighbor datesets D and D' if:

$$\Pr[\mathcal{A}(D) \in \mathcal{Q}] \leq e^{\epsilon} \Pr[\mathcal{A}(D') \in \mathcal{Q}] + \delta \tag{2}$$

Two datasets D, D0 that only differ by one record are called neighbor datasets. The notion of neighboring datasets emphasizes the sensitivity of any individual private data. The parameters (ϵ, δ) denote the privacy budget. It only indicates our confidence level of privacy, given the (ϵ, δ) parameters. The smaller the (ϵ, δ) parameters are, the more confident we become about our algorithm's privacy as $((\epsilon, \delta)$ indicates the privacy loss by definition.

Definition 2. (Gaussian Mechanism) [10]. For a deterministic function f with its ℓ_2 -norm sensitivity as $\Delta_2 f = \max_{\|\mathbf{G}-\mathbf{G}'\|_1=1} \|f(\mathbf{G}) - f(\mathbf{G}')\|_2$, we have:

$$\mathcal{M}_f(\mathbf{G}) \triangleq f(\mathbf{G}) + \mathcal{N}\left(0, \Delta_2 f^2 \sigma^2\right) \tag{3}$$

where $\mathcal{N}\left(0, \Delta_2 f^2 \sigma^2\right)$ is a random variable obeying the Gaussian distribution with mean 0 and standard deviation $\Delta_2 f \sigma$. The randomized mechanism $\mathcal{M}_f(\mathbf{G}) is (\varepsilon, \delta) - DP if \sigma \geq \Delta_2 f \sqrt{2\ln(1.25/\delta)}/\varepsilon and \varepsilon < 1$.

Definition 3. (Rényi divergence) [10]. For two probability distributions P and Q defined over $\mathcal{R}$, the Rényi divergence of order $\alpha > 1$ is

$$D_\alpha(P\|Q) \triangleq \frac{1}{\alpha-1} \log E_{x \sim Q} \left(\frac{P(x)}{Q(x)}\right)^\alpha \tag{4}$$

where $P(x)$ denotes the density of P at x. It motivates exploring a relaxation of DP based on the Rényi divergence.

Definition 4. $((\alpha, \epsilon) - RDP)$ [10]. A randomized mechanism $f : \mathcal{X} \mapsto \mathcal{R}$ is said to have ϵ -Rényi D P of order α, or (α, ϵ) R D P for short, if for any adjacent $X, X' \in \mathcal{X}$ it holds that

$$D_\alpha\left(f(X)\|f(X')\right) \leq \epsilon \tag{5}$$

Compared to (ϵ, δ)-differential privacy, Rényi Differential privacy is a stricter definition of privacy that provides an operational It is a convenient and quantitatively accurate method that can track and calculate the cumulative privacy loss during the execution of independent differential privacy mechanisms.

Proposition 1 (Composition of RDP [10]): If $\mathcal{A} : \mathcal{X} \to \mathcal{U}_1$ is (α, ϵ_1) -RDP and $\mathcal{B} : \mathcal{U}_1 \times \mathcal{X} \to \mathcal{U}_2$ is (α, ϵ_2) -RDP, then the mechanism $(\mathcal{M}_1, \mathcal{M}_2)$, where $\mathcal{M}_1 \sim \mathcal{A}(\mathcal{X})$ and $\mathcal{M}_2 \sim \mathcal{B}(\mathcal{M}_1, \mathcal{X})$, satisfies the $(\alpha, \epsilon_1 + \epsilon_2)$ -RDP conditions.

4   RDPGGAN

This paper draws on the optimization strategy proposed by Carl Yang et al. [11], uses its DPGGAN model as the basic framework, and combines Rényi differential privacy to propose RDP-GGAN for secure graph generation. This model aims to solve the problems of GAN model stability and performance degradation caused by the fixed gradient clipping of the traditional DPGAN algorithm, and takes into account privacy protection and the availability of generated data.

The goal of RDP-GGAN is to securely generate graph data, paying special attention to the privacy issue of edges.

Since edges are necessary to protect the interaction of graph data objects. In this model, the gradients of the generator and discriminator are first noised using Rényi differential privacy to guarantee privacy. Then, the optimization strategy in DPGGAN is used to optimize the GAN model to improve the stability and performance of the model.

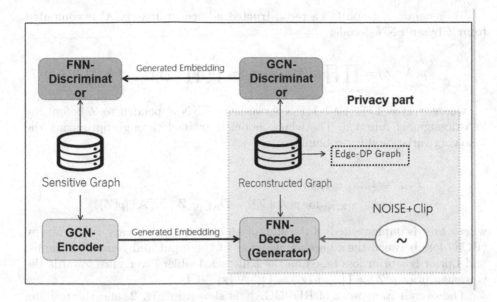

Fig. 2. The overall framework of RDPGGAN

GCNs have shown great potential in computing general-purpose graph representations [19]. In order to protect the privacy of individual links without seriously destroying the global network structure, in consideration of edge privacy, this paper takes advantage of the power and simplicity of GCNs by combining link reconstruction based graph variational autoencoders (GVAE) [20] adapted for our backbone graph generative model.

The GVAE model is an autoencoder model whose autoencoder architecture includes a GCN-based graph encoder and a feedforward neural network (FNN)-based adjacency matrix decoder. Graph encoders use GCNs to learn graph representations that can be used to guide the learning of an adjacency matrix decoder to generate graphs that are topologically similar to the input graph. In addition, the GVAE model also introduces a random latent variable z as a latent representation of the input graph.

$$q(\mathbf{Z} \mid \mathbf{X}, \mathbf{A}) = \prod_{i=1}^{N} q\left(\mathbf{Z}_i \mid \mathbf{X}, \mathbf{A}\right) = \prod_{i}^{N} \mathcal{N}\left(\mathbf{z}_i \mid \mu_i, \operatorname{diag}\left(\sigma_i^2\right)\right) \qquad (6)$$

where $\mu = \mathbf{g}_\mu(\mathbf{X}, \mathbf{A})$ is the matrix of mean vectors μ_i, and $\sigma = \mathbf{g}_\sigma(\mathbf{X}, \mathbf{A})$ is the matrix of standard deviation vectors $\sigma_i \cdot \mathbf{g}_\bullet(\mathbf{X}, \mathbf{A}) = \tilde{\mathbf{A}} \operatorname{ReLU}\left(\tilde{\mathbf{A}} \mathbf{X} \mathbf{W}_0\right) \mathbf{W}_1$ is a two-layer GCN model. $\mathbf{g}_\mu$ and $\mathbf{g}_\sigma$ share the first-layer parameters $\mathbf{W}_0$. $\tilde{\mathbf{A}} = \mathbf{D}^{-\frac{1}{2}} \mathbf{A} \mathbf{D}^{-\frac{1}{2}}$ is the symmetrically normalized adjacency matrix of $\mathbf{G}$, with degree matrix $\mathbf{D}_{ii} = \sum_{j=1}^{N} \mathbf{A}_{ij} \cdot \mathbf{g}_\mu$ and $\mathbf{g}_\sigma$ form the encoder network.

To generate a graph G', a reconstructed adjacency matrix A' is computed from Z by an FNN decoder

$$p(\mathbf{A} \mid \mathbf{Z}) = \prod_{i=1}^{N}\prod_{j=1}^{N} p\left(\mathbf{A}_{ij} \mid \mathbf{z}_i, \mathbf{z}_j\right) = \prod_{i=1}^{N}\prod_{j=1}^{N} \sigma\left(\mathbf{f}\left(\mathbf{z}_i\right)^T \mathbf{f}\left(\mathbf{z}_j\right)\right) \tag{7}$$

where $\sigma(z) = 1/\left(1 + e^{-z}\right)$, $\mathbf{f}$ is a two-layer FNN appended to $\mathbf{Z}$ before the logistic sigmoid function. The whole model is trained through optimizing the following variational lower bound

$$\begin{aligned}
\mathcal{L}_{\text{vae}} &= \mathcal{L}_{\text{rec}} + \mathcal{L}_{\text{prior}} \\
&= \mathbb{E}_{q(\mathbf{Z}|\mathbf{X},\mathbf{A})}[\log p(\mathbf{A} \mid \mathbf{Z})] - D_{\text{KL}}(q(\mathbf{Z} \mid \mathbf{X}, \mathbf{A})\|p(\mathbf{Z}))
\end{aligned} \tag{8}$$

where $Lrec$ is implemented as the sum of an element-wise binary cross entropy (BCE) loss between the adjacency matrices of the input and generated graphs, and $Lprior$ is a prior loss based on the Kullback-Leibler divergence towards the Gaussian prior $p(\mathbf{Z}) = \prod_{i=1}^{N} p(\mathbf{z}_i) = \prod_{i}^{N} \mathcal{N}(\mathbf{z}_i \mid \mathbf{0}, \mathbf{I})$.

The overall framework of RDPGGAN is shown in Fig. 2, and the training process is detailed.

When performing gradient perturbation on the discriminator optimization process, random samples from each batch During the training process, calculate the gradient of the batch of samples g_θ. Using the clipping parameter C Clip the second norm of the gradient, and control the value of the second norm of the gradient in C. Then, set the value C to the gradient sensitivity and add the obedience $(0, (\sigma C)^2 \boldsymbol{I})$ distributed random noise, make the lot satisfy $(\varepsilon_i, \delta) - \text{DP}$. Assuming that the number of training batches is m, then according to the combination mechanism, it can be known that in a single round of the discriminator During training, the whole satisfies it $(\varepsilon_1 + \ldots + \varepsilon_m, \delta) - \text{DP}$ This paper uses Rényi Accountant finds the privacy budget used in each round $\varepsilon_t(\delta)$, and the privacy budget satisfy $\varepsilon_t(\delta) \geq \sum_{i=1}^{m} \varepsilon_i$, That is, in each round of iterative process, the discriminator model as a whole satisfies $(\varepsilon_t(\delta), \delta) - \text{DP}$. According to nature 1 can Therefore, the model satisfies the overall requirement in the optimization training process $(\varepsilon_T(\delta), \delta) - \text{DP}$ and The algorithm also sets the overall privacy budget $mathbf\varepsilon_0$, when the accumulated hidden When the private budget is higher than $mathbf\varepsilon_0$, the algorithm terminates. Assuming that the algorithm terminates at T', it can be obtained that $\varepsilon_0 \leq \varepsilon_{T'}(\delta)$ That is, the algorithm as a whole satisfies (ε_0, δ)-DP.

5   Experimental Evaluations

To enable a direct comparison between the original networks and the generated networks, we utilized two commonly used real-world network datasets: DBLP and IMDB. DBLP comprises 72 networks consisting of author nodes and co-author links, with an average node count of 177.2 and an average link count of 258. IMDB comprises 1500 networks of actor/actress nodes and co-star links, with an average node count of 13 and an average link count of 65.9. To enhance the generalization ability of the model, increase the diversity of data, and reduce the risk of model overfitting, we augmented the dataset and perturbed the graph data to obtain better data. The basic parameter configuration of the experiment is shown in Table 1.

Algorithm 1: algorithm 1

1 Graph data $\mathbf{G}(\mathbf{A}, \mathbf{X})$, clipping parameter C , privacy budget ε , $noisescale\sigma$, total number of nodes N, batch size $B = qN$, learning rate η ,decay ratio γ, maximum number of training epochs T , loss weighing parameters $\lambda_1 and \lambda_2$ Differentially private decoder f . Initialize weights randomly for $\mathbf{g}_\mu, \mathbf{g}_\sigma, \mathbf{f}, \mathbf{g}'$ and $\mathbf{f}'$.

2 **for** $t \in T$ **do**

3 $\quad$ **for** $i = 1, 2, \ldots, bs$ **do**

4 $\quad\quad$ Sample a subgraph $\mathbf{G}_{sub}\,(\mathbf{A}_{sub}, \mathbf{X}_{sub})$ of size B

5 $\quad\quad$ Mean vector: $\mu \leftarrow \mathbf{g}_\mu\,(\mathbf{X}_{sub}, \mathbf{A}_{sub})$ Standard deviation vector:
$\quad\quad\quad \sigma \leftarrow \mathbf{g}_\sigma\,(\mathbf{X}_{sub}, \mathbf{A}_{sub})$

6 $\quad\quad$ Update
$\quad\quad\quad q(\mathbf{Z} \mid \mathbf{X}, \mathbf{A}) \leftarrow \prod_{i=1}^{N} \mathcal{N}\left(\mathbf{z}_i \mid \mu_i, \mathrm{diag}\left(\sigma_i^2\right)\right) Sample \mathbf{z}_i, \mathbf{z}_j \sim q(\mathbf{Z} \mid \mathbf{X}, \mathbf{A})$
$\quad\quad\quad$ Reconstruct adjacent matrix $\mathbf{A}' \leftarrow \sigma\left(\mathbf{f}\,(\mathbf{z}_i)^T, \mathbf{f}\,(\mathbf{z}_j)\right)$

7 $\quad\quad$ $\mathcal{L}_{\mathrm{prior}} = D_{\mathrm{KL}}(q(\mathbf{Z} \mid \mathbf{X}, \mathbf{A})\|p(\mathbf{Z}))$

8 $\quad\quad$ $\mathcal{L}_{\mathrm{gan}} = \log(\mathcal{D}(\mathbf{A})) + \log\left(1 - \mathcal{D}\,(\mathbf{A}')\right)$

9 $\quad\quad$ **for** $node x_i \in \mathbf{G}_{sub}$ **do**

10 $\quad\quad\quad$ $\mid Compute g_\theta\,(x_i) \leftarrow \partial\left(\mathcal{L}_{rec} - \lambda_2 \mathcal{L}_{gan}\right)/\partial x_i$

11 $\quad\quad$ Perturb gradient: $\tilde{g}_\theta \leftarrow \frac{1}{B}\left(\sum_i \bar{g}_\theta\,(x_i) + \mathbf{N}\left(0, \sigma^2 C^2 \mathbf{I}\right)\right)$

12 $\quad\quad$ Clip gradient: $\bar{g}_\theta\,(x_i) \leftarrow g_\theta\,(x_i) / \max\left(1, \frac{\|g_\theta(x_i)\|_2}{C}\right)$

13 $\quad\quad$ Average gradient: $g_\theta \leftarrow \frac{1}{B}\sum_i g_\theta\,(x_i)$

14 $\quad\quad$ Update $\mathbf{g}_\mu, \mathbf{g}_\sigma \overset{\pm}{\longleftarrow} \cdot\nabla_{\mathbf{g}}\left(\mathcal{L}_{rec} + \lambda_1 \mathcal{L}_{\mathrm{prior}}\right)$ Calculate ε about ε_0 using Rényi Accountant

15 $\quad\quad$ **if** $\varepsilon \geq \varepsilon_0$ **then** End Training.

To evaluate the effectiveness of RDPGGAN, the authors utilized a standard set of graph statistics to assess its ability to capture the global network structure. These statistics have been used in previous studies to evaluate the performance of graph generative models, particularly from a global perspective. Additionally, for a more comprehensive comparison, the authors compared the results of RDPGGAN with RDPGVAE, NetGAN, and GraphRNN.

During the comparison, the authors trained all models from scratch for each graph in the dataset. They then used the trained model to generate a graph, which was compared to the original graph using graph statistics to measure the difference between them. To more comprehensively evaluate the global utility of generated graphs, the authors also conducted a study on graph classification, which is a widely used graph-level downstream task. The graph statistics used in this study include LCC (size of the largest connected component), TC (triangle count), CPL (characteristic path length), GINI (Gini index), and REDE (relative edge distribution entropy).

In Tables 2 and 3(Performance evaluation of comparative models on a critical set of graph structure statistics is conducted. The statistic being assessed is the average absolute difference between the networks produced by the various models and the original network in its raw form. Consequently, lower values indicate a more accurate representation of the overall network structure, thereby implying superior utility in terms of global data analysis.), as we gradually increase the privacy budget v, it is evident that the data generated under the RDPGGAN model exhibits the smallest average absolute difference compared to the original network (the average absolute difference is an indicator used to evaluate the difference between the generated network by the different models and the original network. Thus, a smaller value indicates a better performance, which captures the global network structure more accurately and thus obtains better global data utility). This demonstrates the effectiveness of our privacy constraints and the apparent trade-off between privacy and utility. Moreover, under the same privacy budget, RDPGGAN consistently outperforms RDPGVAE, which highlights the superiority of the new design of our GAN framework.

In addition to basic statistical metrics, we also analyze the generated graphs by comparing their degree distributions and motif counts. For the degree distribution, we represent each graph as a 50-dimensional vector (where nodes with degree greater than 50 are grouped together). As for topic counting, we consider all 29 undirected topics with 3–5 nodes and convert each graph into a 29-dimensional vector by matching topics. To evaluate the similarity between the original and generated graphs, we compute the average cosine similarity between them. Furthermore, we evaluate the overall quality of the generated graphs by employing graph classification, specifically, we measure the accuracy of GIN using default hyperparameter configurations. In Table 4, we evaluate performance in terms of degree distribution, motif counts, and GIN accuracy. The graphs generated by RDPGGAN are competitively similar to the original graphs regarding both degree distributions and motif counts, while achieving satisfactory graph classification accuracy. All these results demonstrate the global structure preservation ability of RDPGGAN. Higher values for cosine similarity and GIN accuracy indicate better utility of the resulting graph. Notably, the

Table 1. Experimental parameters

Variable Name	Describe	Default
δ	Privacy Deviation	1e-5
α	RDP Restrain	–
σ	Noise Scale	5
ε	Privacy Budget	–
T	Maximum iteration times	0.001
γ	ratio	0.99

Table 2. Data for each model under DBLP Networks

DBLP Networks					
Models	LCC	TC	CPL	GINI	REDE
GVAE	42.1	444.1	0.9867	0.0442	0.0005
GGAN	41.1	302.1	0.7697	0.0972	0.0105
RDPGVAE($\varepsilon = 0.1$)	6.5	63.1	0.8942	0.1932	0.0505
RDPGVAE($\varepsilon = 1$)	6.1	21.3	0.8553	0.1822	0.0465
RDPGVAE($\varepsilon = 10$)	**4.7**	**3.1**	**0.7133**	**0.1802**	**0.0465**
RDPGGAN($\varepsilon = 0.1$)	28.85	28.9	0.8963	0.1932	0.0465
RDPGGAN($\varepsilon = 1$)	26.46	20.1	0.8773	0.01752	0.0455
RDPGGAN($\varepsilon = 10$)	**11.9**	**11.1**	**0.8553**	**0.1742**	**0.0435**

Table 3. Data for each model under IMDB Networks

IMDB Networks					
Models	LCC	TC	CPL	GINI	REDE
GVAE	9.8	193.3	1.3535	0.0468	0.0264
GGAN	5.6	187.9	0.8615	0.0108	0.0194
RDPGVAE($\varepsilon = 0.1$)	10.6	236.1	0.7715	0.1258	0.0514
RDPGVAE($\varepsilon = 1$)	10.6	222.7	0.6725	0.1238	0.0485
RDPGVAE($\varepsilon - 10$)	**9.6**	**223.3**	**0.5885**	**0.1208**	**0.0454**
RDPGGAN($\varepsilon = 0.1$)	11.4	254.3	0.7435	0.1558	0.0594
RDPGGAN($\varepsilon = 1$)	11.2	233.5	0.6818	0.1278	0.484
RDPGGAN($\varepsilon = 10$)	**9.8**	**209.9**	**0.6025**	**0.1278**	**0.0354**

GIN model trained on the original network achieves 0.3578 and 0.6212 accuracies on the two datasets, respectively. These accuracies serve as upper bounds for all compared graph generative models.

To demonstrate the effectiveness of DPGGAN in ensuring individual link privacy, we perform additional training on each dataset using the model shown in Fig. 3. After training and generation, we align the nodes in the generated network with those in the original network according to the degree distribution. Subsequently, we evaluate the accuracy of individual link predictions by comparing the predicted links in the generated network with those hidden during training in the original network.

As shown in Fig. 3, the performance of the RDPGGAN model generated by link prediction on the two datasets is significantly lower than the prediction performance on the original network. This indicates that even if an attacker can identify nodes in the resulting network, they cannot use the information to accurately deduce whether a link exists between a specific pair of nodes on the original network.

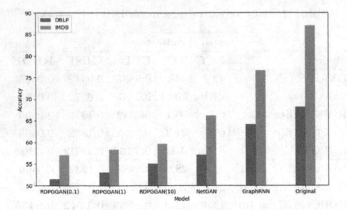

Fig. 3. model accuracy

Table 4. Model performance

DBLP				IMDB		
Models	Degree	Motif	Gin	Degree	Motif	Gin
GVAE	0.2643	0.4085	0.2826	0.5244	0.4118	0.5243
GGAN	0.2208	0.4119	0.3012	0.5396	0.4256	0.5356
NetGAN	0.2823	0.4056	0.3504	0.4652	0.3853	0.4848
GraphRNN	0.2796	0.3601	0.3261	0.4347	0.3702	0.4202
RDPGVAE	0.2836	0.4016	0.3043	0.4693	0.3986	0.5215
RDPGGAN	0.3343	0.4123	0.3478	0.4795	0.4053	0.5454

This performance drop is due to the RDPGGAN generative network being created by randomly perturbing the original network. This generates some incorrect node links or misleading information, which affects the accuracy of link prediction. Therefore, an attacker cannot accurately infer links on the original network from the generated network.

6   Conclusion

In order to solve the privacy problem of graph generation data, the author proposes the framework of RDPGAN based on Rényi differential privacy. Through the experimental tests of DBLP and IMDP datasets, it is verified that the framework proposed in this paper is effective in protecting the global graph structure and protecting the privacy of individual links.

References

1. Ji, S., et al.: Structural data deanonymization: theory and practice. IEEE/ACM Trans. Netw. **24**(6), 3523–3536 (2016)

2. Goodfellow, I., et al.: Generative adversarial nets. In: Proceedings of the Advances in Neural Information Processing Systems, pp. 2672–2680. ACM (2014)
3. Ciresan, D., et al.: Deep neural networks segment neuronal membranes in electron microscopy images. In: Proceedings of the 25th International Conferences on Neural Information Processing Systems, pp. 2843–2851. ACM, New York (2012)
4. Hinton, G., Deng, L., Yu, D., et al.: Deep neural networks for acoustic modeling in speech recognition: The shared views of four research groups. IEEE Signal Process. Mag. **29**(6), 82–97 (2012)
5. Zhu, J.Y., et al.: Unpaired image-to-image translation using cycle-consistent adversarial networks. In: Proceedings of the IEEE International Conference on Computer Vision. Piscataway, pp. 2223–2232. IEEE (2017)
6. Dwork, C., et al.: The algorithmic foundations of differential privacy. Found. Trends® Theor. Comput. Sci. **9**(3–4), 211–407 (2014)
7. Abadi, M., et al.: Deep learning with differential privacy. In: SIGSAC (2016)
8. Xie, L., Lin, K., Wang, S., Wang, F., Zhou, J.: Differentially private generative adversarial network. arXiv preprint: arXiv:1802.06739 (2018)
9. Acs, G., Melis, L., Castelluccia, C., De Cristofaro, E.: Differentially private mixture of generative neural networks. IEEE Trans. Knowl. Data Eng. **31**(6), 1109–1121 (2018)
10. Mironoy, I., Talwar, K., Zhang, L.: Rényi differential privacy of the sampled gaussian mechanism [EB/OL]. [2019-08-28]. https://arxiv.org/pdf/1908.10530.pdf
11. Yang, C., et al.: Secure deep graph generation with link differential privacy. In: International Joint Conference on Artificial Intelligence International Joint Conferences on Artificial Intelligence Organization (2021)
12. Lu, P.H., Wang, P.C., Yu, C.M.: Empirical evaluation on synthetic data generation with generative adversarial network. In: Proceedings of the 9th International Conference on Web Intelligence, Mining and Semantics, Seoul, Korea, pp. 1–6 (2019)
13. Wang, Y.X., Balle, B., Kasiviswanathan, S.P.: Subsampled Rényi differential privacy and analytical moments accountant. In: The 22nd International Conference on Artificial Intelligence and Statistics, pp. 1226–1235. PMLR, Naha, Okinawa, Japan (2019)
14. Torkzadehmahani, R., Kairouz, P., Paten, B.: DP-CGAN: differentially private synthetic data and label generation. In: Proceedings of the IEEE/CVF Conference on Computer Vision and Pattern Recognition Workshops, pp. 98–104. IEEE, Piscataway (2019)
15. Zhang, X., Ji, S., Wang, T.: Differentially private releasing via deep generative model (technical report). [Online] https://arxiv.org/abs/1801.01594.. Accessed 28 Mar 2020
16. Ma, C., et al.: RDP-GAN: a Rényi-differential privacy based generative adversarial network [EB/OL]. [2020-07-04]. https://arxiv.org/pdf/2007.02056.pdf
17. Liu, Y., et al.: PPGAN: privacy-preserving generative adversarial network. In: Proceedings of the 2019 IEEE 25Th International Conference on Parallel and Distributed Systems (ICPADS), pp. 985–989. IEEE, Piscataway (2019)
18. Triastcyn, A., Faltings, B.: Generating differebtially private datasets using GANs [EB/OL]. [2018-03-08]. https://arxiv.org/pdf/1803.03148v1.pdf
19. Maron, H., Ben-Hamu, H., Shamir, N., Lipman, Y.: Invariant and equivariant graph networks. In: ICLR (2019)
20. Kipf, T.N., Welling, M.: Semisupervised classification with graph convolutional networks. In: ICLR (2017)

Clustered Federated Learning Framework with Acceleration Based on Data Similarity

ZhiPeng Gao($\boxtimes$), ZiJian Xiong, Chen Zhao, and FuTeng Feng

Beijing University of Posts and Telecommunications, Beijing, China
{gaozhipeng,xiongzijian,zc_zhaochen,fengfuteng}@bupt.edu.cn

Abstract. Federated Learning is a distributed machine learning framework which allows multiple participants training machine learning model without exchanging their local data. It addresses critical issues such as data privacy in distributed machine learning. In real circumstances, the statistical heterogeneity of data on different devices will cause bad performance of training process. In this paper, we propose FedCSA, a clustered federated learning framework with acceleration algorithm using the similarity of data distribution between federated learning clients. Clients with similar data distribution are clustered and the acceleration algorithm is performed among them to obtain group model, which can maximum the utilization of similarity. The global model is aggregated from group models in each round of training. The empirical evaluation shows that FedCSA outperforms state-of-art approaches on datasets with different non-IID settings.

Keywords: Federated learning · Clustering · Data similarity · Distributed machine learning

1 Introduction

With the increasing computing power of mobile devices, artificial intelligence has become more and more widely used on mobile devices and in the Internet of Things [1–3]. It become possible to jointly model the same target using local data on devices of different users, which can improve training efficiency and data resource utilization. But at the same time, data privacy of user and other security issues need to be guaranteed, which is a challenge for joint training. Federated learning is a framework to address this problem, proposed by McMahan et al. [4] in 2016. In federated learning framework, clients participating in training collaborate to model with each other under the coordination of a central server, and in the process devices are allowed to jointly train models without exchanging local data. Therefore, federated learning can effectively reduce the risk of user privacy data leakage and the cost of taking privacy protection measures.

But at the same time, statistical heterogeneity brings a lot of challenges to federated learning. In fact, the assumption of independent and identically distributed does not hold in data distribution of different clients, which is called non-IID. In federated learning, this causes a larger number of training epochs required to meet the training objectives and make model converge, compared to the IID case. In practical applications, it will cause considerable additional communication overhead and training time [5,6]. Based on FedAvg algorithm [4], Li et al. added a proximal term to the loss function of the client-side training process to limit the deviation of the local update from the global model [7], so that the training process had better stability and alleviated the impact of non-IID. However, the superiority of this algorithm is only manifested in the situation that many clients fail to complete the specified number of rounds of training within the specified time.

Methods to reduce the impact of data heterogeneity by using the similarity information of data held by each client are gradually proposed. The algorithm proposed by Li et al. [8] uses the similarity information of client data to build decision trees to speed up the convergence of federated learning. But this algorithm broadcasts the hash value of each user data among all clients. The degree of privacy protection is weaker than traditional federated learning. Also, this method is only used for the building of decision trees in gradient boosting tree [9]. It is not applicable to other models such as neural networks which are more widely used. Other methods [10–12] cluster clients based on the similarity of data distribution between the clients. The FedAvg algorithm is run inside clusters to obtain the group model, and then the models of different groups are aggregated to obtain the global model. These methods can also speed up the convergence speed of the training process. However, in circumstance of high degree of heterogeneity of client data, there is no obvious advantage over traditional federated learning and the stability of these algorithms declines.

In this paper, we propose a efficient **Fed**erated learning framework with **C**lustering and **S**imilarity based **A**cceleration algorithm (FedCSA). We propose a federated learning acceleration algorithm based on client data similarity. The similarity information of client data is used to identify the data distribution of the client first. Then, the clients with similar data distribution are clustered in a group to train local model and obtain a group model by aggregation. Clients with similar data distributions can be approximated as IID clients, so the impact of non-IID is reduced. Finally, these group models are aggregated to obtain a global model.

At the same time, among clients with similar data distribution, the proportion of data with a certain degree of similarity is increased in the training process. Thus, the statistical heterogeneity of client data is further reduced to improve the convergence speed and accuracy of federated learning process.

The main contributions of this paper are listed below:

- we propose a efficient clustered federated learning framework for non-IID environment.
- we propose an acceleration algorithm based on the similarity of client data distribution to improve the convergence speed and accuracy of federated learning process.

- we experimentally prove that our FedCSA algorithm performs better than the previous state-of-art method such as *FedAvg* in terms of model accuracy and convergence speed under heterogeneous environments.

2   Preliminaries

In this section, we introduce the basic background knowledge on federated learning and locality-sensitive hash.

2.1   Federated Learning

The most widely used federated learning algorithm is FedAvg proposed by McMahan et al. *FedAvg* allows clients to collaboratively train models such as neural networks without sharing their dataset, which provides privacy protection on private data of clients.

The goal of training is to find global model parameters w^* which minimized the global objective function:

$$w^* = argmin_w\{F(w)\} \ where \ F(w) = \sum_{i=1}^{n} p_i F_i(w^i) \tag{1}$$

where $F_i(w)$ is the local empirical loss on the local dataset of client i. w^i are the model parameters of client i. n is the number of total client participating in training process and p_i is the contribution rate of client i where $\sum p_i = 1$.

In each round of the training process, server selects a number of clients to start training. Clients perform stochastic gradient descent (SGD) on their local dataset and send the model parameters to server to obtain global model. The aggregation process of client models is as follows:

$$w_t = \sum_{i=1}^{n} \frac{|D_i|}{|D|} w_t{}^i \tag{2}$$

where D is the total data sample size from all clients participating in this round, and D_i is the local dataset of client i. w_t denotes the global model parameters in round t and $w_t{}^i$ denotes the local model parameters of client i in round t. In FedAvg, the contribution of client is represented by the size of client local dataset.

2.2   Locality-Sensitive Hashing

Locality-Sensitive Hashing [13] (LSH) is a hashing algorithm. Similar raw data has a high probability of remaining adjacent after LSH calculation, so it can be used for data clustering and nearest neighbor search. Locality-sensitive hash functions can also play a good role in privacy-preserving computing. LSH can achieve dimensionality reduction and local matching of data, which can reduce the amount of information carried by the data, enhance its privacy. And the statistical characteristics of the data are not lost. LSH can take advantage of similarity to preserve the relationship between data.

3   Design of Our FedCSA Framework

In this section, we introduce the details of the proposed FedCSA framework.

In Sect. 1, we mentioned that exploiting the similarity of client data distribution can speed up the training process. In the actual federated learning process, the data distribution of the clients participating in each round of training is different. If the data distribution of a selected client is used to accelerate training based on data similarity, the global model obtained in this round will be shifted to this data distribution. However, the data distribution of the client selected in the next round is usually different from the previous round, so the global model will shift in a different direction. In this case, the global model will oscillate between different data distributions, and the use of data similarity will prolong the convergence of the global model and reduce the model accuracy. Therefore, our proposed algorithm groups clients with similar data distribution in advance, and uses the data similarity to speed up the training process within each group, so that the data similarity can be effectively utilized. And we present a similarity-based approach named *Intra-Group Acceleration* to obtain better models within each client group.

The basic steps of FedCSA are similar to traditional federated learning framework. Figure 1 shows the overall framework of FedCSA. The main steps are presented as follows:

Preparation Stage. Each client calculates the similarity feature (LSH) of the local data and sends it to the central server. The central server receives the similarity information, calculates the similarity matrix and sends the similarity matrix to each client. The central server divides the clients into different groups for training according to the similarity of the data distribution of the client.

Training Stage. In each training round, the central server selects a certain number of clients to participate in training. Clients assigned to the same group collaborate to train a model, which is called group model. Within each group, clients receive global model and similarity information from the central server and execute local SGD to train their local models. Intra-Group Acceleration algorithm is used in training process to obtain better model. Then clients send their local models to the central server.

Aggregation Stage. The central server first aggregates the local models of clients by group to obtain group models. Then the server aggregates group models to generate new global model for next training round.

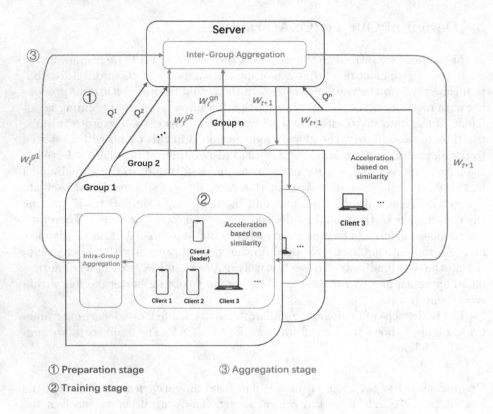

① Preparation stage   ③ Aggregation stage
② Training stage

Fig. 1. Overall Framework of FedCSA.

The training process will last until the global model reaches the target accuracy or converges.

Our proposed FedCSA includes two key parts: *Intra-Group Acceleration* and *Group Aggregation* in clustered federated learning.

3.1 Intra-group Acceleration

Clients in the same group have similar data distribution. On this basis, we further improve the similarity of the data distribution to generate a better group model. We use the data distribution of a client in the group as a benchmark, and build datasets with similar data distribution on each client in the same group.

The *Intra-Group Acceleration* algorithm consists of algorithms in preparation stage and training stage.

Before the training process starts, clients first obtain similarity information and establish similarity matrices. Algorithm 1 describes the process of generating similarity matrix of each client. Each client first compute parameters q_i^m for similarity computing of each local instance x_i^m, and then broadcast them to the server. After receiving the loss information from all clients, server generate similarity matrices by finding the instance $S_{ij}{}^m$ in client c_j with the highest similarity with instance i in dataset D_m of client c_m, and then send them to clients.

The function which computes parameters for similarity computing can be implemented in many different ways. The gradients or loss of client local dataset above the global model can be used as parameters for similarity computing [14], which is widely used in clustered federated learning frameworks [10,11]. And LSH value of each instance in client local dataset can also be used as parameters for similarity computing, which is used in SimFL [8]. In this paper, we choose the LSH method due to privacy requirements.

At the same time, server uses the parameters $q_i{}^m$ of each client to divide clients into groups by clustering algorithm (K-means algorithm is used in this paper).

In each round of training, the selected clients are divided into groups according to the result of clustering. The acceleration algorithm is performed within each group. In training round t, server first select a certain client c_l from the group as leader, the data distribution of which will be used as reference for acceleration.

As the leader, c_l only needs to train the model on its local dataset D_l, and then send the model parameters obtained by training to the central server. For the client c_m that is not the leader, it first queries the similarity matrix S^m to obtain the partial data $D_{sim}{}^m$ in the local dataset which is similar to the data in D_l. The client then checks the proportion of $D_{sim}{}^m$ in its local dataset D_m, which can describe how similar the local dataset is to D_l. If the proportion is less than a threshold α, a data augmentation operation is performed on $D_{sim}{}^m$. In this paper, we use the *AUGMIX* method proposed by Hendrycks et al. [15] for data augmentation. Then, the augmented data $D_{sim}{}^m$ is combined with the dissimilar data in the local dataset to generate new train set D_m. The client trains the model on D_m, and sends the model parameters $w_t{}^m$ obtained in training round t to the central server.

Our proposed *Intra-Group Acceleration* algorithm further improves the similarity of the data distribution between the leader and other members of the group by increasing the proportion of similar data. Since the adverse effects of non-IID can be further reduced, a group model with higher accuracy can be obtained.

3.2 Group Aggregation

After receiving the model update $w_t{}^c$ sent by clients, the server first performs model aggregation for each group. The obtained group model after aggregation is a representative model of the data distribution for this group. For group g_j, in training round t, the aggregation process of the group model is as follows:

$$w_t{}^{g_j} = \sum_{c_m \in g_j} \frac{|D_m|}{|D|} w_t{}^m \qquad (3)$$

Algorithm 1: Client Similarity Matrix Preparation

Input: Dataset D_m of client m

Output: Similarity matrix S^m of client m

1 **for** *each client $c_m \in C$* **do**
2 | **for** *each instance $x_i^m \in D_m$* **do**
3 | | $q_i^m \leftarrow ComputeSimilarityParameters(x_i^m)$;
4 | Send Q^m to server;

5 **Server Execution:**
6 **for** *each client c_m in C* **do**
7 | **for** *each instance $x_i^m \in D_m$* **do**
8 | | **for** $c_j \in C$ **do**
9 | | | **if** $j \neq m$ **then**
10 | | | | Find instance S_{ij}^m which has the highest similarity with x_i^m;
11 | | | **else**
12 | | | | $S_{ij}^m \leftarrow x_i^m$;

13 | **return** S^m;

---

where $|D|$ is the total size of datasets in group g_j, and $w_t^{g_j}$ is the group model after aggregation. The intra-group aggregation process is basically the same as the model aggregation process of traditional federated learning in (2).

After group model aggregation is complete, server aggregates the group models into the global model. In training round t, the aggregation process of the global model is as follows:

$$w_t = \frac{1}{|G|} \sum_{g_j \in G} w_t^{g_j} \tag{4}$$

where G is the set of all groups, $|G|$ is the number of groups, and w_t is the global model in training round t. When there is no member in a group participating in this round of training, we use the latest group model of this group in the aggregation process.

The global model is the arithmetic mean of the group models. In our proposed algorithm, data distribution of each group is considered to be of equal importance.

The whole process of FedCSA is shown in detail in Algorithm 2. Our proposed method reduces the discrepancy between the global optimization objective and the optimization objective of groups with different data distribution. Therefore, global model with higher prediction accuracy can be obtained, and the convergence speed of training process of federated learning can be improved.

3.3 Privacy Analysis

If the number of LSH functions is smaller than the number of training data dimensions, the privacy in FedCSA can be guaranteed.

Algorithm 2: Clustered Federated Learning with Similarity Acceleration

---

Input: Cluster numbers $n_{clusters}$, Initial global model w_0, number of total training round T

Output: global model w_T IN training round T

1 $G = Cluster(C, n_{clusters})$;
2 **for** *each round* $t = 1, ..., T$ **do**
3   Server broadcasts global model w_{t-1} to all clients;
4   Select K clients $\in C$;
5   **for** *each group* $g_j \in G$ **do**
6   select leader client $c_l \in g_j$;
7   **for** *each client* $c_m \in g_j$ **do**
8              $w_t{}^m \leftarrow w_{t-1}$;
9              $w_t{}^m \leftarrow ClientUpdate(c_l)$;
10         $w_t{}^{g_j} \leftarrow \sum_{c_m \in g_j} \frac{|D_m|}{|D|} w_t{}^m$;
11     $w_t \leftarrow \frac{1}{|G|} \sum_{g_j \in G} w_t{}^{g_j}$;
12 **return** w_T;
13 **ClientUpdate** (c_l):
14   **if** $m \neq l$ **then**
15         $D_{sim}{}^m \leftarrow \{S_{im}{}^l, x_i \in D_l\}$;
16   **if** $\frac{|D_{sim}{}^m|}{|D_m|} < \alpha$ **then**
17             $D_m \leftarrow (D_m \setminus D_{sim}{}^m) \cup DataAugumentation(D_{sim}{}^m)$;
18   **for** *each batch* $b \in D_m$ **do**
19         $w_t{}^m \leftarrow w_t{}^m - \eta \nabla loss(w_t{}^m, b)$;
20   **return** $w_t{}^m$;

It can be proved by basic mathematical theories. We define, $L_1, L_2, ..., L_m$ are the LSH functions. And we define that $d_1, d_2, ..., d_n$ are the data dimensions. For one data instance, we obtain that:

$$\begin{cases} L_1(d_1, \ d_2, \ ..., \ d_n) = l_1 \\ L_2(d_1, \ d_2, \ ..., \ d_n) = l_2 \\ ... \\ L_m(d_1, \ d_2, \ ..., \ d_n) = l_m \end{cases} \tag{5}$$

Since $m < n$, the system of equations in (5) is indeterminate. An indeterminate system usually has infinitely many solutions. Therefore, it is difficult to obtain the original data. The data privacy of using LSH can be guaranteed to some extent. However, from the perspective of the data transmission, the data privacy protection of this method is not as good as traditional federated learning approaches.

4   Experiment and Result

4.1   Experiment Settings

In our experiments, we compare our FedCSA algorithm with *FedAvg* [4], *FedProx* [7] and *FedSim* [10]. FedAvg and FedProx are traditional federated learning frameworks and FedSim is a clustered federated learning framework.

We conduct the experiments on a device with an Intel Xeon Processor (Cascadelake) CPU and 2 Nvidia Tesla V100S 32 GB GPUs.

Dataset. We use 2 standard datasets, MNIST and CIFAR-10, in the experiments. MNIST is a handwritten digit recognition dataset with 10 categories ranging from 0 to 9, with a total of 60,000 images. The training set contains 50,000 images, and the test set contains 10,000 images. Each instance of data is a 28×28 single-channel picture. CIFAR-10 is a real-world object recognition dataset containing 10 categories, with a total of 60,000 images. The training set contains 50,000 images, and the test set contains 10,000 images. Each instance of data is a 3-channel color image with size of $32 \times 32 \times 3$.

Non-IID Settings. In our experiments, we simulate the situation of data heterogeneity on class labels. The distributions of class labels of different clients are different. We use the following method to divide the original dataset into non-IID datasets. First we select one class label as the main class label among all classes in the dataset. The proportion of the selected main class label in the dataset allocated to the client is θ, and the remaining data with a proportion of $1 - \theta$ are randomly selected from the data of three class labels. The three classes are also randomly selected from other classes. Obviously, in this data division method, the value of θ determines the degree of heterogeneity of the client dataset. The larger the value of θ, the higher the degree of heterogeneity of the client data distribution. In the experiments on MNIST dataset, we set the value of θ to 0.8, 0.9, 0.95. In the experiments on CIFAR-10 dataset, we set the value of θ to 0.7, 0.8, 0.9.

Hyperparameters Settings. For the MNIST and CIFAR-10 datasets, we uniformly use LeNet5 model as the training model, SGD as the optimizer, and the learning rate is set to 0.01 without learning rate decay. For FedSim and FedCSA, we set the number of clusters to 5. And in each experiment, we set the value of α to 0.8 for FedCSA.

Real Environment Simulations. The availability of clients in real-world environments may not be guaranteed throughout the training process. For both MNIST and CIFAR10, we set the number of clients to 50. At the beginning of each round of training, 10 clients are randomly selected from total the 50 clients to participate in the training.

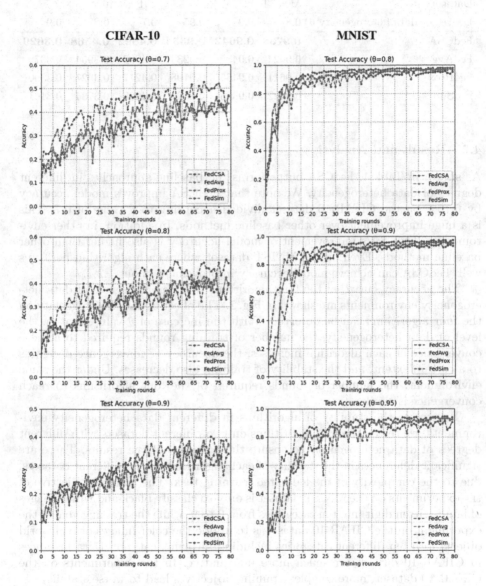

Fig. 2. Test accuracy curves on MNIST and CIFAR-10 ($\theta = 0.8$, 0.9, 0.95 for MNIST and $\theta = 0.7$, 0.8, 0.9 for CIFAR-10)

Table 1. The model test accuracy performance of FedCSA, FedAvg and FedProx on two federated datasets with different degree of data heterogeneity.

Dataset	MNIST			CIFAR-10		
Degree of data heterogeneity θ	0.8	0.9	0.95	0.7	0.8	0.9
FedCSA	**0.9738**	**0.9642**	**0.9334**	**0.4892**	**0.4568**	**0.3629**
FedAvg	0.9624	0.9485	0.9233	0.4202	0.3954	0.3379
FedProx	0.9611	0.9465	0.9198	0.4217	0.3927	0.3390
FedSim	0.9422	0.9305	0.8823	0.4008	0.3734	0.3161

4.2   Result and Analysis

As shown in Table 1, FedCSA outperforms the baseline approaches in different degrees of data heterogeneity. We note that FedCSA improves model accuracy by 6.75% on the CIFAR-10 dataset divided according to θ value of 0.7. This is a huge improvement over other baseline methods. In addition, in other environments, FedCSA's improvement of model accuracy is also higher than other baseline methods. Even on the MNIST dataset where each algorithm performs well, FedCSA can achieve model accuracy improvement about 1.2%.

The accuracy curves of the four algorithms in different degrees of data heterogeneity environments are shown in Fig. 2. The effect of data heterogeneity on the four algorithms can be observed. With the increase of θ value, that is, the level of data heterogeneity, the number of training rounds required to achieve convergence of each algorithm increases, the accuracy of convergence decreases to a certain extent, and the stability of training also decreases. Under the same environment, the number of rounds required of the four algorithms to reach convergence is similar.

To explore the stability of algorithms in different data heterogeneity environments, we tested the four algorithms on two different datasets with different degrees of data heterogeneity to ensure the accuracy of the results. There are significant differences in the stability of each algorithm on two different datasets, due to the complexity of the federated learning task. The training objective of the experiments on the MNIST dataset is only to classify black and white images of handwritten digit into ten categories from 0 to 9, while the training goal of the experiment on the CIFAR-10 dataset is to divide the color images of real world objects into ten different categories including airplane, cat, etc. Data instances in CIFAR-10 dataset are much more informative. In the experiments on the CIFAR-10 dataset, more complex training objectives lead to worse stability.

In the experiments on the MNIST dataset, all four methods suffer to some extent under different degrees of data heterogeneity ($\theta = 0.8$ to 0.95), since larger heterogeneity leads to worse and steeper convergence. Under relatively low degree of data heterogeneity ($\theta = 0.7$), the jitter amplitudes of the accuracy curves of FedCSA, FedAvg, and FedProx are similar and smaller than which of the accuracy curve of FedSim. At a relatively high level of data heterogeneity

($\theta = 0.95$), the jitter amplitude of the accuracy curve of FedCSA is slightly larger than that of FedAvg and FedProx, but still smaller than that of FedSim. The experimental results on the CIFAR-10 dataset are similar to that. This suggests that FedCSA is more robust than FedSim and the stability of FedCSA is acceptable compared to popular federated learning algorithms.

5   Conclusion

To deal with the problem that the accuracy of the model decreases and the training convergence speed becomes slower under the heterogeneous environment in federated learning, we propose a clustered federated learning framework based on data similarity. By grouping clients for training, the influence of large differences in the data distribution of different clients is reduced. FedCSA further construct similar data distribution within client group, thereby increasing the accuracy of the group model. Therefore, it can aggregate better group models to obtain global model with better performance. We evaluated the proposed framework on two datasets with different data heterogeneity settings. In the experimental results, FedCSA achieves up to a 6.5% improvement of model accuracy on non-IID dataset. Our evaluation has demonstrated that our FedCSA framework is superior to the previous state-of-the-art framework, and is robust enough in heterogeneous environments.

Acknowledgments. This work is supported by the General Program of National Natural Science Foundation of China (62072049).

References

1. Mukhopadhyay, S.C., Tyagi, S.K.S., Suryadevara, N.K., Piuri, V., Scotti, F., Zeadally, S.: Artificial intelligence-based sensors for next generation IoT applications: a review. IEEE Sens. J. **21**(22), 24920–24932 (2021)
2. Shah, R., Chircu, A.: IoT and AI in healthcare: a systematic literature review. Issues Inf. Syst. **19**(3) (2018)
3. Kankanhalli, A., Charalabidis, Y., Mellouli, S.: A research agenda, IoT and AI for smart government (2019)
4. McMahan, B., Moore, E., Ramage, D., Hampson, S., y Arcas, B.A.: Communication-efficient learning of deep networks from decentralized data. In: Artificial Intelligence and Statistics, pp. 1273–1282. PMLR (2017)
5. Li, X., Huang, K., Yang, W., Wang, S., Zhang, Z.: On the convergence of FedAvg on non-IID data. arXiv preprint arXiv:1907.02189 (2019)
6. Li, Q., Diao, Y., Chen, Q., He, B.: Federated learning on non-IID data silos: an experimental study. In: 2022 IEEE 38th International Conference on Data Engineering (ICDE), pp. 965–978. IEEE (2022)
7. Li, T., Sahu, A.K., Zaheer, M., Sanjabi, M., Talwalkar, A., Smith, V.: Federated optimization in heterogeneous networks. Proc. Mach. Learn. Syst. **2**, 429–450 (2020)

8. Li, Q., Wen, Z., He, B.: Practical federated gradient boosting decision trees. In: Proceedings of the AAAI Conference on Artificial Intelligence, vol. 34, pp. 4642–4649 (2020)

9. Chen, T., Guestrin, C.: XGBoost: a scalable tree boosting system. In: Proceedings of the 22nd ACM SIGKDD International Conference on Knowledge Discovery and Data Mining, pp. 785–794 (2016)

10. Palihawadana, C., Wiratunga, N., Wijekoon, A., Kalutarage, H.: FedSim: similarity guided model aggregation for federated learning. Neurocomputing **483**, 432–445 (2022)

11. Duan, M., et al.: FedGroup: efficient federated learning via decomposed similarity-based clustering. In 2021 IEEE International Conference on Parallel & Distributed Processing with Applications, Big Data & Cloud Computing, Sustainable Computing & Communications, Social Computing & Networking (ISPA/BDCloud/SocialCom/SustainCom), pp. 228–237. IEEE (2021)

12. Tian, P., Liao, W., Wei, Yu., Blasch, E.: WSCC: a weight-similarity-based client clustering approach for non-IID federated learning. IEEE Internet Things J. **9**(20), 20243–20256 (2022)

13. Paulevé, L., Jégou, H., Amsaleg, L.: Locality sensitive hashing: a comparison of hash function types and querying mechanisms. Pattern Recogn. Lett. **31**(11), 1348–1358 (2010)

14. Wang, Y., Wolfrath, J., Sreekumar, N., Kumar, D., Chandra, A.: Accelerated training via device similarity in federated learning. In: Proceedings of the 4th International Workshop on Edge Systems, Analytics and Networking, pp. 31–36 (2021)

15. Hendrycks, D., Mu, N., Cubuk, E.D., Zoph, B., Gilmer, J., Lakshminarayanan, B.: AugMix: a simple data processing method to improve robustness and uncertainty. arXiv preprint arXiv:1912.02781 (2019)

An Anonymous Authentication Scheme with Low Overhead for Cross-Domain IoT

Long Fan, Jianfeng Guan$^{(\boxtimes)}$, Kexian Liu, and Pengcheng Wang

School of Computer Science (National Pilot Software Engineering School), Beijing University of Posts and Telecommunications, Beijing, China
{fanlong,jfguan,kxliu,wpc1021}@bupt.edu.cn

Abstract. In a decentralized Internet of things (IoT) environment, it is inevitable that devices from different administrative domains communicate and collaborate with each other, yet there is often a lack of trust among them. In this case, it is necessary to design a reliable authentication scheme to ensure the secure cross-domain access of devices. However, existing cross-domain authentication schemes suffer from a number of issues that have not been fully considered. On the one hand, as one of techniques commonly used for cross-domain authentication, blockchain has limitations in storage capacity and throughput speed. On the other hand, users' privacy and sensitive information are not strictly protected. In this paper, we propose a certificateless cross-domain authentication scheme that combines low overhead and anonymity, named CALA. The cryptographic theory of CALA is Certificateless Public Key Cryptography (CL-PKC) mechanism, which relies on a semi-trusted third party and gets rid of the key escrow problem. To reduce the storage overhead and enable light-weight data verification, we design a storage structure named Merkle Hash Tree Grid Combination (MHTGC) for data management. To preserve the privacy of users, we propose an anonymous authentication protocol based on Zero-Knowledge Proof (ZKP) algorithm. The security analysis and experimental results demonstrate the effectiveness of our scheme.

Keywords: Cross-domain authentication · Certificateless signature · Privacy protection · Blockchain · Storage optimization

1 Introduction

In recent years, researches related to Internet of Things (IoT), such as smart healthcare [29], smart home [20], intelligent transportation [23] and industrial Internet [22], have been emerging in an endless stream. Taking the Industrial Internet of Things (IIoT) as an example, it is difficult for a standalone administrative domain to produce a product that satisfies customers due to the complexity of product manufacturing process [22]. In order to access the services or resources unavailable in its own trust domain, communication and collaboration between devices from different administrative domains become inevitable, consequently. In the environment of IoT, especially in the decentralized environment [31,33], it is urgent to construct trust contracts between different administrative domains which lack trust relationship. Unlike access within a single administrative domain, cross-domain access is facing more security risks, such as unauthorized operations across domains and users' privacy disclosure.

To enable secure communication between IoT devices, it is permitted to authorize legitimate users access resources or services in other domains. A more reliable cross-domain authentication mechanism is therefore essential. Currently, most cross-domain authentication mechanisms are based on Public Key Infrastructure (PKI) [8]. PKI based schemes often rely on a trusted third party named Certificate Authority (CA), which is prone to single point of failure. Besides, CA needs to manage numerous public key certificates, which brings a great burden to the CA [18]. Authentication mechanisms based on Identity-Based Cryptography (IBC) take the users' valid identities as public keys (also known as Identity-Based Public Key Cryptography, ID-PKC) [21], so there is no problem of digital certificate management. However, private keys of all devices are generated by Private Key Generator (PKG), which means that private keys are escrowed by a third party. In contrast, another kind of authentication mechanism based on Certificateless Public Key Cryptography (CL-PKC) gets rid of certificate management and key escrow problems [3].

Traditional cross-domain authentication mechanisms rely on trusted third party, which brings many potential risks, such as single point of failure, high computational overhead, and low flexibility, etc. [26]. The distributed consensus mechanism of blockchain inspires researchers to design cross-domain authentication schemes and construct trust relationship between different administrative domains by utilizing it [4,12,22]. However, blockchain has storage resource constraints, and its query and write delays can slow down the execution of the entire authentication scheme [25,28].

To tackle the above challenges, we propose a cross-domain authentication scheme with low overhead and anonymity (CALA) for IoT environment in this paper. This scheme achieves low computational and storage overhead without exposing the privacy of devices. Our contributions can be summarized as follows:

(1) We propose a certificateless cross-domain authentication scheme that combines low overhead and anonymity (CALA), which supports authentication between IoT devices from different administrative domains. By utilizing Zero-Knowledge Proof (ZKP) algorithm, users' privacy can be protected in the process of cross-domain authentication.
(2) We design an efficient data management mechanism. Specifically, we adopt a data storage structure named Merkle Hash Tree Grid Combination (MHTGC) for storing and processing most of the data. As an extension of Merkle Hash Tree (MHT), Merkle Hash Grid (MHG) is leveraged not only to reduce storage overhead but also to reduce computational overhead in data verification.
(3) We conduct a theoretical analysis to prove the high security of our scheme. Simulation experiments are also conducted, and the experimental results demonstrate that our scheme has lower storage and computational overhead.

The remainder of this paper is organized as follows. In Sect. 2, we review the existing authentication schemes. The preliminaries of this work are introduced in Sect. 3. Section 4 gives the overview of the proposed scheme. Then, Sect. 5 describes of the design details. Section 6 provides a theoretical analysis of the proposed solution in this paper. In Sect. 7, we conduct experiment and evaluate the performance. Finally, Sect. 8 concludes this paper.

2   Related Work

In recent years, many intra-domain and cross-domain device authentication schemes have been proposed. Intra-domain authentication schemes are mainly based on traditional authentication mechanisms, while cross-domain authentication schemes need to introduce new technologies to construct trust between different administrative domains. At the same time, due to the improvement of privacy protection awareness, many papers focus on anonymous authentication schemes. Du et al. proposed a cross-domain authentication scheme based on ZKP to protect the privacy of the devices [9], but this scheme made many ideal assumptions in the communication process, which is quite different from the real environment and needs further improvement. Moreover, existing literature pays little attention to the problem of resource constraints in blockchain. These are also the major concerns of our work.

Intra-domain authentication can be divided into symmetric cryptography and asymmetric cryptography. The advantage of symmetric cryptography is the high speed of implementation [2]. However, the distribution and management of keys are expensive, and it cannot realize the function of non-repudiation. Asymmetric cryptosystem includes PKI based mechanism [8], IBC based mechanism [21] and CL-PKC based mechanism [3].

The mechanism based on CL-PKC alleviates the certificate management problem of PKI and the key escrow problem of IBC. In CL-PKC, a semi-trusted third party named Key Generation Center (KGC) generates partial private key for users. The legitimate user then decides on a secret value to generate public key and the full private key derived from the partial private key. That is, KGC cannot obtain the full private key, and thus there is no key escrow problem. Due to the advantages above, encryption and signature schemes based on CL-PKC have been proposed in quantity. He et al. proposed a certificateless public key authentication and encryption scheme based on keyword search and deployed it to the IIoT [11], which is considered to be secure and efficient. Karati et al. proposed a secure and light-weight certificateless signature scheme for IIoT [13]. Ma et al. proposed a new scheme of outsourcing semi-trusted cloud revocation agent based on bilinear pairing [17], which solves the identity revocation problem in CL-PKC and realizes the uniqueness of public key and reliable revocation flexibility under low computing and communication costs. Tomar proposed to apply a blockchain-based certificateless authentication scheme to intelligent transportation systems [23], which achieves secure authentication with low computational and storage costs. In general, the CL-PKC based authentication mechanisms have the advantages of certificateless management, light-weight system, low communication overhead and strong non-repudiation.

For the cross-domain authentication issue, it needs to be addressed how to construct mutual recognition and trust of identity between different administrative domains [30]. The existing solutions can be divided into centralized and decentralized schemes. Centralized solutions include third-party bridge pattern [16], cross-authentication pattern [6], key exchange pattern [15], and path proof pattern [5]. These four patterns have the disadvantages of single point of failure, high computational overhead, low scalability and low efficiency, respectively. Decentralized solutions are mainly based on blockchain technology, and such cross-domain authentication solutions mainly construct trust between different administrative domains through distributed ledgers and consensus.

Shen et al. proposed a blockchain assisted cross-domain identity authentication scheme [22], in which specific parameter information of different domains is stored off-blockchain, while domain identifier, unified resource identifier (URI) and hash value calculated from real data are stored on the chain. Tong et al. proposed a cross-domain authentication scheme under the condition of limited computing resources [24], allowing each domain to adopt different authentication mechanisms, and adopting a robust identity management scheme to protect device privacy. The identity authentication scheme based on blockchain has alleviated the problems of data tampering, single point of failure and the difficulty in public key revocation in traditional schemes. Meanwhile, considering the limitations of blockchain storage resources and throughput speed, Chen et al. decouples the control layer and storage layer and designs the Multiple Merkle Hash Tree (MMHT) to improve response speed [7]. In this paper, we make further improvements to the work of Chen et al.

3   Preliminaries

This section briefly discusses the background knowledge for the proposed scheme.

3.1   MHT and MHG

Merkle Hash Tree (MHT) has been widely used in file systems and P2P systems. Specifically, MHT is a data structure commonly used in data storage of blockchain technology, which can ensure data security and verify the integrity of data. In an MHT, each leaf node stores the hash of one data block, and each branch node stores the hash of its child nodes. The left of Fig. 1 illustrates an example of MHT, in which Node1 stores the hash of data block L1, and Node5 stores the hash of the concatenation of the values stored in Node1 and Node2. As a result, any modification to the data block will change the hash values on the path from the corresponding leaf node to the root node. This feature gives MHT the advantage of fast data integrity verification, and thus it has become a fundamental component of blockchain technology.

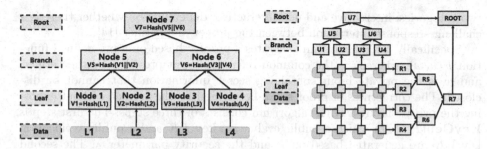

Fig. 1. Example of Merkle Hash Tree and Merkle Hash Grid.

Merkle Hash Grid (MHG) is an extension of MHT [19], which arranges the data blocks into a two-dimensional grid. The hash values are maintained in two MHTs in the horizontal and vertical directions, respectively. The right of Fig. 1 illustrates an example of MHG, in which each leaf node of the sub-tree on the upper side stores the hash of all data blocks in the corresponding column, and each leaf node of the sub-tree on the right side stores the hash of all data blocks in the corresponding row. In this way, the number of leaf nodes stored in each MHT is reduced from N to $\sqrt{N}$, and thus the storage overhead is significantly reduced to a minimum of 50%.

To prove this, suppose that N denotes number of leaf nodes contained in each MHT or MHG structure. Then, the number of additional hash values that an MHT needs to store can be estimated as the total number of N leaf nodes and the branch nodes at each depth:

$$N + (1 + 2 + 4 + \cdots + 2^{\lceil \log_2 N \rceil - 1}) \approx 2N - 1. \tag{1}$$

For an MHG, it contains two sub-trees with approximately $\sqrt{N}$ leaf nodes each, N leaf nodes associated with the data, and one additional root node. Thus, the number of hash values that an MHG needs to store can be estimated as:

$$N + 1 + 2 \times (1 + 2 + \cdots + 2^{\lceil \log_2 \sqrt{N} \rceil - 1}) \approx N + 4\sqrt{N} - 1. \tag{2}$$

When $N \to +\infty$, the limit of the proportion of additional storage overhead required by MHG to MHT is 0.5:

$$\lim_{N \to +\infty} \frac{N + 4\sqrt{N} - 1}{2N - 1} = \frac{1}{2}. \tag{3}$$

3.2 Zero-Knowledge Proof

Zero-Knowledge Proof (ZKP) is first proposed by S. Gadasser, S. Moali and C. Radof [10], which is a cryptographic protocol running between the prover and verifier. The prover makes verifier believe a certain statement, provided that no useful information is disclosed. Generally, ZKP algorithm can be divided into two categories: interactive and non-interactive, depending on whether there is a challenge-response interaction between the prover and verifier [14].

Specifically, ZKP algorithm generates a proof π based on a pre-defined function $F(u, w)$, where u is the common reference input provided by the verifier, and w is the private input containing secret information that cannot be disclosed. The correctness of proof π can be verified by the verifier without exposing the private input w. ZKP algorithm consists of three steps: The first step is **KeyGen**(F, λ), in which a public evaluation key EK and a public verification key VK are generated based on F and the security parameter λ. The second step is **Compute**(EK, u, w), in which the prover computes the proof π and the expected output y. The third step is **Verify**(VK, u, π, y), in which the verifier checks whether the proof π can make function F obtain the expected output y.

The ZKP algorithm satisfies the following three properties:

- *Correctness:* For the validation function F and any inputs u, w, the proof (y, π) can be obtained through **Compute**(EK, u, w). If the user identity is valid, then **Verify**$(VK, u, \pi, y) = 1$ always holds. Otherwise, the output of function **Verify**(VK, u, π, y) is 0.
- *Soundness:* For the prover, it is extremely difficult to generate forged identity information or evidence that satisfies **Verify**$(VK, u, \pi, y) = 1$. When a malicious adversary generates a zero-knowledge proof using an invalid identity or evidence, the output of **Verify**(VK, u, π, y) is always 0. Besides, the probability that an adversary generates correct evidence by repeated attempts is negligible.
- *Zero-Knowledge:* For the verifier, it is extremely difficult to obtain any part of the private input w based on the proof π, the expected output y, the function F, and the common reference input u. The process of computing ZKP is completed locally, and the private input is not leaked to the verifier. Therefore, our cross-domain authentication scheme is zero-knowledge and the privacy of devices can be protected well.

4   Proposed Scheme

4.1   System Overview

In the proposed authentication scheme, the architecture consists of three layers, namely entity layer, blockchain layer and storage layer. The overview of our authentication architecture is illustrated in Fig. 2.

- *Entity Layer:* The entity layer consists of IoT devices and KGC. IoT devices have sensing and processing capabilities, and their main task is to provide specific services. KGC is unique in an administrative domain, which is a semi-trusted third party responsible for generating partial private keys for IoT devices in its domain. Specifically, an IoT device sends a registration request to KGC. KGC generates the partial private key according to the identity information submitted. After receiving the partial private key, IoT device selects a secret value to generate its own public key and full private key derived from the partial private key.

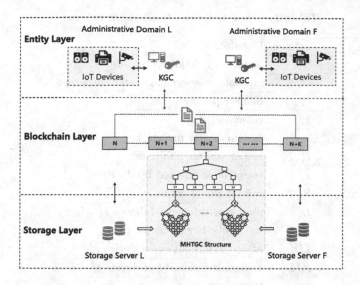

Fig. 2. Overview of our authentication architecture.

- *Blockchain Layer:* The blockchain layer is actually a global distributed ledger. The information stored in the blockchain is written by KGCs from different administrative domains. The blockchain collects and binds these information quickly. The data shared by the storage layer must pass integrity and consistency verification. Considering that the blockchain has the limitation of storage capacity and throughput speed, only the minimum information is stored in the blockchain.
- *Storage Layer:* The content maintained in the storage layer includes the public key list of IoT devices, system public parameters, blockchain addresses and verification time. The blockchain address refers to the address of the verification value bound to data, and the verification time refers to the latest time of consistency and integrity verification. Specifically, MHT is still adopted in the blockchain layer, which does not change the original data storage structure of blockchain. However, using MHG in the storage layer greatly reduces the storage overhead. In a MHTGC, the leaf nodes of the upper MHTs actually store the same value as the root nodes of the corresponding lower MHGs.

4.2 Threat Model

In our model, KGCs from different administrative domains are considered to have mutual trust, but there is a lack of trust between IoT devices. We assume that an adversary has the following capabilities:

- The adversary can illegally monitor, collect, forge, replay and tamper with communication messages.

Table 1. List of Notations Used in our Scheme

Symbol	Description
G_1	additive cyclic group of order q
G_2	multiplicative cyclic group of order q
$e(\cdot,\cdot)$	bilinear pairing
P	generator of G_1
k	system master private key
P_{pub}	system master public key
x_i^f	secret value selected by device e_i^f
S_i^f	private key of device e_i^f
D_i^f	partial private key of device e_i^f
P_i^f	full public key of device e_i^f
ID_i^f	identity information of device e_i^f
AID_i^f	anonymous identity of device e_i^f
t_i	timestamp
EK	evaluation key
VK	verification key
$H_i(\cdot)$	cryptographic hash functions

- The adversary can illegally read the data in the blockchain or storage server and try to impersonate a legitimate user. However, the user's real identity information will never be exposed.
- The adversary can attack the nodes in blockchain, but cannot control more than 1/3 of nodes.

5 Design Detail

We use CL-PKC mechanism to deploy CALA. In the CL-PKC mechanism, KGC only generates the partial private key, and the full private key of the device is generated locally that will never be transmitted on the Internet. Table 1 describes the symbols used in our design scheme.

5.1 Initialization Phase

Before registration, each administrative domain needs to initialize the system. KGC is the executor of system initialization.

(1) The parameters generated by KGC are: a large prime number q as order, additive group $(G_1, +)$ and multiplicative group $(G_2, \cdot)$ of order q, a generator P of G_1, bilinear pair $e : G_1 \times G_1 \to G_2$.

(2) KGC selects a random number $k \in Z_q^*$ as the master private key of system, and the public key can be calculated by $P_{pub} = kP$. Thus the master key pair of the system is (k, P_{pub}). k is kept secretly by KGC, while P_{pub} is written into blockchain and exposed to IoT devices in its administrative domain.

(3) The following four hash functions are adopted, $H_1 : \{0,1\}^* \rightarrow G_1$, $H_2 : \{0,1\}^* \times \mathcal{T} \rightarrow Z_q^*$, $H_3 : \{0,1\}^* \times G_2 \rightarrow Z_q^*$, $H_4 : \{0,1\}^* \times Z_q^* \rightarrow \{0,1\}^n$. Here n is the bit-length of messages, and $\mathcal{T}$ is the set of binary strings representing the timestamp.

To sum up, the system parameters to be disclosed are: q, G_1, G_2, e, P, P_{pub}, H_1, H_2, H_3, H_4.

5.2   Device Registration

Device A submits a registration request and its real identity information ID_A to KGC through a secure channel, where $ID_A \in \{0,1\}^*$. KGC checks whether device A is registered repeatedly and is valid. If A is a legitimate user, KGC selects a random number $r \in Z_q^*$, generates anonymous identity $AID_A = H_4(ID_A, r)$ and calculates the partial private key $D_A = kH_1(ID_A) \in G_1$. AID_A and D_A are transmitted to device A. AID_A will be regenerated by KGC after expiration.

After receiving the partial private key, device A verifies the correctness of D_A by checking $e(D_A, P) = e(H_1(ID_A), P_{pub})$. Device A selects the secret value x_A based on security parameters and ID_A, where $x_A \in Z_q^*$. The full private key $S_A = x_A D_A \in G_1$ can be obtained if secret value x_A and partial private key D_A are set as input. The public key P_A of A can be calculated by $P_A = x_A P_{pub} \in G_1$. Note that the generation of the full private key S_A and public key P_A of device A is not necessarily chronological. That is, the signature key pair (S_A, P_A) of device A is generated locally. P_A is written into the storage server by KGC along with AID_A, while S_A is always kept locally secretly.

Eventually, KGC updates and maintains the list of legitimate users.

5.3   Intra-domain Authentication

When device A applies for services or resources from device B in the same administrative domain, it needs to complete authentication and get authorization. Device A sends an authentication request to device B. If device A is not in the legitimate user list, it should complete device registration first.

The authentication process is initialized by device A. Device A applies its own private key to sign message $M \in \mathcal{M}$, where message space $\mathcal{M} = \{0,1\}^n$. The signing algorithm used in this process can be described as follows:

(1) Generate a random number $r \in Z_q^*$.
(2) Compute integer $\alpha = H_2(M, t)$, here t is the timestamp when message M is generated.

(3) Compute element $Q = S_A + \alpha D_A$ in G_1.
(4) Compute element $W = e(Q, P)^r$ in G_2.
(5) Compute integer $u = H_3(M, W)$.
(6) Compute element $V = rH_1(ID_A)$ in G_1.
(7) Output $C = \langle u, V \rangle$ as the signature on message M.

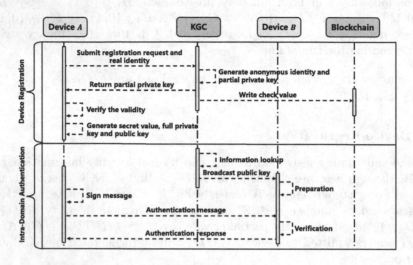

Fig. 3. The process of device registration and intra-domain authentication.

$C = \langle u, V \rangle$ is the signature of device A on message M, and device A sends the signed message $Token_{AB} = t||M||C$ to device B. Device B first checks the validity of timestamp t. If the difference between t and the current timestamp of system exceeds a certain threshold, no further process will be performed. If the timestamp is valid, device B will verify the signature of A with the following verifying algorithm:

(1) Compute integer $\alpha = H_2(M, t)$.
(2) Compute element $R = P_A + \alpha P_{pub}$ in G_1.
(3) Compute element $W' = e(V, R)$ in G_2.
(4) Compute integer $u' = H_3(M, W')$.
(5) If $u' = u$ holds, the verification succeed. Otherwise, the verification fails.

If the above steps are correct, the authentication succeeds. Device B will send a response to Device A to confirm the success of authentication. Then, device B will allocate relevant resources and services to device A within the expiration time T. The process of device registration and local authentication is illustrated in Fig. 3.

After A is authenticated by B, the two parties exchange roles, that is, A becomes the verifier and B becomes the claimant. Device B sends authentication request and signed message $Token_{BA} = t||M||C$ to device A. Once B is authenticated by A with the same algorithm, the two parties complete mutual authentication.

5.4   Cross-Domain Authentication

For the privacy protection in the process of cross-domain authentication, we use ZKP algorithm to realize the anonymity of the authentication mechanism. Specifically, the ZKP algorithm generates evidence π based on the predefined function $F(u, w)$, where u is the common reference input provided by the verifier and w is the private input provided by the prover. The verifier can confirm the correctness of π without exposing w.

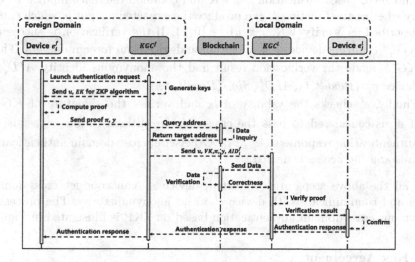

Fig. 4. The process of anonymous cross-domain authentication based on ZKP.

Suppose that a foreign domain IoT device e_i^f sends a cross-domain access request to a local domain IoT device e_j^l. At this time, it is necessary for device e_j^l to verify the identity of device e_i^f. The specific process of applying ZKP algorithm to cross-domain authentication can be described as follows:

(1) Device e_i^f of foreign domain launches a cross-domain authentication request $(Request_1, AID_i^f, t_1, Sig_{S_i^f}(ID_i^f, t_1))$.

(2) KGC^f always monitors various requests from its administrative domain. When KGC^f receives the cross-domain authentication request sent by e_i^f, KGC^f first checks the validity of the timestamp t_1. If t_1 is valid, KGC^f leverages the public key P_i^f for signature verifying. Afterwards, KGC^f generates evaluation key EK and verification key VK based on predefined function F and security parameters λ. This process can be described as **KeyGen**$(F, \lambda) \to (EK, VK)$. Then, KGC^f sends evaluation key EK and common reference input u to device e_i^f, $(EK, u, t_2, Sig_{k^f}(EK, t_2))$.

(3) Device e_i^f uses evaluation key EK and common reference input u to prove its legality without revealing its real identity ID_i^f. This process can be

described as $\mathbf{Compute}(EK, u, w) \rightarrow (\pi, y)$, where y is the expected output of function F, and the evidence π is used to prove the correctness of the result. Device e_i^f sends the message (AID_i^f, π, y, t_3) to KGC^f.

(4) After checking the validity of timestamp t_3, KGC^f adds the verification key VK and common reference input u to the original message. KGC^f signs the message and sends $(VK, u, AID_i^f, \pi, y, t_4, Sig_{k^f}(t_4))$ to KGC^l.

(5) KGC^l receives the message from KGC^f and checks the validity of timestamp t_4. Then, it uses the public key of KGC^f to verifies the signature. And KGC^l uses verification key VK and common reference input u to verify whether the function F can produce the expected output y, which can be described as $\mathbf{Verify}(VK, u, \pi, y) \rightarrow \{0,1\}$. If the verification is successful, KGC^l confirms device e_i^f is a legitimate device from foreign domain. Then, KGC^l sends the verification result and the anonymous identity AID_i^f to device e_j^l, $(Result, t_5, AID_i^f, Sig_{k^l}(Result, t_5))$.

(6) Finally, e_j^l checks the validity of t_5 and verifies the signature of KGC^l. If e_i^f is considered to pass the cross-domain authentication, e_j^l sends the authentication response to e_i^f at t_6. Otherwise, cross-domain authentication fails and the access is denied.

If all the above steps are completed, device e_i^f can conduct cross-domain access and communicate with device e_j^l in an anonymous way. The process of anonymous cross-domain authentication based on ZKP is illustrated in Fig. 4.

5.5 Key Agreement

The key agreement mechanism is adopted to achieve secure communication between devices. Here we assume that the two devices share the same elliptic curve equation $E(F_p)$. For two devices e_i and e_j, the process of key agreement is described as follows: First, device e_i generates a random number $r_i \in Z_q^*$, and calculates a public value $R_i = r_i P_i + r_i P_j$, where P_i, P_j are the public keys of the devices. Then, e_i sends R_i and timestamp t to device e_j. After receiving the message from e_i, device e_j checks the validity of the timestamp t, and generates another random number $r_j \in Z_q^*$. Then, the public value $R_j = r_j P_i + r_j P_j$ is calculated by e_j and sent to e_i. At this moment, both e_i and e_j independently calculate $K_{ij} = r_j R_i = r_i R_j$ using the random number generated by themselves and the public value provided by the other. Finally, the session key sk can be derived based on the following formula:

$$sk = H_2(K_{ij}\|R_i\|R_j, t) \tag{4}$$

5.6 Data Correctness Verification

Most of the information is stored in the storage layer, which includes system public parameters, domain master public key, public keys and anonymous identities of IoT devices, etc. The data stored in storage servers is considered to be at risk of being corrupted or maliciously tampered. Blockchain is leveraged to verify the correctness of the data. However, if the hash value of each data is stored separately in the blockchain, it will significantly increase the storage overhead and the communication overhead of the blockchain layer. Inspired by Ref. [7], which uses an MMHT structure to manage the data, we design an MHTGC structure to achieve fast correctness verification of large amounts of data.

Table 2. Comparison of Security Attributes

Scheme	Ref. [7]	Ref. [13]	Ref. [22]	Ref. [27]	Ref. [32]	**Ours**
Mutual Authentication	✗	✗	✓	✓	✗	✓
Key Agreement	✗	✗	✓	✗	✗	✓
Anonymity	✓	✗	✓	✓	✓	✓
Unlinkability	✓	✓	✗	✓	✓	✓
No Trusted Third Party	✗	✓	✗	✗	✓	✓
Cross-Domain	✓	✗	✓	✓	✗	✓
Common Attacks Resistance	✓	✓	✓	✓	✓	✓

The lower layer of our MHTGC structure is located at the storage layer and consists of MHG with original data as leaf nodes. The upper layer of our MHTGC structure is MHT located at the blockchain layer, with the root nodes of MHG in the lower layer as leaf nodes. When the storage layer collects N new data, an MHG containing N leaf nodes is created to manage the data. Then, the storage layer sends the root node of the MHG to the blockchain layer. After receiving the root node, the blockchain layer needs to verify the correctness of the root of the MHG. Specifically, the blockchain randomly selects b leaf nodes from the MHG and sends them to the storage layer. Then, the storage layer generates a Merkle proof $\pi = \sigma \backslash \rho$ about these b leaf nodes, where ρ is the set of nodes on the paths from the leaf nodes to the root node, σ is the set consisting of the sibling nodes of each node in ρ, and $\backslash$ denotes the set difference operation. Finally, the blockchain layer verifies the correctness of the root node. If the root node is correct, it will be written into the blockchain and an address will be returned. Otherwise, the root node will be discarded.

6 Theoretical Analysis of CALA

We compare the security attributes between our scheme and existing relevant schemes. The results are listed in Table 2, where "✓" means the corresponding security attribute is satisfied, and "✗" means the attribute is not satisfied.

- *Mutual Authentication:* In our scheme, both devices participating in authentication are required to verify the identity of each other, and to verify the reliability of received messages. For both intra-domain and cross-domain authentication, such mutual trust is necessary so that attackers cannot impersonate a legal device to pass the authentication.

- *Key Agreement:* Attackers have the ability to intercept messages sent by devices through eavesdropping. Therefore, participants in the authentication process need to negotiate a shared key $sk = H_2(K_{ij}||R_i||R_j, t)$ to further ensure communication security and avoid the leakage of sensitive information. Since our scheme is designed on the elliptic curve discrete logarithm problem (ECDLP), it is impossible for an adversary to derive the negotiated session keys in conventional polynomial time.

- *Anonymity:* Participants use anonymous identities instead of real identities for communication. In our scheme, no adversary can speculate the real identity ID through anonymous identity AID, because the hash function $AID_i^f = H_4(ID_i^f, r)$ has a strong collision resistance characteristic. Therefore, even if adversaries illegally access the information stored in blockchain, they cannot know the real identity of other devices.

- *Unlinkability:* Even if the attacker illegally accesses the blockchain and continuously obtains anonymous identity, the attacker cannot induce the user's real identity. Since the anonymous identity in CALA is temporary, there is no mathematical relationship between anonymous identities generated by the device at different periods. In addition, it is not feasible for an attacker to obtain the private key of the device by intercepting session messages.

- *No Trusted Third Party:* In our design, KGC only generates partial private key, while the full private key is generated by the device and always stored locally. In other words, KGC is a semi-trusted third party, which is different from CA in PKI and PKG in IBC. Our scheme is considered that there is no trusted third party.

- *Cross-Domain:* There is often a lack of trust between different administrative domains. Compared with intra-domain authentication, cross-domain authentication faces greater challenges. In this scheme, IoT devices from different administrative domains construct mutual trust without a trusted third party involved.

- *Common Attacks Resistance:* (i) In our scheme, the receiver checks the validity of the timestamp at the first moment of receiving the interactive message. At the same time, the sender signs the message containing timestamp with private key, which ensures that the timestamp is not tampered with. That is, if attackers **replay** the messages intercepted, the subsequent operation will be interrupted because the messages have expired. (ii) During the authentication process, only legitimate devices have anonymous identities, and their anonymous identities and public keys will be written to blockchain by KGC. Even if attackers obtain anonymous identities and public keys, the corresponding private keys cannot be leaked. Thus, it is impossible for **impersonation attacks** to succeed. (iii) The scheme we designed binds the devices' anonymous identities to public keys. The same device cannot be registered repeatedly. If the content of message is tampered by an attacker, the message will not be verified by the receiver. Our solution can effectively prevent **man-in-the-middle attacks**. (iv) The anonymous identities of devices have an expiration time, thus devices need to reapply for anonymous identities from KGC if identities expire. If a device has any malicious behavior, KGC will reject the request and record the device in the revocation list. Therefore, the scheme we designed can effectively resist **internal attacks**.

Table 3. Time Consumption in Signature Generation and Verification

Scheme Name	Signing	Sign.Time	Verifying	Verify.Time
Ref. [7]	T_{exp}	1.34 ms	T_{exp}	1.36 ms
Ref. [13]	$2T_s + T_e$	0.85 ms	$T_p + 2T_e$	2.65 ms
Ref. [22]	$T_p + T_s + T_e$	2.93 ms	$2T_p + T_s + T_e + 2T_a$	5.42 ms
Ref. [27]	$6T_s + 2T_a$	1.97 ms	$8T_s + 6T_a$	2.59 ms
Ref. [32]	T_s	0.41 ms	$T_p + T_s + T_a$	5.30 ms
Ours	$T_p + 2T_s + T_e + T_a$	3.19 ms	$T_p + T_s + T_a$	2.81 ms

7 Experiment and Evaluation

7.1 Experimental Settings

We implement blockchain on the FISCO BCOS platform which adopts PBFT as the consensus protocol [1]. Blockchain communicates with other entities through the Web3 library. SHA-256 is chosen as the hash algorithm for the system. Among the various ZKP algorithms, we choose the Groth16 algorithm to assist the experiment and complete the authentication process. We simulate two administrative domains in the experiments. The operations of KGC are executed in a laptop with Intel(R) Core(TM) i5-11300H CPU @3.1 GHz, 16.0 GB RAM, and Window 11 operating system. The operations of IoT devices are executed in a virtual machine with 4 GB memory and Ubuntu 16.04 LTS system, hosted on the desktop using VMware Workstation 15 Pro.

7.2 Performance Analysis

Time Consumption in Authentication. The comparison of the time consumption between our scheme and other relevant schemes is shown in Table 3. Here, T_p denotes the time of a bilinear pairing operation, T_s denotes the time of a scalar point multiplication operation in G_1, T_a denotes the time of a point addition operation, T_e denotes the time of an exponentiation operation, T_v denotes the time of a modular inversion operation, and T_{exp} denotes the time of an exponentiation operation in RSA algorithm. We calculate the time consumption of different authentication schemes in signing and verifying based on the average of 1,000 runs with various inputs. The results are illustrated in Table 3.

To sign a given message M, our authentication scheme requires one bilinear pairing operation, two scalar point multiplications, one point addition, and one exponentiation operation. Due to the introduction of timestamp t and partial private key DA, the computational complexity of the signing process is slightly increased. Compared with Ref. [22], our scheme only spends an additional 0.26ms in signing to achieve a trade-off between security and efficiency. Nevertheless, the process of our signature verifying is more efficient than Ref. [22], Ref. [32], and the time consumption is close to that of Ref. [13] Ref. [27].

Table 4. Comparison of Signature Length

Scheme	Ref. [7]	Ref. [13]	Ref. [22]	Ref. [27]	Ref. [32]	**Ours**
Length	$|Z_r|$	$|G_1| + |G_2|$	$|G_1| + |Z_q^*|$	$3|G_1| + 4|Z_q^*|$	$|G_1| + |Z_q^*|$	$|G_1| + |Z_q^*|$
Size	128B	128B	96B	256B	96B	96B

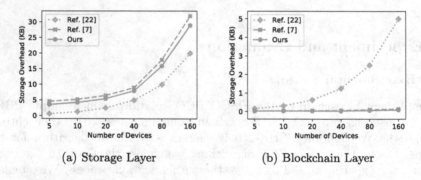

(a) Storage Layer   (b) Blockchain Layer

Fig. 5. The comparison of storage overhead of different schemes.

Signature Length. We compare the length of signatures generated by different authentication schemes. In our experiments, an elliptic curve with 512-b group is utilized, which is equal to 1024-b RSA security level [13]. The results are listed in Table 4, where $|Z_r|$ represents the length of the modulus adopted in RSA algorithm, $|G_1|$, $|G_2|$, $|Z_q^*|$ represent the length of the elements in each set, respectively. The length of the signature generated by our scheme is $|G_1| + |Z_q| =$ 96 Bytes, which is equal to that of Ref. [22], Ref. [32], and is smaller than that of Ref. [7], Ref. [13] and Ref. [27]. The signature length affects the communication overhead, which gives our scheme an advantage of communication.

Storage Overhead. The storage overhead is only compared with Ref. [7] and Ref. [22], because both of them design a storage layer. However, due to the fact that each scheme stores different data fields in the storage layer and some fields (e.g., domain names, etc.) are variable in length, it is impossible to obtain accurate storage overhead. Thus, in our experiment, we assume that storing relevant information (e.g., public keys, etc.) for each device (or user) consumes a total of 128 bytes of memory in the storage layer. Meanwhile, we assume that the sizes of the hash values for data verification stored in the blockchain are all 32 bytes. The number of leaves in each MHT and MHG is set to 64.

Figure 5 illustrates the storage overhead of different schemes at the storage layer and blockchain layer, respectively. In the storage layer, Ref. [22] has the lowest storage overhead, because it does not require additional space for data management. The storage overhead of our scheme is lower than that of Ref. [7]. However, in the blockchain layer, the storage overhead of Ref. [22] is significantly higher than that of other schemes, since it does not adopt integrated data verification. In practice, due to the limited storage capacity and throughput speed of the blockchain, writing excessive amounts of data into the blockchain should be avoided.

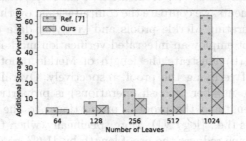

Fig. 6. The comparison of additional storage overhead for data management.

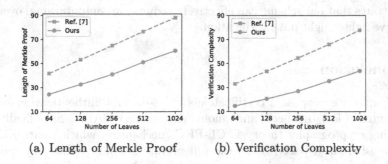

(a) Length of Merkle Proof   (b) Verification Complexity

Fig. 7. The computational overhead of generating and verifying Merkle proof.

To further compare the storage overhead of our scheme with Ref. [7], the additional storage overhead used for data management is illustrated in Fig. 6. Note that Ref. [22] does not adopt integrated data management, so it is not comparable with our model in this experiment. It can be observed that the additional storage overhead is proportional to the number of leaf nodes. With the increase of the number of leaf nodes, the storage overhead that our scheme can reduce is significantly larger. Specifically, when the number of leaf nodes is 256, the number of additional stored hash values of Ref. [7] is 511 (15.97 KB). In contrast, the number of additional stored hash values of ours is only 319 (9.97 KB). The storage overhead is reduced by 37.6%. When the number of leaf nodes is 1024, the number of additional stored hash values of Ref. [7] and ours are 2047 (63.97 KB) and 1151 (35.97 KB), respectively. At this time, the storage overhead is reduced by 43.7%. Theoretically, when the number of leaf nodes is sufficiently large, the storage overhead of our scheme is reduced to 50% of that of Ref. [7], which has been proved in Sect. 3.1.

Computational Overhead in Data Verification. In the process of data verification, Merkle proofs are need to generate for several randomly assigned leaf nodes. At the same time, fast verification of the correctness is also required. In this set of experiments, we compare the computational overhead of our scheme and Ref. [7] in generating Merkle proofs and verifying data correctness, since other schemes do not employ an integrated verification mechanism.

Figure 7(a) and (b) illustrate the length of Merkle proof and the computational overhead of verifying the proof, respectively. Overall, the verification complexity (i.e., the number of hash operations) is proportional to the proof length, while the length of the proof is proportional to the logarithm of the number of leaf nodes (i.e., $\lceil \log_2 N \rceil$). More specifically, when the number of leaf nodes is 64, our scheme reduces the proof length by 41.6% compared to Ref. [7], and the verification complexity is reduced by 55.6%. As the number of leaf nodes increases, there is a slight decrease in the percentage of the reduced complexity by our scheme. However, even when $N = 1,024$, our scheme is still able to reduce the proof length by 31.2% and the verification complexity by 43.6%. This fully demonstrates that our scheme can effectively reduce the computational overhead to achieve light-weight data verification.

8   Conclusion

In this paper, we propose a certificateless cross-domain authentication scheme that combines low overhead and anonymity, named CALA. Specifically, the underlying cryptography theory is CL-PKC mechanism, which treats KGC as a semi-trusted third party. Since the full private key and public key are generated locally, there is no key escrow problem. Due to the limited resources of the blockchain, we design an MHTGC structure to achieve fast correctness verification while reducing a large amount of computational and storage overhead. Compared with MHT, MHG can significantly reduce the length of Merkle proofs as well as the complexity of verification. To address the issue of lack of trust among IoT devices during cross-domain authentication, we implemented anonymous authentication leveraging ZKP algorithm to protect the sensitive identity information of users. Both security analysis and experimental results demonstrate the effectiveness of CALA.

Acknowledgements. This work was supported by the National Key R&D Program of China under Grant No. 2022YFB3102304 and in part by National Natural Science Foundation of China Grants (62225105, 62001057).

References

1. FISCO BCOS open source community. https://www.fisco.com.cn/en/about_41.html. Accessed 8 Jun 2023
2. Abdullaziz, O.I., Wang, L., Chen, Y.: HiAuth: hidden authentication for protecting software defined networks. IEEE Trans. Netw. Serv. Manag. **16**(2), 618–631 (2019)
3. Al-Riyami, S.S., Paterson, K.G.: Certificateless public key cryptography. In: Laih, C.-S. (ed.) ASIACRYPT 2003. LNCS, vol. 2894, pp. 452–473. Springer, Heidelberg (2003). https://doi.org/10.1007/978-3-540-40061-5_29
4. Ali, G., et al.: xDBAuth: blockchain based cross domain authentication and authorization framework for internet of things. IEEE Access **8**, 58800–58816 (2020)
5. Andersen, M.P., et al.: WAVE: a decentralized authorization framework with transitive delegation. In: 28th USENIX Security Symposium, USENIX Security, Santa Clara, CA, USA, 14–16 August 2019, pp. 1375–1392. USENIX Association (2019)
6. Bai, Q.H., Zheng, Y., Zhao, L., Chun, H., Cheng, C.: Research on mechanism of PKI trust model. Appl. Mech. Mater. **536–537**, 694–697 (2014)
7. Chen, J., Zhan, Z., He, K., Du, R., Wang, D., Liu, F.: XAuth: efficient privacy-preserving cross-domain authentication. IEEE Trans. Dependable Secur. Comput. **19**(5), 3301–3311 (2022)
8. Diffie, W., Hellman, M.E.: New directions in cryptography. IEEE Trans. Inf. Theor. **22**(6), 644–654 (1976)
9. Du, R., Li, X., Liu, Y.: A cross-domain authentication scheme based on zero-knowledge proof. In: Lai, Y., Wang, T., Jiang, M., Xu, G., Liang, W., Castiglione, A. (eds.) Algorithms and Architectures for Parallel Processing, ICA3PP 2021. LNCS, vol. 13156, pp. 647–664. Springer, Cham (2022). https://doi.org/10.1007/978-3-030-95388-1_43
10. Goldwasser, S., Micali, S., Rackoff, C.: The knowledge complexity of interactive proof systems. SIAM J. Comput. **18**(1), 186–208 (1989)
11. He, D., Ma, M., Zeadally, S., Kumar, N., Liang, K.: Certificateless public key authenticated encryption with keyword search for industrial internet of things. IEEE Trans. Ind. Inform. **14**(8), 3618–3627 (2018)
12. Huang, C., et al.: Blockchain-assisted transparent cross-domain authorization and authentication for smart city. IEEE Internet Things J. **9**(18), 17194–17209 (2022)
13. Karati, A., Islam, S.H., Karuppiah, M.: Provably secure and lightweight certificateless signature scheme for IIoT environments. IEEE Trans. Ind. Inf. **14**(8), 3701–3711 (2018)
14. Lesavre, L., Varin, P., Mell, P., Davidson, M., Shook, J.M.: A taxonomic approach to understanding emerging blockchain identity management systems. CoRR abs/1908.00929 (2019)
15. Liu, X., Ma, W.: CDAKA: a provably-secure heterogeneous cross-domain authenticated key agreement protocol with symptoms-matching in TMIS. J. Med. Syst. **42**(8), 135:1–135:15 (2018)

16. Liu, Y., Yang, Z.: The research and design of the proxy for certificate validation based on distributed cross-certification. In: 5th International Conference on Applied Computing and Information Technology, 4th International Conference on Computational Science/Intelligence and Applied Informatics, 2nd International Conference on Big Data, Cloud Computing, Data Science & Engineering, ACIT/CSII/BCD, Hamamatsu, Japan, 9–13 July 2017, pp. 135–140. IEEE (2017). https://doi.org/10.1109/ACIT-CSII-BCD.2017.18
17. Ma, M., Shi, G., Shi, X., Su, M., Li, F.: Revocable certificateless public key encryption with outsourced semi-trusted cloud revocation agent. IEEE Access **8**, 148157–148168 (2020)
18. Matsumoto, S., Reischuk, R.M.: IKP: turning a PKI around with decentralized automated incentives. In: 2017 IEEE Symposium on Security and Privacy, SP 2017, San Jose, CA, USA, 22–26 May 2017, pp. 410–426. IEEE Computer Society (2017). https://doi.org/10.1109/SP.2017.57
19. Pâris, J., Schwarz, T.J.E.: Merkle hash grids instead of Merkle trees. In: 28th International Symposium on Modeling, Analysis, and Simulation of Computer and Telecommunication Systems, MASCOTS, Nice, France, 17–19 November 2020, pp. 1–8. IEEE (2020). https://doi.org/10.1109/MASCOTS50786.2020.9285942
20. Rathore, M.M., Bentafat, E., Bakiras, S.: Smart home security: a distributed identity-based security protocol for authentication and key exchange. In: 2019 28th International Conference on Computer Communication and Networks (ICCCN), pp. 1–9 (2019). https://doi.org/10.1109/ICCCN.2019.8847034
21. Shamir, A.: Identity-based cryptosystems and signature schemes. In: Blakley, G.R., Chaum, D. (eds.) CRYPTO 1984. LNCS, vol. 196, pp. 47–53. Springer, Heidelberg (1985). https://doi.org/10.1007/3-540-39568-7_5
22. Shen, M., et al.: Blockchain-assisted secure device authentication for cross-domain industrial IoT. IEEE J. Sel. Areas Commun. **38**(5), 942–954 (2020)
23. Tomar, A., Tripathi, S.: BCAV: blockchain-based certificateless authentication system for vehicular network. Peer-to-Peer Netw. Appl. **15**, 1733–1756 (2022). https://doi.org/10.1007/s12083-022-01319-2
24. Tong, F., Chen, X., Wang, K., Zhang, Y.: CCAP: a complete cross-domain authentication based on blockchain for Internet of Things. IEEE Trans. Inf. Forensics Secur. **17**, 3789–3800 (2022)
25. Wang, S., Ma, Z., Liu, J., Luo, S.: Research and implementation of cross-chain security access and identity authentication scheme of blockchain. Netinfo Secur. **22**, 61–72 (2022)
26. Wei, S., Li, S., Wang, J.: A cross-domain authentication protocol by identity-based cryptography on consortium blockchain. Chin. J. Comput. **44**, 908–920 (2021)
27. Xi, N., Li, W., Jing, L., Ma, J.: ZAMA: a ZKP-based anonymous mutual authentication scheme for the IoV. IEEE Internet Things J. **9**(22), 22903–22913 (2022)
28. Xu, R., Chen, Y., Li, X., Blasch, E.: A secure dynamic edge resource federation architecture for cross-domain IoT systems. In: 31st International Conference on Computer Communications and Networks, ICCCN, Honolulu, HI, USA, July 2022, pp. 1–7. IEEE (2022). https://doi.org/10.1109/ICCCN54977.2022.9868843
29. Xue, L., Huang, H., Xiao, F., Wang, W.: A cross-domain authentication scheme based on cooperative blockchains functioning with revocation for medical consortiums. IEEE Trans. Netw. Serv. Manage. **19**(3), 2409–2420 (2022). https://doi.org/10.1109/TNSM.2022.3146929
30. Yang, D., Cheng, Z., Zhang, W., Zhang, H., Shen, X.: Burst-aware time-triggered flow scheduling with enhanced multi-CQF in time-sensitive networks. IEEE/ACM Trans. Netw. (2023). https://doi.org/10.1109/TNET.2023.3264583

31. Yang, D., et al.: DetFed: dynamic resource scheduling for deterministic federated learning over time-sensitive networks. IEEE Trans. Mob. Comput. (2023). https://doi.org/10.1109/TMC.2023.3303017
32. Zhang, L., Xu, J.: Anonymous authentication scheme based on trust and blockchain in VANETs. In: Lai, Y., Wang, T., Jiang, M., Xu, G., Liang, W., Castiglione, A. (eds.) Algorithms and Architectures for Parallel Processing, ICA3PP 2021, Part II. LNCS, vol. 13156, pp. 473–488. Springer, Cham (2022). https://doi.org/10.1007/978-3-030-95388-1_31
33. Zhang, W., et al.: Optimizing federated learning in distributed industrial IoT: a multi-agent approach. IEEE J. Sel. Areas Commun. **39**(12), 3688–3703 (2021). https://doi.org/10.1109/JSAC.2021.3118352

UAV-Assisted Data Collection and Transmission Using Petal Algorithm in Wireless Sensor Networks

Xueqiang Li[iD], Ming Tao[(✉)][iD], and Shuling Yang[iD]

School of Computer Science and Technology, Dongguan University of Technology, Dongguan 523808, People's Republic of China
{lixq,taom,shulingyang}@dgut.edu.cn

Abstract. With advancements in unmanned aerial vehicle (UAV) technology, the utilization of UAVs for data collection and transmission has become widespread in wireless sensor networks (WSNs). In this paper, the energy consumption of UAVs, the integrity of data collection and full coverage and so on are taken into account. Consequently, a dynamic UAV data collection model is formulated, with the objectives of minimizing the number of UAVs, reducing their flight distances, and optimizing service quality within WSNs. To address this model, the data collection nodes are initially determined using the Kmeans algorithm, followed the petal algorithm is proposed to search for the optimal flight route of UAVs. Finally, experimental comparisons were conducted, involving four test problems with different scales of sensors and five classic path planning algorithms, in comparison with the algorithm proposed in this paper. The results consistently demonstrate that the proposed algorithm yields better solution outcomes, effectively addressing the challenge of the UAV-assisted data collection.

Keywords: Wireless sensor networks · the UAV-assisted data collection · Path planning

1 Introduction

With the development of big data technology, urban and industrial intelligence has become a trend in national, societal, and enterprise development [1]. In this trend, the utilization of UAVs for data collection and transmission, based on the Internet of Things (IoT) architecture, has been widely applied in various domains such as disaster management, agricultural protection, environmental monitoring, and surveillance. It is estimated that by 2030, the number of sensors will be more than 100 trillion worldwide [2]. Hence, the task of collecting substantial data from sensors, characterized by low energy consumption, minimal latency, and high reliability, presents a substantial challenge. Over the years, scholars have been actively researching the issues in this field [3,4].

Z. Tari et al. (Eds.): ICA3PP 2023, LNCS 14493, pp. 114–125, 2024.
https://doi.org/10.1007/978-981-97-0862-8_8

Conventionally, data collection in WSNs is achieved through multi-hop transmissions scheme, where data collected by sensors is transmitted hop by hop through intermediate sensors and ultimately access the system to the internet [5,6]. Under this scheme, the distance between adjacent sensors must be within the communication range. Furthermore, the issue of energy imbalance of sensors during data transmission is known to result the occurrence of "energy holes" around the receiver, reducing network lifetime [7]. Additionally, sensors near the receiver typically experience better Quality of Service (QoS), while those farther away tend to experience relatively poorer QoS. This QoS imbalance experienced by sensors at different positions adversely affects the timeliness and reliability of data transmission [8]. Therefore, it is crucial to balance QoS among sensors. Moreover, it is also required that all data collected by sensors should be covered by the UAVs. Otherwise, data from uncovered sensors will be discarded [9].

In order to address the data collection and transmission issues within traditional WSNs, the UAV-assisted data collection and transmission have gradually been proposed, due to the high mobility of UAVs. By directly collecting all or a subset of data from sensors, UAVs can effectively reduce the number of transmission hops in WSN, thereby reducing network energy consumption of sensors and improving data transmission reliability of WSNs [10,11]. In recent years, a substantial amount of research work has been conducted by scholars in the field of the UAV-assisted data collection and transmission.

Miao et al. aimed to achieve both time and energy efficiency in data collection by incorporating metrics such as the concurrent uploaded data volume, the number of neighbors, and the moving tour length of the sink into a single metric [12]. Zhou et al. proposed a three-phase energy-balanced heuristic algorithm for scheduling mobile sinks and prolonging the network lifetime. They employed a clustering algorithm inspired by the k-dimensional tree algorithm to assign uniformly divided grid cells to clusters, ensuring similar energy consumption in each cluster [13]. Chang et al. introduced a novel tree-based power-saving scheme to reduce the data transmission distances of the sensors [14]. Tao et al. introduce a new concept of 'great full-coverage subgraph' to generate candidate area for UAV deployment and transform the investigated problem into a traditional K-center problem and build a multi-objective joint optimization problem with multiple constraints [15].

Although the proposed strategies were demonstrated demonstrate high effectiveness in data collection within specific monitoring regions, most approach still possesses certain limitations that warrant further discussion. Moreover, none of the previous studies have considered the dynamic utilization of UAVs for data collection.

The issue of the UAV-assisted data collection and transmission faces the challenge of planning the flight path of the UAVs, which is similar to the vehicle routing problem (VRP) and falls under the category of NP problems. Currently, there exist several well-known deterministic algorithms. Among them are the nearest neighbor algorithm proposed by Rosenkrantz et al. [16], the savings algorithm introduced by Clarke and Wright [17], the insertion algorithm proposed by Mole and Jameson [18], the sweep algorithm was presented by Gillett and Miller [19], the Split algorithm developed by Beasley [20], as well as various randomly generated methods. Additionally, some deterministic algorithms leverage effective inequalities and combine techniques from operations research, such as branch and bound or integer programming, to continuously optimize the current path. For instance, Perboli et al. [21] solve the vehicle routes problems through the application of effective inequality constraints and branch and bound methods. Baldacci R [22] proposed a deterministic algorithm based on integer programming to optimize vehicle routes.

The current path is optimized and adjusted by the aforementioned algorithms based on distance, angles, or inequality information. Local information of the problem is primarily focused on by them without the consideration of the global optimality characteristics of the optimal solution. By taking inspiration from previous achievements and utilizing UAVs as mobile sinks, the problem of the UAV-assisted data collection and transmission is modeled, and the petal algorithm is proposed to solve this problem.

2 Mathematical Model for the UAV-Assisted Data Collection in WSNs

The WSN model is designed to facilitate data collection and transmission from sensors in the region to the data center through UAVs. The data can be transmitted either through intermediate sensors along the communication links or directly to the UAV. Subsequently, the UAVs transfer the collected data to the data center.

2.1 WSN Model and Fundamental Concepts

Assuming M UAVs denoted as $U = \{u_1, u_2, \cdots, u_M\}$ are deployed for dynamic data collection and transmission among N sensors $S = \{s_1, s_2, \cdots, s_N\}$ in a square area of size R^2. The entire WSN can be represented by an undirected graph $G(V, E)$, where $V = (W, C, S)$ consists of one UAV depot $W = \{c_0\}$, K collection nodes $C = \{c_1, c_2, \cdots, c_K\}$ of UAV and N sensors S. The sets $C = \{c_0, c_1, \cdots, c_K\}$ and $E = \{(v_i, v_j)|v_i, v_j \notin W \times C\}$ represent the edges between sensors, the edges between sensors and collection nodes, and the edges between different collection nodes, respectively.

To establish a reasonable data collection and transmission model for UAVs in WSN, the following concepts are defined as follows:

Definition 1 (Neighbor Nodes). *Assuming the communication radius between sensors is denoted as r and the communication radius of UAVs is denoted as R, for any two nodes v_i, v_j within the collection area, their Euclidean distance is denoted as d_{ij}. v_i and v_j are considered neighbor nodes if one of the following conditions is met.*

(1) When $(v_i, v_j) \in S$, $d_{ij} \leq r$.
(2) When $(v_i, v_j) \in C \times S$, $d_{ij} \leq R$.

Definition 2 (Path). *For a given undirected graph G, let there be an alternating sequence $\Gamma = v_0 e_1 v_1 e_2 \cdots e_l v_l$ that contains the nodes and edges of G. For any $i = 1, 2, \cdots, l$, the adjacent nodes v_i, v_j in the sequence, if they are neighbors of each other, Γ is referred to as a path.*

Definition 3 (Reachable Nodes). *For any two nodes v_i, v_j in an undirected graph G, if there exists a path that connects v_i to v_j, then v_i and v_j are said to be reachable.*

Definition 4 (Hop Count). *For any two nodes v_i, v_j in an undirected graph G, if v_i, v_j are reachable from each other, and the number of edges connecting v_i and v_j is l, then the hop count between v_i and v_j is $H(v_i, v_j) = l$. If v_i and v_j are not reachable, the hop count $H(v_i, v_j) = \infty$.*

As shown in Fig. 1, the data of 30 sensors are collected by two UAVs, and all sensors are served by and only one collection node. One of the routes is that the UAV u_1 departs from the depot and passes through the three collection nodes c_1, c_2 and c_3 to collect data and then returns. The other route is that U_2 set off and return to the depot after passing through c_4 to c_7. During the process of data collection, except for sensors s_5, s_{11} and s_{15}, which need to be transferred by s_4, s_9 and S_{18} respectively, the data of other sensors can be collected directly by the UAVs, therefore, the hop count of s_5, s_{11} and s_{15} is 2, and the hop count of other sensors is 1.

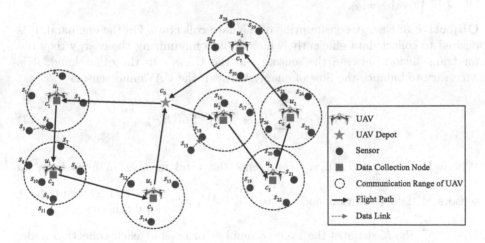

Fig. 1. Optimization of the UAV-assisted data collection.

2.2 Mathematical Model of the UAV-Assisted Data Collection

In WSNs, when utilizing UAVs for data collection from sensors, several considerations need to be considered. Firstly, due to the expensive nature of UAVs, there is a need to reduce their usage quantity. Secondly, considering the battery energy of UAVs and the timeliness of data collection in different scenarios, it is necessary to minimize the flying distance of UAVs while satisfying energy constraints during flight, hovering, and data collection. Thirdly, taking into account the battery capacity of sensors and the network's lifespan, it is necessary to balance QoS and the hops of data collection.

Objective 1: Cost of UAV usage. Considering the expensive prices of UAVs, the first objective is to minimize the number of UAVs used for data collection.

$$min\ f_1 = M \tag{1}$$

where M represents the number of UAVs.

Objective 2: UAV flight path planning problem. Considering the data transmission efficiency of UAVs and the energy consumption requirements for their normal operation during flight, data collection, transmission, and return to the UAV depot, the second objective is to minimize the total UAV flight distance.

$$min\ f_2 = \sum_{m=1}^{M} \sum_{(i,j) \in C^2} d_{ij} t_{ij}^m \tag{2}$$

where d_{ij} represents the Euclidean distance between collection node c_i and c_j, $t_{ij}^m = \begin{cases} 1 \text{ the UAV } u_m \text{ flies from } c_i \text{ to } c_j, \text{ and } i \neq j. \\ 0 \text{ otherwise.} \end{cases}$

Objective 3: Energy consumption of the data collection. On the one hand, it is desired to collect data efficiently, which implies minimizing the energy required for transmission between the sensors and the UAVs. On the other hand, it is necessary to balance the flow of energy between the UAVs and sensors.

$$min\ f_3 = \sum_{k=1}^{K} H_k + p \cdot e_1 \cdot Std(H_C) \tag{3}$$

Where $H_k = \sum_{n=1}^{N} H(c_k, s_n) x_k^n$ represents the total number of hops from all sensors at data collection node c_k to the UAV, $x_k^n = \begin{cases} 1 \text{ } s_n \text{ is served by } c_k. \\ 0 \text{ otherwise.} \end{cases}$,

$\overline{H}_C = \sum_{k=1}^{K} H_k / K$ denotes the average number of hops at each collection node,

$Std(H_C) = \sqrt{\sum_{k=1}^{K} (H_k - \overline{H})^2}$ represents the standard deviation of hop counts at each collection node, and $p(0 < p < 0.1)$ is introduced to balance the trade-off between data collection efficiency and energy balance.

Additionally, several constraints need to be satisfied for the UAV-assisted sensor data collection, as follows:

Constraint 1: Each UAV departs from the UAV depot, follows a predetermined planned path to collect and transmit data from multiple collection nodes, and returns to the UAV depot.

$$\sum_{k=1}^{K} t_{0k}^m = \sum_{k=1}^{K} t_{k0}^m, \forall m \in 1, 2, \cdots, M. \tag{4}$$

Constraint 2: The energy consumption including data transmission, hovering and flight of the UAV, must be less than or equal to the battery capacity of the UAV.

$$\max_{m \in 1, 2, \cdots, M} E_m \leq E. \tag{5}$$

Where $E_m = q_1 \cdot e_2 \sum_{n=1}^{N} y_m^n v_n + q_2 \cdot e_3 \sum_{k=1}^{K} z_{mk} H_k + e_4 \sum_{m=1}^{M} \sum_{(i,j) \in (W \cup C)^2} d_{ij} t_{ij}^m$.

$y_m^n = x_k^n \cdot z_{mk}$, $y_m^n = \begin{cases} 1 \text{ the data from } s_n \text{ is collected by } u_m. \\ 0 \text{ otherwise.} \end{cases}$ · $z_{mk} = \sum_{i=0}^{K} t_{ik}^m$,

$y_{mk} = \begin{cases} 1 \text{ the UAV } u_m \text{ collects data at collection node } c_k. \\ 0 \text{ otherwise.} \end{cases}$

E represents the maximum energy consumption that a UAV can provide for the service. E_m represents the total energy consumption of the m-th UAV during flight, hovering, and data collection. v_n represents the amount of data that collected from the n-th sensor. Assuming that the energy consumption of the data collection is positively correlated with the amount of data from the serviced sensor, with a correlation coefficient of q_1, and the hover duration of UAV at each collection node is proportional to the total hops count, with a correlation coefficient of q_2. e_2 represents the energy required per unit of data during data transmission. e_3 represents the energy consumption of the UAV per unit hop during each stop at a collection node. e_4 represents the energy consumption per unit distance during the UAV flight.

Constraint 3: Considering the practicality of UAV recharging and maintenance, each UAV is limited to performing at most one complete cycle of flight mission.

$$\sum_{k=1}^{K} t_{0k}^m \leq 1, \forall m \in 1, 2, \cdots, M. \tag{6}$$

Constraint 4: To improve the efficiency of data collection, it is required that the data of each sensor is collected only from one collection node.

$$\sum_{k=1}^{K} x_k^n = 1, \forall n \in 1, 2, \cdots, N. \tag{7}$$

Constraint 5: To ensure data integrity, the collection nodes must cover all the sensors in the area, meaning that each sensor is served by only one UAV when it passing through the collection node.

$$\sum_{m=1}^{M} t_{ij}^m = \sum_{m=1}^{M} t_{hi}^m = 1, \forall i \in C, \ j, h \in W \cup C. \tag{8}$$

3 Petal Algorithm for the UAV-Assisted Data Collection

In WSN, assuming the collection nodes and the UAV depot are determined, the UAV-assisted data collection can be considered as a two-layer optimization problem. The first layer involves the assignment of sensors, determining which collection node should be responsible for collecting data from each sensor. The second layer focuses on UAV path planning, determining the optimal route for the UAV to collect data from the assigned collection nodes. Therefore, this problem falls into the category of NP-hard problems that combine two optimization aspects. To address this challenge, a two-stage approach is employed in this paper to solve the problem, and the petal algorithm is proposed for the path planning of UAV.

3.1 Analysis of the UAV-Assisted Data Collection in WSNs

Regarding the first-layer optimization problem, the positions and quantities of collection nodes not only affect the UAV flight routes in Objective 2, but also impact the data collection efficiency and flow balance in Objective 3, because the position of the collection point will affect the affiliation of the sensors. When the positions of collection nodes are determined, the sensors that serviced by each collection node can be identified by distance and hop count between sensors and collection nodes. Thus, the amount of data and hop count required for data collection and transmission at each collection node can be calculated.

Considering the complexity and difficulty of the problem, we first utilize the position information of sensors and set the number and positions of multiple collection nodes using the Kmeans clustering method, while satisfying the full coverage constraint of sensors as the optimization objective. Then, we discuss the optimization of UAV quantity and flight routes using the proposed petal model algorithm under different positions and quantities of collection nodes. Based on this approach, we propose an algorithm based on clustering and the petal model to solve the dynamic UAV data collection problem in WSNs. The algorithm can be divided into two steps to solve the problem:

Based on the above analysis, the position of the collection nodes will first be determined by Kmeans clustering method, which needs to meet the requirements of full sensor coverage, while making the objective 3 as small as possible. Then, the petal algorithm is proposed to optimize the UAV quantity and flight routes under different positions and quantities of collection nodes.

3.2 Petal Algorithm for the UAV-Assisted Data Collection in WSNs

The petal model is inspired by the non-crossing loop characteristics exhibited by optimal paths in VRP. It adopts the concept of segmentation to generate initial paths. The basic principle is as follows: firstly, the starting and ending points of each path are determined based on angle and distance information; then, the nodes to be inserted are pre-classified according to the "along the way" principle; finally, each node is inserted into the corresponding class of paths. The pseudopod of the petal algorithm is as follow.

Algorithm 1. Pseudocode of Petal Algorithm

Input: the set of sensor $S = \{s_1, s_2, \cdots, s_N\}$; the UAV depot $W = \{c_0\}$; the battery
   capacity E of UAV;
Output: the data collection nodes and the flight routes of the UAVs;
 1: Generate hop count matrix H. The minimum hops between any two sensors can
    be calculated by the floyd algorithm.
 2: Set $K = 1$.
 3: Obtain K collection nodes through the Kmeans algorithm;
 4: **while** the sensors in each cluster are unreachable **do**
 5:     $K = K + 1$.
 6:     Obtain K collection nodes through the Kmeans algorithm;
 7: **end while**
 8: **for** $k = 1$ to K **do**
 9:     Calculate the energy consumption E_k for data collection at the k-th collection
    node according to Equation (5);
10: **end for**
11: Calculate the radian of each collection node starting at the UAV and ending at the
    collection node.
12: Take a random radian as the starting radian.
13: Initial an empty route start with the UAV depot.
14: **while** exist an uninserted collection node **do**
15:     **if** the constraint is violated **then**
16:         Reinitialize an new empty route start with the UAV depot;
17:     **else**
18:         Insert the uninserted collection node into the route one by one in a coun-
    terclockwise direction;
19:     **end if**
20: **end while**
    /*Assume that the sum of the generated paths is T */
21: **for** $t = 1$ to T **do**
22:     Select the collection node farthest from the UAV depot in t-th route;
23:     Removes all nodes from the route except the selected collection nodes;
24: **end for**
25: Calculate the minimum cost when the deleted point is inserted into the T routes.
26: Reinsert the deleted into one of the T routes in turn according to the order of cost
    from large to small with the minimum cost;

---

4   Experimental Simulation

4.1   Test Problem and Parameter Description

In order to evaluate the performance of the proposed algorithm, the UAV-assisted data collection and transmission problem are experimented at $N = 500, 1000, 1500, 2000$ of sensors. The sensors were randomly distributed in an area $(0, \sqrt{N})^2$, and the communication radius $R = 3$ of UAV. The communication radius $r = 1$ of sensor. The relevant parameter values in Sect. 2 are set as follows: $p = 1$, $q_1 = 0.01$, $q_2 = 0.1$, $e_1, e_2, e_3, e_4 = 1$, $E = 150$.

All the algorithms proposed and compared in this paper were implemented using Matlab 2016b. Each algorithm was independently run 50 times to ensure reliable results.

4.2   Comparative Analysis of Experimental Results

(1) Membership Relationship between Sensors and Collection nodes In order to address the issue of sensor coverage, the Kmeans algorithm was employed in this paper to determine the relationship between collection nodes and coverage rates, as illustrated in Fig. 2(a). Obviously, as the number of collection nodes increases, the coverage of collection points also increases until full coverage. Additionally, an example of the flight path maps of the UAVs on Test1 is presented in Fig. 2(b).

(2) Path Planning for UAV Data Collection In the scenario of fixed collection nodes, a comparative experiment was conducted between the method based on the petal model and five classical path planning methods. The test results on four different data sets are presented in Table 1 and Table 2. The bolded data in the tables represent the best results obtained among the six algorithms. The algorithm proposed in this paper can obtain a shorter flight path while using a smaller number of UAVs. In addition, since the collection

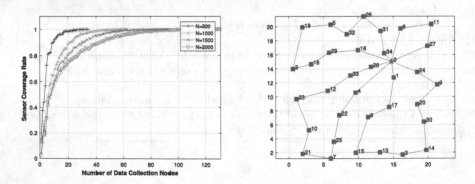

Fig. 2. (a) Relationship between coverage and number of collection points. (b) an example of the flight path maps of the UAVs on Test1.

nodes of all algorithms on different problems are obtained by the kmeans algorithm, the results of objective 3 are the same. The comparison results for Objective 1–3 are shown in Fig. 3.

Table 1. The average value of six algorithms on Objective 1

Test problem	Sweep algorithm	Savings algorithm	Nearest Neighbor algorithm	Insertion algorithm	Split algorithm	Petal algorithm
Test 1	4	4	4	4	4.42	4
Test 2	11.60	8	8	8	12.32	8
Test 3	15.78	11	12	12	16.48	11
Test 4	23.4	16	16	17	24.94	15

Table 2. The minimum value of six algorithms on Objective 2.

Test problem	Sweep algorithm	Savings algorithm	Nearest Neighbor algorithm	Insertion algorithm	Split algorithm	Petal algorithm
Test 1	220.94	162.51	199.22	196.83	218.97	**145.67**
Test 2	805.86	492.51	511.02	521.21	801.98	**458.68**
Test 3	1148.76	669.00	798.82	748.86	1136.51	**621.47**
Test 4	1825.81	999.27	1073.14	1135.95	1693.01	**949.33**

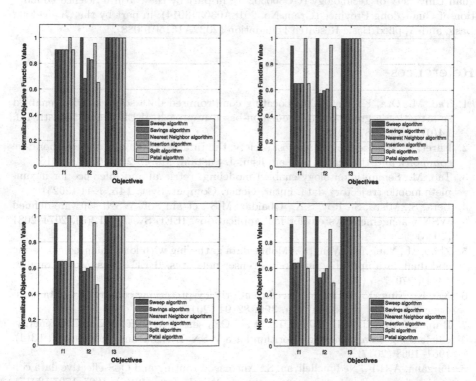

Fig. 3. The normalized objective function values obtained by six algorithms on four test problems.

5   Conclusion

This paper aims to minimize the number of UAVs, shorten their flying distances, and optimize the service quality of the sensor network in WSNs. Under the constraints of UAV energy consumption, integrity of sensor data collection, and full coverage, we establish and solve the data collection and transmission problem based on a dynamic UAV data collection and transmission model. Experimental results demonstrate that the petal algorithm proposed in this paper, achieve superior results in UAV path planning compared to five traditional path planning algorithms on four test problems. The solutions affirm the effectiveness of the proposed model and algorithms in this paper.

Acknowledgements. This work was supported in part by the Guangdong Key Construction Discipline Research Ability Enhancement Project (Grant No. 2021ZDJS086); in part by the Guangdong University Key Project (Grant No. 2019KZDXM012); in part by the Natural Science Foundation of Guangdong Province (Grant No. 2021A1515010656); in part by Guangdong Basic and Applied Basic Research Foundation (2022B1515120059); in part by the research team project of Dongguan University of Technology (Grant No. TDY-B2019009); in part by the PhD Start-Up Fund of Dongguan University of Technology (GC300502-3); in part by the Natural Science Foundation of Guangdong Province (Grant No. 2018A030313014); in part by the Guangdong Basic and Applied Basic Research Foundation (2022A1515010088).

References

1. Tao, M., Ota, K., Dong, M.: Locating compromised data sources in IoT-enabled smart cities: a great alternative-region-based approach. IEEE Trans. Industr. Inf. **14**(6), 2579–2587 (2018)
2. Ehret, M.: The zero marginal cost society: the Internet of Things, the collaborative commons, and the eclipse of capitalism. J. Sustain. Mobil. **2**(2), 67–70 (2015)
3. Tao, M.: Semantic ontology enabled modeling, retrieval and inference for incomplete mobile trajectory data. Futur. Gener. Comput. Syst. **145**, 1–11 (2023)
4. Bera, S., Misra, S., Roy, S.K., Obaidat, M.S., et al.: Soft-WSN: software-defined WSN management system for IoT applications. IEEE Syst. J. **12**(3), 2074–2081 (2018)
5. Zhao, M., Yang, Y., Wang, C.: Mobile data gathering with load balanced clustering and dual data uploading in wireless sensor networks. IEEE Trans. Mob. Comput. **14**(4), 770–785 (2015)
6. Xie, K., Ning, X., Wang, X.: An efficient privacy-preserving compressive data gathering scheme in WSNs. Inf. Sci. **390**, 82–94 (2017)
7. Rani, S., Ahmed, S.H., Talwar, R., et al.: Can sensors collect big data? An energy-efficient big data gathering algorithm for a WSN. IEEE Trans. Industr. Inf. **13**(4), 1961–1968 (2017)
8. Farzana, A.H.F., Neduncheliyan, S.: Ant-based routing and QoS-effective data collection for mobile wireless sensor network. Wirel. Netw. **23**(6), 1697–1707 (2017). https://doi.org/10.1007/s11276-016-1239-6
9. Joshi, Y.K., Younis, M.: Restoring connectivity in a resource constrained WSN. J. Netw. Comput. Appl. **66**, 151–165 (2016)

10. Wu, Q., Liu, L., Zhang, R.: Fundamental tradeoffs in communication and trajectory design for UAV enabled wireless network. IEEE Wirel. Commun. **26**(1), 36–34 (2019)
11. Tunca, C., Isik, S., Donmez, M.Y., et al.: Distributed mobile sink routing for wireless sensor networks: a survey. IEEE Commun. Surv. Tut. **16**(2), 877–897 (2014)
12. Miao, Y., Sun, Z., Wang, N., et al.: Time efficient data collection with mobile sink and vMIMO technique in wireless sensor networks. IEEE Syst. J. **12**(1), 639–647 (2018)
13. Zhou, Z., Du, C., Shu, L.: An energy-balanced heuristic for mobile sink scheduling in hybrid WSNs. IEEE Trans. Industr. Inf. **12**(1), 28–40 (2016)
14. Chang, J.Y., Shen, T.H.: An efficient tree-based power saving scheme for wireless sensor networks with mobile sink. IEEE Sens. J. **16**(20), 7545–7557 (2016)
15. Tao, M., Li, X.Q., Yuan, H.Q., Wei, W.H.: UAV-aided trustworthy data collection in federated-WSN-enabled IoT applications. Inf. Sci. **532**, 155–169 (2020)
16. Rosenkrantz, D.J., Steams, R.E., Lewis, P.M.: An analysis of several heuristics for the traveling salesman problem. SIAM J. Comput. **6**(3), 563–581 (1977)
17. Clarke, G., Wright, J.W.: Scheduling of vehicles from a central depot to a number of delivery points. Oper. Res. **12**, 568–581 (1964)
18. Mole, R.H., Jameson, S.R.: A sequential route-building algorithm employing a generalized savings criterion. Oper. Res. Q. **27**(2), 503–511 (1976)
19. Gillett, B.E., Miller, L.R.: A heuristic algorithm for the vehicle-dispatch problem. Oper. Res. **22**(2), 340–349 (1974)
20. Beasley, J.: Route first-cluster second methods for vehicle routing. Omega **11**(4), 403–408 (1983)
21. Perboli, G., Tadei, R., Vigo, D.: The two-echelon capacitated vehicle routing problem: models and math-based heuristics. Transp. Sci. **45**(3), 364–380 (2011)
22. Baldacci, R., Mingozzi, A., Roberti, R., Calvo, R.W.: An exact algorithm for the two-echelon capacitated vehicle routing problem. Oper. Res. **61**(2), 298–314 (2013)
23. Li, X., Tao, M.: Location planning of UAVs for WSNs data collection based on adaptive search algorithm. In: Chen, X., Yan, H., Yan, Q., Zhang, X. (eds.) ML4CS 2020. LNCS, vol. 12487, pp. 214–223. Springer, Cham (2020). https://doi.org/10.1007/978-3-030-62460-6_19

DeletePop: A DLT Execution Time Predictor Based on Comprehensive Modeling

Yongzhe He[1,2], Yueyuan Zhou[1], En Shao[1,2,3(✉)], Guangming Tan[1,2],
and Ninghui Sun[1,2]

[1] State Key Lab of Processors, Institute of Computing Technology, CAS,
Beijing 100190, China
{heyongzhe22z,zhouyueyuan,shaoen,tgm,snh}@ict.ac.cn
[2] University of Chinese Academy of Sciences, Beijing 100049, China
[3] Nanjing Institute of InforSuperBahn, Nanjing 211100, China

Abstract. The modeling and simulation of Deep Learning Training
(DLT) are challenging problems. Due to the intricate parallel patterns,
existing modelings and simulations do not consider enough factors that
influence the training, which brings inaccuracy for the prediction of DLT
time. To address these rising challenges, we propose *DeletePop*, a **D**eep
Learning **T**raining **E**xecution time **P**redictor based on comprehensive
modeling at the **Op**erator level. It systematically abstracts the process
of DLT by dividing it into computation, memory access, and communication three parts. *DeletePop* could predict the Job Execution Time
(JET) according to the operator dataset obtained from the homogeneous
network. Finally, we integrate the *DeletePop* into a Job Scheduling Simulator (JSS) *DLTSim* to make support more efficient scheduling. Although
the implementation of *DeletePop* is based on the TensorFlow framework,
the theoretical model could adapt to any other frameworks that use static
graphs. *DeletePop* achieves up to 90% accuracy for Homogeneous Networks, and we also provide the theoretical manners to add support for
Heterogeneous Networks.

Keywords: Deep Learning Training · Job Execution Time · Modeling
and Simulation · Job Scheduling Simulator

1 Introduction

Large-scale Deep Learning Training (DLT) jobs require a significant amount of
time and are typically executed in a distributed cluster environment. However,
the diverse neural network topologies used in DLT jobs can lead to decoupling
from large-scale clusters. The DLT software stack lacks optimized solutions for
job load, distributed training methods, and scheduling strategies, resulting in
significantly lower parallel efficiency than its theoretical performance.

In order to improve parallel efficiency and resource utilization, researchers
need to choose or even propose suitable scheduling algorithms. However, the

Z. Tari et al. (Eds.): ICA3PP 2023, LNCS 14493, pp. 126–145, 2024.
https://doi.org/10.1007/978-981-97-0862-8_9

adjustment of scheduling algorithms by just-in-time optimizations is difficult. To achieve this goal, it is imperative to utilize JSS for preemptive verification of the efficacy of various DLT scheduling algorithms.

The execution time of DLT jobs would influence the resource location decision of scheduling algorithms. Existing JSSs cannot achieve satisfactory results unless keeping the execution time prediction of DLT jobs at an accurate level. To extend the application scope of job scheduling and solve the accuracy issue of execution time prediction, we firstly explore the execution mode of deep learning jobs to find the factors that affect program performance. For example, the degree of parallelism is a kind of execution mode.

Some research works have modeled the training process of deep learning and provided support for the prediction of JET. However, most of them have only modeled and predicted the computation part of the job, ignoring the memory access and communication parts which account for the same large proportion. [3] presents that when the Bert-Large model is trained in a distributed manner, its communication overhead accounts for up to 76.8% of the total training time. At the same time, the proportion of memory access intensive operators in BERT, Transformer, CRNN, and other networks also exceeds 50% [33].

Unfortunately, the existing JSSs lack the support for predicting JET, they have narrow application scopes and poor scalability. Job scheduling simulators need a fine-grained model for scheduling algorithms and cluster environment configuration. On the one hand, a large number of scheduling algorithms are based on the execution time of jobs [6], such as longest job priority scheduling [1,16], deadline-constrained scheduling [12], etc. On the other hand, different cluster environment configurations will also affect the execution time of jobs. It is necessary to realize the accurate execution time of DLT jobs in advance when using a JSS.

In fact, the predictions of JET used in simulators are often different from the real execution time, reasoning from the difficulty of gaining the exact execution time. This dilemma worsens the credibility and effectiveness of the scheduling simulator. Many JSSs require researchers to make statistics of the execution time frequently and provide it to the simulator as input. Such a modeling method brings great difficulty to time prediction. It is hard to determine the job's execution time in advance with any cluster scale. Researchers face difficulty to confirm the execution time of different jobs when running on different cluster environments with different GPU counts.

In order to solve the above issue, we propose *DeletePop*, a DLT execution time predictor based on comprehensive modeling at the operator level. We made the first observation that the execution time of each operator in different parallel DLT jobs is definite on specific hardware, according to the parameters of the DLT model. Besides the execution time on computation, we also model and predict the DLT job's execution time on each process's memory access and communication. In addition, we take each job's operator as the granularity in the whole execution prediction.

The development of *DeletePop* is based in *DLTSim*. The architecture of *DeletePop* is shown in Fig. 1. A DLT job always contains many epochs, each epoch is formed by many steps. During a single step, the behavior of program could be split into three parts, they are computation, communication, and memory access respectively. All behaviors are finished by plenty of operators. *DeletePop* takes a job as input, then predicts its JET without running a job. Then *DeletePop* uses a JSS to optimize the scheduling algorithm according to the predicted JET. By using *DeletePop*, researchers could quickly determine a JET of large-scale DLT job, and further find a optimal scheduling algorithm with a specific cluster and workload.

The main contributions of this paper are as follows:

1. It provides a fine-grained modeling method of DLT job's process by three kinds of resources: computing, memory access, and communication.
2. It provides a time prediction method for DLT job's operator by the prior knowledge of similar neural networks. The time prediction method of DLT operator can reach an average accuracy of 90%.
3. It implements *DeletePop*, a JSS for distributed DLT job by fine-grained modeling and time prediction method, which is adaptable to TensorFlow framework.

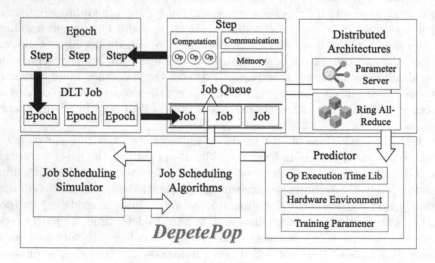

Fig. 1. Architecture of *DeletePop*

2   Background and Current Challenges

This chapter introduces the background of a JSS. Firstly, we introduce some distributed strategies of parallel training. Then we introduce some common scheduling algorithms, job schedulers, and JSSs. Finally, the shortcomings of the existing methods for predicting JET are discussed.

2.1 Distributed Strategy for Deep Learning Training

The most common parallelism mode is data parallelism [13], which places model copies on each device so that each device processes a subset of the training data and updates network parameters between model copies at the end of each iteration. Data parallelism is effective for computationally intensive operators with a small number of parameters, but not for operators with a large number of parameters. Another common parallel mode is model parallelism [5], which assigns disjoint subsets of a model to each device respectively. Model parallelism can avoid parameter synchronization, but it needs data transmission between devices. Model parallelism can effectively solve the problem of large computing graphs. Although model parallelism solves the problem that a single device cannot accommodate the entire large-scale computing graph, there are strong dependencies among computing graphs among devices, which have a certain impact on the parallel efficiency. There are also studies using hybrid parallelism [19,25], that is, both data parallelism and model parallelism, which combines the advantages of the two parallel modes to achieve better results than the single parallel mode. In addition, there are some works that use pipeline parallelism, which divides the training work into several sub-parts and takes the form of pipeline to parallel the sub-parts. For example, PipeDream [17] comprehensively adopts intra-batch pipeline parallelism and inter-batch pipeline parallelism. Compared with the traditional work that only uses intra-batch pipeline parallelism, PipeDream achieved a speed increase of 5.3× on model training.

In the data parallel mode, distributed architectures can be further selected. The most widely used distributed architectures include PS (Parameter Server) [16] and All-Reduce [20,22]. In PS architecture, data and workload are distributed on worker nodes, while server nodes maintain global shared parameters, which are expressed as dense or sparse vectors and matrices. However, PS architecture will bring a lot of communication. All-reduce effectively improves the problem of massive communication. Ring All-Reduce [9,12] is a commonly used All-Reduce architecture. Ring All-Reduce organizes devices into a Ring structure, and each device only needs to communicate with its two neighboring devices. Ring All-Reduce reduces the amount of data in communication, it has good scalability but poor fault tolerance. Besides the scheduling algorithm and the hardware architecture, the greatest difference of DLT jobs comes from the parallel mode. To avoid mistakes and lower the cost, JSS is essential.

2.2 Job Scheduling Simulator

Each scheduling algorithm has its own design principles and target orientation, which drive scheduling decisions through fairness, capacity guarantee, resource availability and other factors. It is very important to test and evaluate the scheduling algorithm completely before deploying it to the production cluster. However, testing and evaluating the scheduling algorithm in the real cluster consumes a lot of computing resources and time and energy, which brings difficulties to the research of scheduling algorithms.

The scheduling simulator can predict the performance of the scheduling algorithm under some specific workloads and reduce the consumption of resources and time during testing and evaluation. For example, the YARN scheduling simulator [28] is a tool that can load applications on machines and simulate large-scale YARN machines. It also combines the YARN scheduling simulator with the distributed network simulator MaxiNet to enable it to conduct research on joint job and flow scheduling. The Slurm simulator [24] can perform several months of workload simulations for large HPC systems in a few days to investigate configuration strategies for large cluster splitting and node sharing, evaluate their potential benefits, and provide information on the possible benefits of deploying such configurations. CQsim [30] is a Trace-based scheduling simulator that is used to evaluate various scheduling policies and test the performance of the task power-aware scheduling mechanism proposed in the paper under different scales. Yang et al. [31] combined inference characteristics with serverless paradigm in order to make universal servers meet the low latency and high throughput requirements of machine learning services, then proposed INFless, the first specific serverless platform in the field of machine learning, and tested the scheduling logic under large-scale clusters on the simulator. Evaluate the effectiveness of controller algorithms and the overhead in INFless. Robson et al. [21] developed a clock-cycle level accurate performance simulator to simulate thread startup, scheduling, and communication between host and accelerator for ProSE, an accelerator for protein design and discovery. Dadu et al. [4] proposed TaskStream, a task execution model for accelerators, in order to make reconfigurable accelerators such as CGRAs suitable for common task-parallel workloads, and developed a simulator for its accelerator Delta to provide simulation for small, individually scheduled program units of work. *DLTSim* [29] is a GPU cluster scheduling simulator with fault generator, it could simulate the execution of large-scale jobs under different scheduling policies. Due to its excellent modularization and flexibility, we develop *DeletePop* based on *DLTSim*.

The above functions of the JSS can be summarized in Table 1.

2.3   Major Challenges of Modeling and Simulation of Deep Learning Training Job

Using the simulator, we can quickly test and evaluate scheduling algorithm. On the one hand, the use of scheduling simulators needs to know the expected completion time of the program. On the other hand, many scheduling algorithms rely on the execution time of operation. Therefore, in an era when the scale of DLT operations has increased dramatically, how to effectively forecast the program execution time became a problem to be solved. There are three major challenges when modeling and predicting the execution time of DLT job.

Challenge I: Current modeling of DLT jobs does not consider the cost of Non-computing and lacks support for different architectures. Kaufman et al. [11] used Graph Neural Network to predict the execution time of programs on TPU by means of machine learning. Their performance model achieved an average error of 3.7% and an average correlation coefficient of 0.8, which is slightly better than

Table 1. The Summarization of Job Scheduling Simulators

Job Scheduling Simulator	Function	ⓐ	ⓑ	ⓒ*
YARN Scheduling Simulator	Load applications on the machine to simulate a large-scale YARN cluster	√	×	×
YARN with MaxiNet	Research on joint job and flow scheduling	√	×	×
Slurm Simulator	Perform workload simulations for large HPC systems in a short period of time	√	×	×
CQsim	A Trace-based scheduling simulator for evaluating various scheduling policies	√	√	×
INFless Simulator	Scheduling simulations for INFless, the first serverless platform in the ML domain	×	√	×
PreSE Simulator	Accurate performance simulations for ProSE, an accelerator for protein design and discovery	×	√	×
Delta Simulator	Make CGRAs applicable to TaskStream, the universally loaded task execution model	×	√	×
DeletePop (ours)	Provide predicted execution time for DNN jobs and then simulate the whole time under different scheduling policies	√	√	√

* ⓐ: Support for large-scale clusters.
ⓑ: Design for specific types of jobs.
ⓒ: Consider the difference between theoretical and practical parallelism patterns.

the analysis model. However, this method is based on machine learning and lacks sorting out the job execution process and distributed architecture. Arafa et al. [2] built a Trace-driven performance model "PPT-GPU" for GPGPU, which has the advantages of portability, extensibility, comprehensive, verifiability, open source, and so on. Compared with the actual NVIDIA Volta hardware, in the scale of clock cycle level, the average absolute errors for the predicted execution time, occupancy, L1 and L2 cache hit rates are 15%, 10%, 4% and 16%, respectively. The method can obtain an accurate calculation for operators in GPGPUs, but did not consider distributed communication overhead in the job. At the same time, using this method needs to obtain the program running trace, which takes several hours when predicting. This is unacceptable in many application scenarios.

Challenge II: Existing prediction methods have limited support for mainstream deep learning frameworks. Jia et al. [10] built a deep learning framework FlexFlow by themselves, which is mainly used to search for the optimal distributed strategy of a task. The test results of FlexFlow on two GPU clusters with six real DNN benchmarks significantly outperform the state-of-the-art parallelization methods. The prediction accuracy of the simulator for the program execution time is also more than 80%. This method is based on the operator of deep learning jobs, which can get the prediction result quickly. At the same time, the distributed architecture is considered. However, this method can only provide the prediction of execution time for programs based on FlexFlow framework, and its support for mainstream deep learning frameworks (such as TensorFlow) is limited.

Challenge III: Simulation Relies on Actual Execution. TensorBoard [7] is a built-in JET prediction tool of TensorFlow. The time required for training can be predicted by the average time of the training steps. This method can obtain more accurate time prediction, but there are still three disadvantages: first, it must be practical training to know the expected time consumption, and if we want more accurate results, we need to train enough steps; Second, the hardware conditions of large-scale training or reasoning are scarce, so it is difficult for researchers to obtain the data in the real environment. Third, the prediction of the execution time cannot adapt to different training parameters. The above three disadvantages show that the time prediction tool helps researchers a little.

The advantages and applicability coming of the above modeling and simulation methods are summarized in Table 2, where A_1, A_2 and A_3 stand for the applicability of *Challenge I, II and III* in Sect. 2.3 individually.

Table 2. Advantages and Applicability of Existing JET Prediction Methods

JET Prediction Methods	Advantages	A_1	A_2	A_3
TPU Performance Model	It achieves high accuracy on TPU			✓
PPT-GPU	Portable, extensible and high accuracy			✓
FlexFlow	Fast, automatically find the optimal policy	✓		✓
TensorBoard	Accurate, out of the box	✓	✓	
DeletePop (**ours**)	Accurate, fast, extensible	✓	✓	✓

3 Cost Model Design

A cost model is used to predict the JET. In this Section, we will introduce the design process and method of the cost model in *DeletePop*. Firstly, the execution process of DLT job is divided into three parts: computation, memory access and communication, which are modeled respectively. Then the execution time of the operator is tested and statistically analyzed. Finally, the factors affecting the data parallel execution mode are explored.

3.1 Modeling of the DLT Process

In our cost model, the process of DLT job is composed of three events: computation, memory access and communication. The main influencing factors in the computation part include the type and quantity of operators, the dependence between operators, training parameters, and parallel mode. As for memory access, the most important influencing factors are training parameters, data sets, and bandwidth. While using multi-device or distributed framework, the number of parameters and devices, and communication mode (bandwidth) influence communication the most.

The set of events Δ contains computation, memory access and communication, which are denoted by cp, ma and cm individually. Then the execution time T of DLT job can be predicted by Formula 1, where ζ represents the number of steps in the training. The modeling details of three events are explained in the following three sections.

$$T = \zeta \cdot T_{step} = \zeta \cdot \sum_{\delta \in \Delta} T_\delta, \text{ where } \Delta = \{cp, ma, cm\} \tag{1}$$

3.1.1 The Cost Model of Computation

For a single GPU, our model defines the computation time of DLT as the sum of operator execution time t_{op}. For multiple GPUs, it is determined according to the single GPU time and the dependence between each other. For the performance reason, most of the TensorFlow programs would run in the form of static graph, so our model also uses static graph to describe the computation. The computation graph of TensorFlow is composed of operators, tensors, data flows and control flows between them. Within the area of time prediction, tensors can also be regarded as a special kind of operator. When executing a graph, TensorFlow will analyze its dependencies from the top down in order to obtain an adaptive order, then submit it to the specific device to execute. In Nvidia GPU, operators are organized in the form of stream. Operators in the same stream are executed sequentially, while different streams can be executed in parallel [32]. However, TensorFlow only creates one GPU stream for computation [14], which means only one operator will be executed in the GPU at a time. The analysis above explains why we predict the computation time by the sum of operator execution time.

The prediction of computation time T_{cp} is shown in Eq. 2. The Ψ in the equation represents the set of all operators. For each kind of operator op, ϑ_{op} and f_{op} represent the count in a step and time predict function individually, where f_{op} is a function of batch_size (B) and image_size (I). Besides, the Ω of device τ represents its normalized factor of computation capacity. The computation capacity of GPU is defined as the performance of floating-point computing. We regard the GPU as standardization where we measured the execution time of operators, so the normalized factor Ω of another device τ is its floating-point computing performance divided by the performance of standard GPU.

$$T_{cp} = \sum_{op \in \Psi} t_{op}, \text{ where } t_{op} = \frac{\vartheta_{op} f_{op}(B, I)}{\Omega_\tau} \tag{2}$$

3.1.2 The Cost Model of Memory Access

For memory access in training tasks, it is defined as the process of transferring training data from CPU memory to GPU memory and then to the register. It is comprehensible that the amount of data transferred is proportional to the batch size and the area of the image, which is proportional to the square of the image size. Taking AlexNet network and CIFAR-10 dataset as an example, except for the transmission of a small number of startup parameters before the training starts (about tens of bytes), the transmission of the training dataset accounts for most of the memory access time between CPU and GPU.

Data of memory access is transferred through the *PCIe* bus between CPU and GPU, through *HBM2* between global memory and shared memory, and through *Reg2Reg* between different registers. We denote the set of links as Γ, which includes the above three memory access ways. Considering the total amount of data *AoD* and the bandwidth *BD* of links in set Γ, the predicted memory access time can be obtained by Formula 3. While the bandwidth of *HBM2* and *Reg2Reg* is several orders of magnitude faster than *PCIe*, the memory access time mainly depends on the PCIe transferring time. PCIe is a full-duplex bus [15,27], it can simultaneously open the transmission in both directions, and the transmission rate does not affect each other. Besides, during each batch, several GB of data needs to be transmitted usually. Therefore, the processing time of data packets in PCIe protocol can be ignored and the bandwidth of the bus can be fully used for data transmission.

$$T_{ma} = \sum_{l \in \Gamma} \frac{AoD}{BD_l} \approx \frac{AoD}{BD_{PCIe}} \tag{3}$$

3.1.3   The Cost Model of Communication

For communication in the training task, it is defined as parameter synchronization between GPUs. Depending on the limitations of hardware conditions, GPU synchronization can be implemented through point-to-point communication buses such as NVLink or PCIe bus between CPU and GPU. Either way, it can be abstracted as the collaboration between data volume and bandwidth. The bandwidth is determined by the hardware parameters, and the amount of data is determined by the model. At the same time, the total amount of communication data required by the whole job is also related to the distributed framework [18]. To be specific, if K is the amount of data needed to be transmitted by a single device and N is the number of devices participating in synchronization, then the total number of parameters that need to be communicated can be calculated by Formula 4, where ϕ is equal to 1 when using Ring All-Reduce and 0 when using Parameter Server.

Similar to the modeling of memory access time, synchronous communication usually needs to transmit a large amount of parameter data, so the utilization of bandwidth can be considered close to the ideal situation. By considering the parameter data amount *PaM* with the bandwidth of chip to chip BD_{c2c} (like PCIe and NVLink), the predicted communication time can be calculated by Formula 4.

$$T_{cm} = \frac{PaM}{BD_{c2c}} = \frac{2K \cdot (1 - \phi/N)}{BD_{c2c}}, \; where \; \phi = \begin{cases} 1, \text{ for RA} \\ 0, \text{ for PS} \end{cases} \tag{4}$$

3.2   The Predicting Algorithm of the Cost Model

The type of general operator in recent popular DNNs is limited. Therefore, as long as we build an execution time library for most operators, the prediction is feasible for common DNNs. After profiling and analyzing eight well-known deep learning networks: ResNet-50, AlexNet, DenseNet-121, DenseNet-169, DenseNet-201, Bert-Base, Seq2Seq and VGG-16, we found that although the structure of different networks is very different, the operators that compose the network are almost the same.

Table 3. The Top 5 Most Frequently Used Operators in DNNs

Operator's Name	Frequency of Occurrence
ReadVariableOp	27%
Identity	16%
Const	10%
VarHandleOp	6%
Switch	5%
Others	16%

Besides, it is feasible and extensible to predict the execution time based on operators. Furthermore, the ten operators with the highest frequency in the above network are shown in Table 3. It can be seen that the operator with the highest frequency is ReadVariableOp, that is, reading the value of a variable, which accounts for up to 27% of the total 95 operators.

The basic idea to predict the operator execution time is using the prior knowledge of operators in the same kind of network model. As the Algorithm 1 shows, we first train a DNN model in a real machine, then we can obtain the execution time of operators. Repeat this process until there is enough data. After that, we would process the data, and generate a suitable function for the operator's execution time and batch size and image size. This function stands for the predicted operator's execution time. Now we have the prediction of each operator, so we can use it to predict the execution time of this operator in other DNNs. For example, the operator's prior knowledge of DenseNet-169 could be used to predict the operator execution time of DenseNet-121 and DenseNet-201. For most operators, their execution time is significantly correlated with Batch Size and Image Size.

4   Simulator Design and Implementation

In this chapter, we will introduce the design and implementation of *DeletePop*. Firstly, the overall architecture of the simulator is introduced. Then we will

---

Algorithm 1. The Predicting Algorithm of Operator's Execution Time

---

Input: A machine, a DNN and a range <L, H>.
Output: Time prediction function F of the operator.
 1: **repeat**
 2:     Assign a pair of (B, I) in the range <L, H>.
 3:     Trace a training.
 4:     Gain the execution time T of each operator.
 5:     Process data.
 6:     Store the relationship r between T and (B, I)
 7: **until** Range <L, H> is traversed.
 8:
 9: **while** Operators still not finished **do**
10:     **repeat**
11:         **if** A relationship r is abnormal **then**
12:             Kick it out.
13:         **end if**
14:     **until** All r in this operators' R finished.
15:     **repeat**
16:         Assign fitting parameters.
17:         Fit function by processed R.
18:     **until** Find the best function.
19: **end while**

---

introduce the function and principle of the cost model in *DeletePop*. Finally, a case study is conducted to help understand its workflow.

DeletePop is based on *DLTSim*, which is a GPU cluster scheduling simulator with fault generator. *DLTSim* contains five basic parts, namely custom workload, custom cluster environment, custom scheduler, error generator and original data generator. The core component of *DeletePop* is the cost model, which is able to predict the JET. The function and principle of our cost model are shown in Fig. 2. By a series of processes in the figure, the cost model can obtain the computation time, communication time and memory access time respectively. Then it provides the JET to simulators, which allows simulators to optimize the job scheduling algorithms.

We use a case study to help understand the workflow of the JSS. After the cost model is added to the simulator, for a DLT job, the predicted execution time of memory access, computation and communication can be obtained respectively through some profiling and analysis tools. Combining these three parts, we can gain the predicted JET. Each job in the queue needs to be labeled with predicted JET before scheduling. Lastly the workload and predicted time are provided to the JSS, so that the JSS can select different scheduling algorithms to generate the execution job queue and calculate the total scheduling time.

5   Evaluation

In this chapter, we test and analyze the simulator, and carry out three experiments in sequence, namely, Prediction of Step Time, Prediction of Epoch Time,

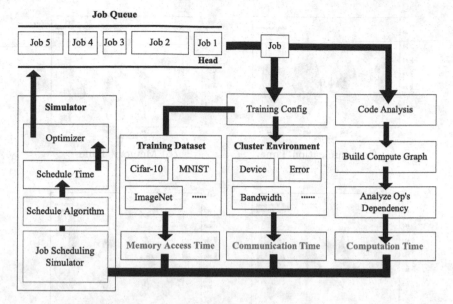

Fig. 2. The Position of Job Execution Time Prediction Model in Workflow

and Performance Simulation Experiment. The Prediction of Step Time is evaluated in three parts, they are calculation, memory access and communication, as well as the total time of a single Step. As for Prediction of Epoch Time, the influence of multi-device parallel training on JET is further considered, and the model is modified according to the experimental results. In the Performance Simulation Experiment, the JET prediction model is combined with the scheduling simulator, and the performance of the model in the simulator is compared and analyzed.

During our experiment, on software, TensorFlow-GPU 1.14.0 and CUDA 10.0 were used in the experiment, and on hardware, we conducted DLT and experiments on a single Intel(R) Xeon(R) Silver 4110 CPU @2.10 GHz CPU and three Tesla V100 (32 GB) GPUs.

5.1 Prediction of Step Time (The Atomic Unit)

To ensure our modeling and prediction are reliable and comprehensive, we first predict the Step Time. We know that a DLT is composed of many epochs, and each Epoch is composed of many steps. Step is the minimum repeated unit in DLT, in other words, the atomic unit. We divided a Step into three parts: calculation, memory access and communication, predicted the execution time of these three parts and the whole Step respectively, and analyzed their accuracy.

We choose DenseNet-169 to reveal this experiment. In the experiment, at each Batch Size, the influence of different Image Sizes on training is tested and the execution time is obtained, as shown in Fig. 3. Each strip in the figure represents the experimental results of a fixed Batch Size, and the width of the strip represents the number of data points located around the execution time. The left

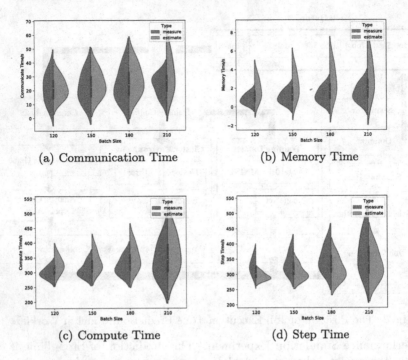

(a) Communication Time   (b) Memory Time

(c) Compute Time   (d) Step Time

Fig. 3. The Result Compare for Communication Time, Memory Access Time, Computing Time and Total Time of a Single Step

strip represents the actual test results, and the right one represents the results predicted by the model. The combination of the two can be used to compare and analyze the accuracy of the forecast. It can be seen that except for the memory access time shown in Fig. 3(b), the prediction results of the rest parts and the overall execution time of Step are accurate, and the average accuracy of the overall time prediction of a single Step achieves 90.63%. However, in this experiment, the memory access time accounted for only a few milliseconds among hundreds of milliseconds of Step, its impact on the whole process could be ignored.

Although there is little difference between the predicted and actual storage access time, the prediction accuracy of Step time is relatively high on the whole, and subsequent experiments can be carried out on this basis.

5.2   Prediction of Epoch Time

To enlarge the area and credibility of prediction, we test the accuracy of *DeletePop* on epoch time. A DLT consists of several Epochs, and each Epoch takes the same behaviors, so we can just analyze one Epoch. Based on 5.1, this experiment further considers the cooperation and dependence of each device during distributed parallel training, as well as the influence of different distributed strategies on execution mode and execution time. This experiment is formed by two small experiments, about Homogeneous Networks and Heterogeneous

Networks respectively. Two small experiments progressively explain the model's advantages and disadvantages from the perspectives of specificity and universality, as well as address its shortcomings in explanation or solution.

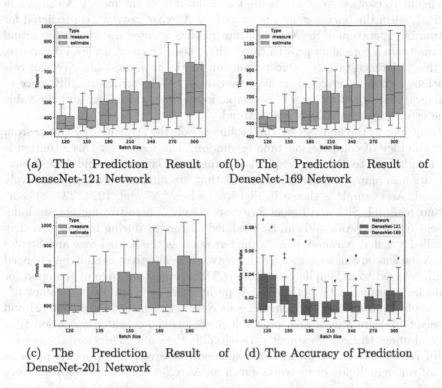

(a) The Prediction Result of DenseNet-121 Network

(b) The Prediction Result of DenseNet-169 Network

(c) The Prediction Result of DenseNet-201 Network

(d) The Accuracy of Prediction

Fig. 4. The Result Compare of DenseNet Networks

5.2.1 The Prediction of Homogeneous Networks

This experiment aims to use the operator database obtained from a certain network to predict the JET of the Homogeneous Networks. Homogeneous Networks refer to networks with similar network structures, layers, and parameters. Otherwise, we call them Heterogeneous Networks. For example, DenseNet-121, DenseNet-169, and DenseNet-201 belong to the Homogeneous Networks, while AlexNet and Resnet-50 belong to Heterogeneous Networks. The difference between the Homogeneous Networks mainly comes from the different repetition times of some layers inside, which also means that they often have similar execution patterns and characteristics.

A large-scale experiment is carried out for different Batch Sizes and different networks, whose results are shown in Fig. 4. We can see that the accuracy of prediction is basically above 90%. The prediction of the model for the Homogeneous Networks achieves very accurate results, which shows high specificity.

5.2.2   The Prediction of Heterogeneous Networks

This experiment will explore the prediction results of execution time for Heterogeneous Networks, so as to discuss the scalability of the model. As shown in Fig. 5(a), when the operator data measured by AlexNet network is predicted for the training operation of ResNet-50, although the expected result and the actual data maintain a consistent growth trend, there is still a constant level difference, and this difference takes a large proportion in the training time. We call this situation "structural error", because it is caused by the structural difference of the neural network. The error is accompanied by every training, and the value of the error keeps a relatively stable trend.

According to the analysis by the profiling tool, the error mainly comes from the inaccuracy of computation time prediction. The reason for this situation is that in addition to the two "explicit parameters" Batch Size and Image Size, there are also many "implicit parameters" that are difficult to be quantitatively analyzed. An example is shown in Fig. 5(b), where "N" and "(224, 224, 3)" correspond to Batch Size and Image Size respectively. These two factors are independent of the network and can be modified as needed during training, so they are called "explicit parameters". Parameters such as the Kernel Size and Strides of convolutions are determined by the network itself, they cannot be changed at will, referred to as "implicit parameters." The execution time of the operator should be determined by both "explicit parameters" and "implicit parameters". The "implicit parameters" of Homogeneous Networks (such as DenseNet-121 and DenseNet-169) are almost the same, which means the operator execution time obtained under the same "explicit parameter" is the same. Therefore the prediction of Homogeneous Networks could succeed. However, the "implicit parameters" of different kinds of networks (such as AlexNet and ResNet-50) are very different. Even if the "explicit parameters" are kept consistent, more "implicit parameters" have changed, so there is a big difference in operator execution time between these networks.

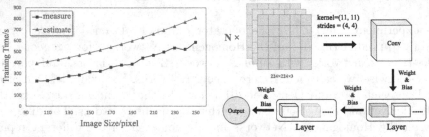

(a) The "Structure Error" in ResNet-50   (b) The Reason of "Structure Error"

Fig. 5. The "Structure Error" and Its Reason

The above analysis shows the effective range of time prediction, it can only work for Homogeneous Networks. The main reason is that *DeletePop* does not take into account the"implicit parameters". However, this conclusion also means that as long as the "implicit parameters" of the operator are further considered, *DeletePop* will be applicable to Heterogeneous Networks.

5.3 Performance Simulation Experiment

This experiment reveals the overall performance of our simulator. We combine Experiment 5.1 and Experiment 5.2, integrating *DeletePop* into the workflow of DLTsim. Then we compare the performance of the actual measured JET and the predicted JET under different scheduling algorithms.

The workload contains 66 jobs during this experiment, each of them is a DNN training task, coming from AlexNet, DenseNet-121, DenseNet-169 and DenseNet-201. In addition, the execution time, resource usage and training epochs of each job are different. At the same time, we also change the available resources (the number of GPUs) of the whole cluster, to test the scalability and consistency. Results are shown in Fig. 6 and Table 4. It can be seen that no matter for First Come First Service (FCFS), Smallest Resource First (SRF), Shortest Job First (SJF), or Smallest Area First (SAF), compared with the

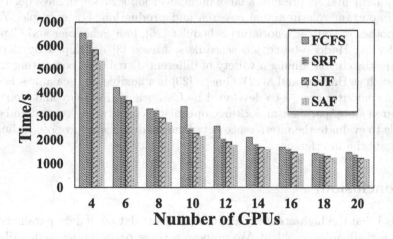

Fig. 6. Scheduling Time Under Different Algorithms

Table 4. The Average Accuracy of Prediction Under Different Scheduling Algorithms

Scheduling Algorithm	Average Accuracy
First Come First Service (FCFS)	96.90%
Smallest Resource First (SRF)	94.18%
Shortest Job First (SJF)	96.35%
Smallest Area First (SAF)	96.10%

actual JET, the average accuracy rate of the predicted JET in the scheduling simulator can reach more than 90%. More importantly, it maintains high accuracy as the number of GPUs in the cluster changes.

6   Related Work

Some current JET predictor does not offer a comprehensive model [11] or only consider partial behavior of DLT [2]. Some are not widely applicable [10] for open-source frameworks like TensorFlow, PyTorch, etc.

There are many scheduling algorithms. Some algorithms have nothing to do with the execution time of jobs. such as First-Come-First-Serve (FCFS) Algorithm, Time Slice Rotation Algorithm, Multilevel Feedback Queue Algorithm [9], and The Lottery Scheduling Algorithm [32]. Some algorithms are based on the execution time of the job [6], such as the Longest Job First Algorithm [1,16], Deadline Constrained Scheduling Algorithm [12], The Shortest Remaining Time First Algorithm [25] and so on.

There are some work studies about job schedulers. Job scheduler is an integration of scheduling algorithms. It adapts and modifies various algorithms according to the working environment of the scheduler. Users can choose different algorithms to schedule jobs in the cluster according to research work and business requirements. At present, a large number of job schedulers have been playing an important role in actual research and production. For example, YARN (Yet Another Resource Negotiator) Scheduler [26], Fair Scheduler and Capacity Scheduler are Hadoop-based job schedulers. Mesos [8] developed by Berkeley can share the cluster among a variety of different distributed computing frameworks (such as Hadoop and MPI). Omega [23] is a flexible and scalable scheduler for large computer groups co-developed by Cambridge, Berkeley and Google. It uses shared state, parallelism, lock-free optimistic concurrency control and other methods to evaluate the interference between schedulers and propose techniques to mitigate the interference.

7   Conclusion

We reveal that the high-accuracy modeling and simulation of data parallelism for DLT is a challenging problem. We propose a three parts model to describe the process of DLT and offer an abstraction for the leverage of parallel pattern. We develop a JET predictor *DeletePop* and integrate it into *DLTSim*. Results show that *DeletePop* achieves an average accuracy of over 90% while predicting the JET of Homogeneous Networks. We believe *DeletePop* would benefit researchers a lot in the field of scheduling, model designing, parallelism optimization, etc.

There are also some limitations of *DeletePop*. Firstly, *DeletePop* does not consider the overlap of different parts, which is a trent of optimization in the future. Secondly, although the idea of *DeletePop* is extensible theoretically, our experiment mainly focus on the neural networks in the area of computer vision, while lacking the further research of other area, like NLP. Besides, the applicability of *DeletePop* on heterogeneous devices is not satisfied, it is necessary to design a more fine-grained cost model. These problems are still waiting to be overcome in the future.

Acknowledgment. This work was sponsored in part by NKRDP (2021YFB0300800), and in part by NSFC (62102396), Beijing Nova Program (Z211100002121143, 20220484217), Youth Innovation Promotion Association of Chinese Academy of Sciences (2021099). Pilot for Major Scientific Research Facility of Jiangsu Province of China (NO. BM2021800).

References

1. Aida, K.: Effect of job size characteristics on job scheduling performance. In: Feitelson, D.G., Rudolph, L. (eds.) JSSPP 2000. LNCS, vol. 1911, pp. 1–17. Springer, Heidelberg (2000). https://doi.org/10.1007/3-540-39997-6_1
2. Arafa, Y., et al.: Hybrid, scalable, trace-driven performance modeling of GPGPUs. In: Proceedings of the International Conference for High Performance Computing, Networking, Storage and Analysis, SC 2021, pp. 1–15. Association for Computing Machinery. https://doi.org/10.1145/3458817.3476221
3. Bai, Y., et al.: Gradient compression supercharged high-performance data parallel DNN training. In: Proceedings of the ACM SIGOPS 28th Symposium on Operating Systems Principles CD-ROM, pp. 359–375. ACM (2021). https://doi.org/10.1145/3477132.3483553
4. Dadu, V., Nowatzki, T.: TaskStream: accelerating task-parallel workloads by recovering program structure. In: Proceedings of the 27th ACM International Conference on Architectural Support for Programming Languages and Operating Systems, pp. 1–13. ACM (2022). https://doi.org/10.1145/3503222.3507706
5. Dean, J., et al.: Large scale distributed deep networks. In: Advances in Neural Information Processing Systems, vol. 25. Curran Associates Inc. (2012)
6. Gautam, J.V., Prajapati, H.B., Dabhi, V.K., Chaudhary, S.: A survey on job scheduling algorithms in big data processing. In: 2015 IEEE International Conference on Electrical, Computer and Communication Technologies (ICECCT), pp. 1–11 (2015). https://doi.org/10.1109/ICECCT.2015.7226035
7. Goldsborough, P.: A Tour of TensorFlow (2016)
8. Hindman, B., et al.: Mesos: a platform for fine-grained resource sharing in the data center. In: Proceedings of the 8th USENIX Conference on Networked Systems Design and Implementation, NSDI 2011, pp. 295–308. USENIX Association (2011)
9. Javed, M.H., Ibrahim, K.Z., Lu, X.: Performance analysis of deep learning workloads using roofline trajectories. CCF Trans. High Perform. Comput. **1**(3), 224–239 (2019). https://doi.org/10.1007/s42514-019-00018-4
10. Jia, Z., Zaharia, M., Aiken, A.: Beyond data and model parallelism for deep neural networks, vol. 1, pp. 1–13 (2019)
11. Kaufman, S., et al.: A learned performance model for tensor processing units 3, 387–400
12. Kim, Y., Choi, H., Lee, J., Kim, J.-S., Jei, H., Roh, H.: Towards an optimized distributed deep learning framework for a heterogeneous multi-GPU cluster. Cluster Comput. **23**(3), 2287–2300 (2020). https://doi.org/10.1007/s10586-020-03144-9

13. Krizhevsky, A., Sutskever, I., Hinton, G.E.: ImageNet classification with deep convolutional neural networks. In: Advances in Neural Information Processing Systems, vol. 25. Curran Associates Inc. (2012)

14. Kwon, W., Yu, G.I., Jeong, E., Chun, B.G.: Nimble: lightweight and parallel GPU task scheduling for deep learning, vol. 33, pp. 8343–8354 (2020)

15. Li, A., et al.: Evaluating modern GPU interconnect: PCIe, NVLink, NV-SLI, NVSwitch and GPUDirect. IEEE Trans. Parallel Distrib. Syst. **31**(1), 94–110 (2020). https://doi.org/10.1109/TPDS.2019.2928289

16. Li, M.: Scaling distributed machine learning with the parameter server. In: Proceedings of the 2014 International Conference on Big Data Science and Computing, BigDataScience 2014, p. 1. Association for Computing Machinery (2014). https://doi.org/10.1145/2640087.2644155

17. Narayanan, D., et al.: PipeDream: generalized pipeline parallelism for DNN training. In: Proceedings of the 27th ACM Symposium on Operating Systems Principles, SOSP 2019, pp. 1–15. Association for Computing Machinery (2019). https://doi.org/10.1145/3341301.3359646

18. Ouyang, S., Dong, D., Xu, Y., Xiao, L.: Communication optimization strategies for distributed deep neural network training: a survey. J. Parallel Distrib. Comput. **149**, 52–65 (2021). https://doi.org/10.1016/j.jpdc.2020.11.005

19. Park, J.H., et al.: HetPipe: enabling large DNN training on (Whimpy) heterogeneous GPU clusters through integration of pipelined model parallelism and data parallelism. In: Proceedings of the 2020 USENIX Conference on USENIX Annual Technical Conference, vol. 21, pp. 307–321. USENIX Association (2020)

20. Patarasuk, P., Yuan, X.: Bandwidth optimal all-reduce algorithms for clusters of workstations. J. Parallel Distrib. Comput. **69**(2), 117–124 (2009). https://doi.org/10.1016/j.jpdc.2008.09.002

21. Robson, E., Xu, C., Wills, L.W.: ProSE: the architecture and design of a protein discovery engine. In: Proceedings of the 27th ACM International Conference on Architectural Support for Programming Languages and Operating Systems, pp. 655–668. ACM (2022). https://doi.org/10.1145/3503222.3507722

22. Sanders, P., Mehlhorn, K., Dietzfelbinger, M., Dementiev, R.: Sequential and Parallel Algorithms and Data Structures: The Basic Toolbox. Springer, Cham (2019). https://doi.org/10.1007/978-3-030-25209-0

23. Schwarzkopf, M., Konwinski, A., Abd-El-Malek, M., Wilkes, J.: Omega: flexible, scalable schedulers for large compute clusters. In: Proceedings of the 8th ACM European Conference on Computer Systems, EuroSys 2013, pp. 351–364. Association for Computing Machinery (2013). https://doi.org/10.1145/2465351.2465386

24. Simakov, N.A., et al.: Slurm simulator: improving slurm scheduler performance on large HPC systems by utilization of multiple controllers and node sharing. In: Proceedings of the Practice and Experience on Advanced Research Computing, PEARC 2018, pp. 1–8. Association for Computing Machinery (2018). https://doi.org/10.1145/3219104.3219111

25. Song, L., Mao, J., Zhuo, Y., Qian, X., Li, H., Chen, Y.: HyPar: towards hybrid parallelism for deep learning accelerator array. In: 2019 IEEE International Symposium on High Performance Computer Architecture (HPCA), pp. 56–68. IEEE (2019). https://doi.org/10.1109/HPCA.2019.00027

26. Vavilapalli, V.K., et al.: Apache hadoop YARN: yet another resource negotiator. In: Proceedings of the 4th Annual Symposium on Cloud Computing, SOCC 2013, pp. 1–16. Association for Computing Machinery (2013). https://doi.org/10.1145/2523616.2523633

27. Verma, A., Dahiya, P.K.: PCIe bus: a state-of-the-art-review. IOSR J. VLSI Sig. Process. **7**(4), 24–28 (2017). https://doi.org/10.9790/4200-0704012428
28. Wette, P., Schwabe, A., Splietker, M., Karl, H.: Extending Hadoop's yarn scheduler load simulator with a highly realistic network & traffic model. In: Proceedings of the 2015 1st IEEE Conference on Network Softwarization (NetSoft), pp. 1–2 (2015). https://doi.org/10.1109/NETSOFT.2015.7116169
29. Yang, K., Cao, R., Zhou, Y., Zhang, J., Shao, E., Tan, G.: Deep reinforcement agent for failure-aware job scheduling in high-performance computing. In: 2021 IEEE 27th International Conference on Parallel and Distributed Systems (ICPADS), pp. 442–449 (2021). https://doi.org/10.1109/ICPADS53394.2021.00061
30. Yang, X., et al.: Integrating dynamic pricing of electricity into energy aware scheduling for HPC systems. In: Proceedings of the International Conference on High Performance Computing, Networking, Storage and Analysis, SC 2013, pp. 1–11 (2013). https://doi.org/10.1145/2503210.2503264
31. Yang, Y., et al.: INFless: a native serverless system for low-latency, high-throughput inference. In: Proceedings of the 27th ACM International Conference on Architectural Support for Programming Languages and Operating Systems, pp. 768–781. ACM (2022). https://doi.org/10.1145/3503222.3507709
32. Zhang, P., Fang, J., Yang, C., Huang, C., Tang, T., Wang, Z.: Optimizing streaming parallelism on heterogeneous many-core architectures. IEEE Trans. Parallel Distrib. Syst. **31**(8), 1878–1896 (2020). https://doi.org/10.1109/TPDS.2020.2978045
33. Zheng, Z., et al.: AStitch: enabling a new multi-dimensional optimization space for memory-intensive ML training and inference on modern SIMT architectures. In: Proceedings of the 27th ACM International Conference on Architectural Support for Programming Languages and Operating Systems, pp. 359–373. ACM (2022). https://doi.org/10.1145/3503222.3507723

CFChain: A Crowdfunding Platform that Supports Identity Authentication, Privacy Protection, and Efficient Audit

Yueyue He[1] , Jiageng Chen[2] , and Koji Inoue[1]([✉])

[1] Kyushu University, Fukuoka 8190385, Japan
inoue@ait.kyushu-u.ac.jp
[2] Central China Normal University, Wuhan 430079, Hubei, China

Abstract. Charity crowdfunding is a technique for raising funds that involves collecting modest contributions from a vast number of individuals or groups via established crowdfunding platforms or other digital avenues. The objective is to provide support for charitable organizations, social welfare initiatives, or personal requirements. The widespread adoption of the Internet and the rapid advancement of digital technology have facilitated the global dissemination and promotion of charity crowdfunding. However, crowdfunding platforms have recently experienced a decline in credibility due to various factors such as fraudulent donations, inadequate fund management, and other forms of disorder. The blockchain's decentralization and anti-tampering features exhibit a high degree of compatibility with the requirements of a crowdfunding platform. Most current state-of-the-art techniques do not ensure the non-linkability of user identities in the face of sybil attacks, nor do they offer a streamlined auditing mechanism for crowdsourcing modest donations that simultaneously preserves transactional privacy. This paper presents a novel crowdfunding system called CFChain based on blockchain technology. Initially, the distributed identity and BLS signature are employed to establish a user authentication mechanism, enabling CFChain to withstand sybil attacks while preserving the non-linkability of user identities. Subsequently, a crowdfunding mechanism is constructed utilizing zero-knowledge proofs to facilitate streamlined auditing procedures while safeguarding donations' confidentiality. Additionally, a security analysis of CFChain is presented. The system prototype is subsequently implemented on the Hyperledger Fabric. Empirical evidence indicates that the efficiency of CFChain is viable.

Keywords: Crowdfunding · Blockchain · Distributed Identity ·
Zero-knowledge proofs · Private audit

1 Introduction

According to Statista's compiled data, the global crowdfunding market was valued at USD 13.64 billion in 2022 and is projected to grow at a compound annual

Z. Tari et al. (Eds.): ICA3PP 2023, LNCS 14493, pp. 146–167, 2024.
https://doi.org/10.1007/978-981-97-0862-8_10

growth rate (CAGR) of 11.2% to reach USD 28.9 billion by 2028 [2]. Charitable crowdfunding, which is a distinct type of crowdfunding that emphasizes the philanthropic utilization of funds and social accountability, constitutes 71% of all campaigns on the primary crowdfunding platforms [1]. Currently, charitable crowdfunding initiatives have achieved worldwide fundraising via established crowdfunding platforms, fostering global collaboration in charitable endeavors and offering effective means for charitable entities and individuals to secure financial support.

In recent times, an increasing number of scandals pertaining to crowdfunding platforms have come to light, resulting in a loss of trustworthiness for these platforms. This has, in turn, impeded the progress of charity crowdfunding. According to The New York Times, a couple has been jailed for five years for making up a story about homeless people on the GoFoundMe platform to raise more than $400,000 in donations [22]. A leukemia patient's GoFundMe account was reportedly hacked by con artists who took images of the patient's infant daughter and requested donations for another girl [24]. Furthermore, certain crowdfunding platforms impose substantial commissions, such as Indiegogo, which levies a 5% commission upon the attainment of successful fundraising.

In 2008, Satoshi Nakamoto developed a decentralized electronic currency system called Bitcoin. This system utilizes blockchain technology to document and authenticate all financial transactions [20]. The advent of blockchain technology has the potential to address the aforementioned issues and establish a basis of reliability for charitable crowdfunding with the following advantages. First, blockchain technology maintains an immutable record of all transactions and operations, enabling public scrutiny of the flow of funds and utilization of charitable contributions. This feature ensures transparency and accountability in tracking and verifying the allocation of donations. Second, the utilization of encryption technology and distributed storage is a fundamental aspect of blockchain. This technology ensures that the personal information and donation records of donors are safeguarded on the blockchain, thereby mitigating the possibility of data leakage and abuse. Third, the implementation of blockchain technology has the potential to decentralize the management and oversight of charity crowdfunding systems, thereby obviating the necessity of intermediary entities curtailing intermediary expenses.

Despite the gradual expansion of charity crowdfunding in recent years, charity fundraising remains to represent a significant proportion of overall charitable efforts. This is due to the longstanding history and broad societal impact of charity fundraising activities. Although blockchain can potentially enhance the reliability of charity crowdfunding services, current studies predominantly use it in charity fundraising. Charity fundraising and crowdfunding are distinct in their approaches to generating funds. While fundraising focuses on collecting relatively large amounts from a few sources, crowdfunding aims to gather smaller amounts from a large group of donors [14]. Hence, in the context of utilizing blockchain for charitable crowdfunding, it is imperative to tackle the following ensuing obstacles in a manner that ensures privacy preservation:

User Authentication: Participants in charitable fundraising are often qualified organizations or foundations, while charitable crowdfunding allows any individual or organization to participate. So charitable crowdfunding platforms are vulnerable to sybil attacks [11], which involve using multiple false identities or accounts by one or a few individuals to deceive, manipulate, or harm crowdfunding campaigns. In the user authentification phase, maintaining the anonymity of users' actual identities while simultaneously adhering to regulatory requirements such as Know-Your-Customer (KYC) poses a significant challenge.

Transaction Audit: Charitable fundraising places greater emphasis on securing substantial donations and collaborating with established entities such as corporations or foundations, as opposed to charitable crowdfunding. To protect the privacy of the transaction, certain studies [6] document the encrypted transaction data on the publicly accessible ledger. Crowdfunding for charitable purposes typically incentivizes numerous individuals or entities to engage in fundraising activities by soliciting modest contributions from a multitude of individuals. The efficacy of the system would be notably diminished if auditors were to decrypt and authenticate every donation. Efficiently conducting audits while maintaining transaction privacy poses a significant challenge.

This study presents a blockchain-based system for charity crowdfunding as a solution to the aforementioned challenges. The contributions are as follows.

- This paper analyzes the characteristics of charitable crowdfunding and suggests a blockchain-based charity crowdfunding platform, CFChain, that is not controlled by a third-party organization. Blockchain is utilized to record crowdfunding records so that every donation can be traced and cannot be tampered with.
- This paper uses distributed identity to construct the user authentication strategy. The distributed identity is composed of a unique main identifier and several context-based identifiers. The uniqueness of the main identifier provides sybil-resistance, while the context-based identifiers are used to prevent linkage attacks.
- This paper utilizes Pedersen commitment and BLS signature to construct verifiable credentials so that users can flexibly combine verification materials according to the needs of different contexts while not exposing their main identifier during the authentication process.
- This paper employs non-interactive zero-knowledge proof to construct a crowdfunding scheme, which allows each donor to jointly audit the correctness and compliance of crowdfunding records while providing transaction privacy.
- This paper presents the implementation of a system prototype using Hyperledger Fabric and demonstrates the viability of the proposed scheme through rigorous security and performance analyses.

Paper Structure: The rest of this paper is organized as follows. Related work is introduced in Sect. 2. Section 3 introduces cryptographic primitives that will be used in the scheme. In Sect. 4, the system model, workflow, and operations in CFChain are introduced. Then, it will discuss the security definition in Sect. 5. The details of system operations are introduced in Sect. 6. The security analysis and experimental performance of CFChain are presented in Sect. 7 and Sect. 8, respectively. Finally, the conclusion is in Sect. 9.

2   Related Work

Blockchain-Based Donation Scheme: Blockchain technology is widely used in charitable donations due to its decentralization, transparency, and immutability characteristics [33]. Existing research can be divided into two categories, one of which primarily focuses on currency, while the other centers on assets. The first category of research commonly uses cryptocurrencies to construct donation schemes. Akram and Mayes utilized the pre-existing GMS network and One Time Password (OTP) to build a blockchain-based charity collection scheme that accepts Bitcoin donations in both online and offline environments [17]. Some studies use Ethereum [30] to build a charity platform, establish donation guidelines via smart contracts, and procure on-chain tokens for donations [27,29]. Muhammad et al. defined a novel cryptocurrency based on Ethereum and added an external layer with public-private blockchains to provide a secure charity donation mechanism that is transparent and auditable for governmental agencies [12]. Osman et al. proposed a donation scheme based on Zerocash [28] that can protect user privacy and simultaneously use attribute-based signatures to allow donors to specify beneficiaries' attributes [6]. These schemes, while protecting user privacy through the anonymity of cryptocurrencies, make it difficult to audit donations. The cryptocurrency market exhibits significant volatility, thereby resulting in instability in the valuation of donations. Furthermore, certain jurisdictions impose bans or limitations on the use of cryptocurrencies. The second type of scheme usually uploads donation records to the blockchain, making the donation process transparent. For instance, Mehra et al. proposed a secure and transparent framework based on a consortium chain that connects with several charity organizations to enable direct donations [19]. The Karma platform is based on Hyperledger Fabric [5], which allows charitable organizations to create and manage different charitable projects, accept donations, and record all transaction information on the chain for auditing [25]. Although this type of research mitigates the potential hazards associated with cryptocurrencies, ensuring the confidentiality of user identities remains a challenging task. This paper is trying to develop a blockchain-based crowdfunding platform that aims to facilitate effective auditing and user compliance monitoring while ensuring user identity and transaction privacy.

Distributed Identity: Distributed identity, defined and standardized by the W3C, forms the three pillars of autonomous identity along with verifiable credentials and blockchain. It allows users to create, hold, and manage their identities without relying on traditional centralized authentication authorities. There are several implementations of decentralized identity systems today, such as uPort [16], and WeIdentity [4]. Although distributed identities can provide users with different pseudonyms to prevent identities from being associated, it is difficult to police these identities. CanDID divides user credentials into a master and several context-based credentials, binding the master credential to the user's real identity to resist witch attacks [18]. SmartDID also uses a similar idea to set the master identifier for users to implement user identity supervision [31]. CanDID relies on secure multi-party computation (MPC) and zero-knowledge proof, providing strong privacy but with a large computational overhead. This paper delegates certain authentications to an authoritative entity to maintain a balanced performance of CFChain.

Private Transaction Audit: zkLedger is the first system to simultaneously achieve transaction privacy and auditing, using non-interactive zero-knowledge proof [21]. The transaction information is encrypted and stored in the table structure. The auditor can check the transaction amount's correctness and legality without knowing the transaction's specific content. PAChain also uses zero-knowledge proofs to build transactions, but it is based on the UTXO model and cannot be used in the account model [32]. Based on zkLedger, FabZK improves it to obtain higher verification efficiency and realizes the system prototype on the Hyperledger Fabric [15]. In this paper, CFChain is trying to use zero-knowledge proofs and Pedersen commitments to construct crowdfunding records for auditing while preserving transaction privacy.

3   Preliminaries

3.1   Bilinear Groups

Consider G_1, G_2 and G_T as multiplicative cyclic groups of prime order p. Let e be a bilinear map, $e : G_1 \times G_2 \to G_T$ which has the following properties:

- Bilinearity: For $\forall g_1 \in G_1, g_2 \in G_2$ and all $a, b \in Z_p, e\left(g_1^a, g_2^b\right) = e(g_1, g_2)^{ab}$.
- Non-Degeneracy: $e(g_1, g_2) \neq 1$
- Computability: For all $\forall g_1 \in G_1, g_2 \in G_2, e(g_1, g_2)$ can be computed in a polynomial time.

3.2   BLS Aggregate Signature [7]

Let G_1 and G_2 be bilinear groups, g_1 and g_2 are their respective generators, and the bilinear map $e : G_1 \times G_2 \to G_T$, with target group G_T, are system parameters. H is a full-domain hash function $H : \{0, 1\}^* \to G_2$, viewed as a random oracle.

- $KeyGen$: Pick random $x \xleftarrow{R} Z_p$ as user's secret key sk and compute $v \leftarrow g_1^x \in G_1$ as user's public key pk.
- $Sign(sk, m)$: Input $sk = x \in Z_p$, and a message $m \in \{0, 1\}^*$, compute $H \leftarrow H(m)$, where $H \in G_2$, and the signature is $\sigma \leftarrow H^x \in G_2$.
- $Ver(pk, \sigma, m)$: Input $pk = v \in G_1$, a message m, and a signature σ. Compute $H \leftarrow H(m)$; accept if $e(g_1, \sigma) = e(v, H)$ holds.

- *AggSign*: Assign an index i ranging from 1 to n to each user in the aggregating subset of n users. Each user provides a signature $\sigma_i \in G_2$ on a message $m_i \in \{0,1\}^*$ which m_i must all be distinct. The aggregate signature is $\sigma \leftarrow \prod_{i=1}^{k} \sigma_i \in G_2$.
- *AggVer*: Input an aggregate signature $\sigma \in G_2$ for an aggregating subset of users U, the original messages $m_i \in \{0,1\}^*$ and public keys $v_i \in G_1$ for all users. First, it should be checked that the messages m_i are all distinct. Second, compute $H_i \leftarrow H(m_i)$ for $1 \le i \le n$, and accept if $e(g_1, \sigma) = \prod_{i=1}^{k} e(v_i, H_i)$ holds.

3.3   Commitment and Non-interactive Zero-Knowledge Proofs

The present study integrates Non-interactive zero-knowledge (NIZK) proof [13] and Pedersen commitment [23] to collaboratively devise a proficient auditing mechanism that ensures privacy. Pedersen commitment is used to conceal transaction amounts and user-specific data. Let G be a cyclic group of order q with two random generators of g and h. Then, the Pedersen commitment for a secret integer value $v \in \{0, 1, \ldots, q-1\}$ can be calculated as $C_v = g^v h^r$ with the randomness r. The Pedersen commitment has perfect hiding with the random number r. Zero-knowledge proof (ZKP) is a cryptographic protocol that allows a prover to convince a verifier of the truthfulness of a statement without revealing any actual information about the statement. The utilization of Non-interactive zero-knowledge (NIZK) proof obviates the requirement for immediate interaction, thereby enabling the implementation of privacy-preserving protocols that are more efficient and adaptable. As an illustration, a prover may know the commitment's opening and want to persuade the verifier that the committed value is in some range. The prover can produce a binary string proof Π using NIZK to persuade the verifier without divulging any knowledge about the value.

3.4   Distributed Identities

Distributed identity is a way of identity management. It includes two parts: Decentralised identifier DID and Verifiable credential $Cred$. Each DID corresponds to a DID document, which is used to reveal the user's public information, such as public key, usage contexts, etc. According to the World Wide Web (W3C) definition, $Cred$ includes metadata, claims, and proofs. The metadata contains standard entries, such as issuance and expiration dates. The claim can be written in an attribute-value relationship as $Claim = \{att, val\}$. The proof refers to the digital signature. To the definition of W3C, this paper summarizes the distributed identity into the following four algorithms.

- $DIDGen(pk, context) \rightarrow (DID)$: During DID generation, the DID document incorporates both the public key pk and its corresponding context.
- $CredGen(DID_A, Claim, DID_I, sk_I) \rightarrow (Cred_{DID_A})$: The $Cred$ applicant DID_A will request the $Cred$ issuer DID_I to generate a $Cred$. DID_I will verify the $Claim$ provided by DID_A. Then, DID_I will attach its signature to the $Cred$. And the final $Cred_{DID} = \{Claim, \sigma^I, DID_I\}$.

– $CredVer(DID, Cred_{DID}, c) \rightarrow (1/0)$: The verifier inputs its challenge c (against replay attract), DID and credential $Cred_{DID}$. The algorithm outputs 1 if all verifications passed; otherwise, 0.

4   CFChain Overview

4.1   System Model

The scheme comprises five distinct entities, namely the Identity Authority (IA), Authority Agency (AA), User, Charity Platform (CP), and Blockchain. Table 1 enumerates the primary notations employed in this manuscript.

– User: Users can be divided into donors and beneficiaries. Each user holds a pair of $DIDs$ and uses their DID_{ctx}s to participate in crowdfunding. Users can only have one DID_{main} but many DID_{ctx}s for different contexts.
– IA: Within the framework, there is a high-level IA in every nation, exemplified by the Census Bureau, that possesses knowledge of the user's unique identifier (UI), such as the social security number (SSN). IA is responsible for preventing sybil attacks. It will bind the user's DID_{main} to its UI and check that each user can only have one DID_{ctx} in the same context.
– AA: The entities in question bear the responsibility of verifying $Claims$ and issuing $Creds$ to users within their respective domains while also providing associated services, such as banks, schools, hospitals, etc.
– CP: It will execute predefined smart contracts to publish crowdfunding project (CFP) information and crowdfunding records.
– Blockchain: The blockchain is used to record a digest of user identities and corresponding $Creds$ and to record crowdfunding records.

Table 1. Notations

Symbol	Remark
DIDs	Decentralised identifiers, DIDs $= \{DID_{main}, DID_{ctx}s\}$
DID_{main}	The unique identity for supervision
DID_{ctx}	The pseudonymous identities used in different contexts ctx
C_v	Commitment to value v
Cred	Verifiable credentials
Claim	A statement in format of attribute-value

The system model of CFChain is shown in Fig. 1. The workflow can be divided into the initialization phase (1), user authentication phase (2–6), and crowdfunding phase (7–10).

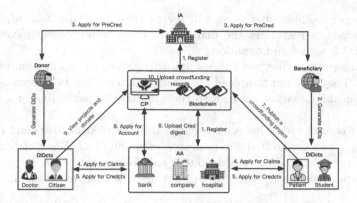

Fig. 1. CFChain architecture overview and workflow.

- Initialization Phase: The system will run the initialization function to set the system parameters. Since IA and AA are public institutions (PI), they only need to generate a publicly available DID. Furthermore, use the services they can provide as attributes to generate $Claim$ and corresponding $Cred$. Ultimately, they will upload the $Cred$'s digest to the blockchain to complete the registration.
- User Authentication Phase: The user first generates a $DIDs$ pair. DID_{main} is considered sensitive and is therefore embedded in the $Claim$s and $Cred$s in the form of commitment to establish their authenticity. The user subsequently submits the $DIDs$ pair to the IA for verification. After that, the IA will generate $PreCred$ for DID_{ctx}. The user generates $Claim$s, procures the corresponding AAs' endorsement, and combines the required $Claim$s into a $Claim_{ctx}$ under the contextual prerequisites. Then, present $Claim_{ctx}$ along with $PreCred$ to the AA accountable for verifying identity in the given context to acquire $Cred_{ctx}$.
- Crowdfunding Phase: Beneficiaries will use their DID_{ctx} to publish crowdfunding projects (CFP) on CP. For example, beneficiaries can simultaneously have student and patient identities and use the patient identity to initiate crowdfunding. After verifying the validity of $Cred_{ctx}$, CP will run a predefined smart contract to publish a CFP for the beneficiary and generate a DID for it to apply for the CFP's account, which is affiliated with the identical bank as the account furnished by the beneficiaries. Donors use their DID_{ctx} to select a CFP for donation, and the bank to which the project account belongs will generate a transaction record and upload it to the blockchain and CP for audit after receiving the donation.

4.2 Threat Model and Design Goals

It is assumed that the system parameters are generated in an honest manner. The present study adopts the subsequent threat model with respect to privacy:

- This paper assumes that users will try to create multiple fake identities to masquerade as real users and manipulate them to obtain certification from IA and AA to defraud donations.
- This paper posits that institutions uploading transactions like banks are dishonest. They attempt to upload fake transaction information, such as misrepresenting the donation value, with the intention of purloining or concealing assets.
- AAs are assumed to be honest-but-curious, they endeavor to acquire the user's DID_{main} and try to find the relevance of different DID_{ctx}s of the user by analyzing crowdfunding records, but it still follows the protocols.
- This paper assumes that the IA will not disclose user information.

Based on the above assumptions, this paper sets the CFChain's privacy and function goals.

- Sybil-resistance: Adversaries cannot use more legitimate user identities than they control.
- Unforgeability: Adversaries cannot fake the identities of honest users or otherwise imitate their behavior unless they have control over their real identities.
- Transaction Privacy: In addition to entities that participated in crowdfunding, no third party is aware of the contents of the transaction.
- Unlinkability: Adversaries cannot discover the linkability of identities used by users in different contexts.
- Private Audit: Users who participate in crowdfunding can audit crowdfunding records, but they cannot know the transaction amount of participants other than themselves.
- Authentication: The system can provide an effective user identity authentication function for screening legitimate users.
- Decentralization: It should not require trusted third parties such as charity organizations for the crowdfunding process.

4.3   System Operations

We define the CFChain operations, $GlobalSetup$, $PIRigister$, $UserRigister$, $PreCredGen$, $ctxCredGen$, $CFPGen$, $Donate$, and $Audit$, as follows.

- $GlobalSetup(1^\lambda) \rightarrow (params)$: This procedure is executed during the initialization of the system. Input a security parameter 1^λ, it outputs system public parameters $params$.
- $PIRigister(params) \rightarrow ((pk_{PI}, sk_{PI}), DID_{PI}, Cred_{PI})$: This operation is run by Public Institutions (PI) such as IA and AA for generation of its public/private key pair (pk_{PI}, sk_{PI}), DID_{PI} and corresponding $Cred_{PI}$.
- $UserRigister(params, contexts) \rightarrow ((pks_U, sks_U), DIDs_U)$: This operation is run by a user U (donor or beneficiary) for generation of their public/private key pairs (pks_U, sks_U) and $DIDs_U = (DID_{main}^U, DIDs_{ctx}^U)$.

- $PreCredGen(DIDs_U, UI_U, DID_{IA}, sk_{IA}) \rightarrow (PreCred_{ctx}^U)$: This operation is instantiated by a two-party protocol between the IA and a user U. U submits his $DIDs_U$ and UI_U to the IA. After verification, IA will use its sk_{IA} to issue $PreCred_{ctx}^U$ for U.
- $ctxCredGen(PreCred_{ctx}^U, Claim_{ctx}^U, \Pi_{link}, sk_{AA}) \rightarrow (Cred_{ctx}^U)$: Instantiated by a two-party protocol between AA and user U. U generates $Claim_{ctx}^U$ according to the requirements of the context, as well as zero-knowledge proof Π_{link}, and sends them to AA for verification along with $PreCred_{ctx}^U$. After verifying the authenticity of $Claim_{ctx}^U$ and the validity of $PreCred_{ctx}^U$, AA will use its sk_{AA} to issue $Cred_{ctx}^U$ to U. Π_{link} is used to prove that $PreCred_{ctx}^U$ and $Claim_{ctx}^U$ belong to the same user.
- $CFPGen(params, DID_B, Cred_B) \rightarrow (DID_{CFP})$: This operation is run by CP. After verifying the validity of beneficiary B's $Cred_B$, CP will run the smart contract to create a crowdfunding project (CFP) and generate DID_{CFP} for it to apply for a project account. For ease of understanding, we use DID here to represent the bank account. DID_{CFP} will transfer the funds to DID_B only after reaching the crowdfunding goal.
- $Donate(DID_{Donor}, DID_{CFP}, sk_{AA}, donation) \rightarrow (record)$: This operation is run by the AA that manages the project account DID_{CFP}. After receiving donation from DID_{Donor}, AA will generate transaction record and upload it to the blockchain.
- $Audit(sk, record) \rightarrow (1/\ 0)$: This operation is run by donor. Donor use his sk to verify the correctness of transaction record. If it is correct, output 1, otherwise, output 0.

5   Security Definitions

In this section, security games are conducted to showcase the security of CFChain. The games in question incorporate interactive protocols that involve a PPT adversary $\mathcal{A}$, and a challenger $\mathcal{C}$. These protocols enable the adversary to access protocol transcripts.

Sybil-Resistance. In this game, the adversary creates as many credentials as are required after initializing n identities. If the adversary generates $>$ n legitimate credentials in the same context, it wins. The process of G_{Sybil} is as follows.

- $\mathcal{C}$ gives to $\mathcal{A}$ the security parameter 1^λ, the system parameters $gparams$ obtained by running $GlobalSetup(1^\lambda)$.
- $\mathcal{A}$ can ask $\mathcal{C}$ to generate IA and AA by running $PIRigister$ operation. $\mathcal{C}$ shares IA's and AA's public keys pk_{IA} and pk_{AA} with $\mathcal{A}$, and provides oracle access $\mathcal{O} = (UserRigister, PreCredGen, ctxCredGen)$ to $\mathcal{A}$.
- $\mathcal{A}$ can generate $Claims$ and run $Creds_{ctx} \leftarrow \mathcal{A}^{\mathcal{O}}(pk_{IA}, pk_{AA}, Claims)$. $\mathcal{A}$ eventually returns $Creds_{ctx}$ to $\mathcal{C}$.
- $\mathcal{C}$ runs algorithm $CredVer$ to check their legality. And $\mathcal{A}$ wins by producing more than n valid $Cred_{ctx}$s.

Definition 1. *Sybil-resistance: CFChain provides Sybil-resistant if, for any stateful PPT adversary $\mathcal{A}$, there exists a negligible function* negl$(\cdot)$ *such that*

$$Pr\left[G_{sybil}\left(\lambda, \mathcal{A}, \mathcal{O}\right) \Longrightarrow 1\right] \leq negl(\lambda)$$

Unforgeability. The unforgeability property of CFChain ensures that the probability that $\mathcal{A}$ can impersonate users such as forge signatures is negligible. Since the verification of the $Cred$ needs to sign the challenge sent by the verifier with the user's private key, we set up a special oracle $\mathcal{O}_{sk_U}$ by referring to the method of [18]. $\mathcal{A}$ can get access to sk_U through $\mathcal{O}_{sk_U}$ that allows calling any algorithm with the user key parameter set to sk_U. The adversary wins by producing a valid signature over a fresh message. The process of G_{UF} is as follows.

- $\mathcal{C}$ gives to $\mathcal{A}$ the security parameter 1^λ, the system parameters *gparams* obtained by running *GlobalSetup*(1^λ).
- $\mathcal{A}$ can ask $\mathcal{C}$ to generate IA, AA, and user by running *PIRigister* and *UserRigister* operations. $\mathcal{C}$ shares their public keys with $\mathcal{A}$, and provides oracle access $\mathcal{O} = (PreCredGen, ctxCredGen, CredVer)$ and $\mathcal{O}_{sk}$ to $\mathcal{A}$.
- $\mathcal{A}$ get several c in calls to $CredVerify$ and corresponding signatures σ by running $c, \sigma \leftarrow \mathcal{A}^{\mathcal{O}, \mathcal{O}_{sk}}(pk_{IA}, pk_{AA}, pk_U)$.
- $\mathcal{A}$ wins by generating a valid signature σ^* on a new challenge c^* which can be verified by $\mathcal{C}$.

Definition 2. *Unforgeability: CFChain offers unforgeability if, for any stateful PPT adversary $\mathcal{A}$, there exists a negligible function* negl$(\cdot)$ *such that*

$$Pr\left[G_{UF}\left(\lambda, \mathcal{A}, \mathcal{O}, \mathcal{O}_{sk}\right) \Longrightarrow 1\right] \leq negl(\lambda)$$

Unlinkability. The unlinkability property of CFChain ensures that the probability that $\mathcal{A}$ can link $DIDs_{ctx}$ used by the user in different contexts is negligible. In this game, $\mathcal{A}$ picks two DID_{main}. $\mathcal{C}$ randomly selects one to generate a $Cred_{ctx}$. If $\mathcal{A}$ cannot guess which DID_{main} is selected, the system guarantees the unlinkability of user identities. Note that, we only includes cryptographic attacks and does not consider other attacks such as network traffic attacks. The process of G_{UL} is as follows.

- $\mathcal{C}$ gives to $\mathcal{A}$ the security parameter 1^λ, the system parameters *gparams* obtained by running *GlobalSetup*(1^λ).
- $\mathcal{A}$ can ask $\mathcal{C}$ to generate IA and AA by running *PIRigister* operation. $\mathcal{C}$ shares IA's and AA's public keys pk_{IA} and pk_{AA} with $\mathcal{A}$, and provides oracle access $\mathcal{O} = (UserRigister, CredVer)$ to $\mathcal{A}$.
- $\mathcal{A}$ runs $DID_{main}^0, DID_{main}^1, DID_{ctx} \leftarrow \mathcal{A}^{\mathcal{O}}(pk_{IA}, pk_{AA})$. Then $\mathcal{A}$ generates $Claim$ for DID_{ctx} and sends them to $\mathcal{C}$.
- $\mathcal{C}$ first checks whether the same DID_{main} exists. Then $\mathcal{C}$ selects a random bit $b \in \{0, 1\}$ and run $PreCredGen$ and $CtxCredGen$ to generate $Cred^{ctx}$ for DID_{main}^b.
- $\mathcal{A}$ finally sends a guess bit $b' \leftarrow \mathcal{A}^{\mathcal{O}}(Cred_{ctx})$. $\mathcal{A}$ wins if $b = b'$.

Definition 3. *Unlinkability: CFChain offers unlinkability if, for any stateful PPT adversary there exists a negligible function* negl(·) *such that,*

$$\left| Pr\left[G_{UL(\lambda,\mathcal{A},\mathcal{O})} \to 1 \right] - \frac{1}{2} \right| \le negl(\lambda)$$

Transaction Privacy. The transaction privacy property of CFChain ensures that the probability that $\mathcal{A}$ can tell the transaction amount from the crowdfunding record is negligible. In this game, $\mathcal{A}$ prepares two transactions, $\mathcal{C}$ randomly selects one to generate a *record*. If the adversary cannot guess which transaction is selected, the system guarantees the transaction privacy. The process of G_{TP} is as follows.

- $\mathcal{C}$ gives to $\mathcal{A}$ the security parameter 1^λ, the system parameters *gparams* obtained by running *GlobalSetup*(1^λ).
- $\mathcal{A}$ can ask $\mathcal{C}$ to generate IA and AA by running *PIRigister* operation. $\mathcal{C}$ shares IA's and AA's public keys pk_{IA} and pk_{AA} with $\mathcal{A}$, and provides oracle access $\mathcal{O} = (UserRigister, PreCredGen, ctsCredGen, Audit)$ to $\mathcal{A}$.
- $\mathcal{A}$ calls $\mathcal{O}$ to register DID_{ctx} for donations. Then, $\mathcal{A}$ sends donations with donation amounts of v_1 and v_2 to $\mathcal{C}$.
- $\mathcal{C}$ selects a random bit $b \in \{0,1\}$ and run *Donate* to generate $record^{v_b}$.
- $\mathcal{A}$ finally sends a guess bit $b' \leftarrow \mathcal{A}^{\mathcal{O}}(record^{v_b})$. $\mathcal{A}$ wins if $b = b'$.

Definition 4. *Transaction privacy: CFChain offers transaction privacy if, for any stateful PPT adversary there exists a negligible function* negl(·) *such that,*

$$\left| Pr\left[G_{TP(\lambda,\mathcal{A},\mathcal{O})} \to 1 \right] - \frac{1}{2} \right| \le negl(\lambda)$$

6 The CFChain Scheme

GlobalSetup and *PIRigister* are run in the initialization phase, *UserRigister*, *PreCredGen* and *ctxCredGen* are run in the user authentication phase, *CFPGen, Donate,* and *Audit* are run in the crowdfunding phase. The details are as follows.

In *GlobalSetup*, on input a security parameter 1^λ, the algorithm chooses bilinear groups G_1, G_2, and G_T of prime order p, $e : G_1 \times G_2 \to G_T$ is a bilinear map. It randomly picks generators $g, h \in G_1$. Suppose $H : \{0,1\}^* \to G_2$ is a collision resistant hash function. It outputs $gparams = (g, h, e, H, G_1, G_2, G_T)$.

PIRigister operation is obtained by PI via execution of *KeyGen, DIDGen,* and *CredGen*. It will first run *KeyGen* to get the public/private key pair (pk_{PI}, sk_{PI}) where $sk_U \in \mathcal{Z}_p$, and $pk_{PI} = h^{sk_{PI}}$. Then PI generates DID_{PI} by running $DIDGen(pk_{PI}, context)$. Next, PI will choose the service it can provide as an attribute to generate $Claim_{PI}$, and run algorithm $CredGen(DID_{PI}, Claim_{PI}, sk_{PI})$ to issue a $Cred_{PI}$ for itself.

Likewise, *UserRigister* is obtained by users via execution of *KeyGen* and *DIDGen*. The user U will generate the required public-private key pairs

(pks_U, sks_U) as needed. And choose one of them to generate DID_{main}^U by running $DIDGen(pk_U, main)$, and use the rest to generate corresponding DID_U^{ctx} according to different contexts by running $DIDGen(pk_U, context)$. The difference is that users cannot issue credentials for themselves.

Further, for $PreCredGen$, the user U submits his $DIDs$ and UI_U to the IA for verification. If $(UI_U, DID_{main}^U) \notin$ the table $MIssueT$, IA will add (UI_U, DID_{main}^U) to $MIssueT$. Else, refuse the user's request. If $(DID_{main}^U, ctx,) \notin$ the table $CIssueT$, IA will add $(DID_{main}^U, ctx, DID_{ctx}^U)$ to $CIssueT$ and generates $Claim = (att, C_v)$ where v is DID_U^{main}. Else, refuse the user's request. Then, IA runs $CredGen$ to issue $PreCred_{ctx}^U$ for U and attach its signature $\sigma^{IA} = H(PreCred_{ctx}^U)^{sk_{IA}}$ on it. $PreCred_{ctx}^U = \{DID_{ctx}^U, DID_{IA}, Claim, ctx, \sigma^{IA}\}$.

The details of $ctxCredGen$ are shown in Protocol 1. Initially, the user U generates $Claims$ and authenticates them with the corresponding AAs. The user then combines the $Claims$ according to the context requirements, executing $AggSign$ to produce an aggregate signature and finally getting $Claim_{ctx}$. To ensure the validity of the $Claim_{ctx}$, namely, it belongs to the same user as $PreCred_U$, it has employed the utilization of DID_{main}^U as a linking attribute within the $PreCred_U$, drawing inspiration from [31]. $Claim_{ctx}$ contains the commitment value of the linking attribute. We use a non-interactive variant of the Chaum-Pedersen zero-knowledge proof [9] Π_{link} to prove that two commitments have the same commitment value. U submits $Claims_{ctx}^U$, $PreCred_U$, and Π_{link} to the contextual AA for authentication. Upon successful authentication, AA will run $CredGen$ to generate DID_U^{ctx}.

Protocol 1 ctxCred Generation

Input: User U inputs $PreCred_{ctx}^U$, $Claims$. AA inputs its sk_{AA}

Output: AA outputs success and U gets $Cred_{ctx}^U$ (or) fail

1. U run $AggSign(Claims)$ to get $Claim_{ctx}$. U generates $\Pi_{link} =$ ZK-POK$\{r, v, r', v' : C_v = g^v h^r$

$\wedge C_{v'} = g^{v'} h^{r'} \wedge v = v'\}$

where v and v' represent the value of U's DID_{main}^U in $Claim_{ctx}^U$ and $PreCred_{ctx}^U$ respectively

(r, r' represent corresponding commitment witness).

2. Randomly select $b_1, b_2, b_3 \in \mathcal{Z}_p$ and compute $c_1 = g^{b_1} h^{b_2}, c_2 = g^{b_1} h^{b_3}$

3. Compute $\beta = H(g, h, C_v, C_{v'}, c_1, c_2)$, $s_1 = v \cdot \beta + r, s_2 = r \cdot \beta + b_2, s_3 = r' \cdot \beta + b_3$

4. U sends $\{\Pi_{link} = (c_1, c_2, s_1, s_2, s_3), Claim_{ctx}^U, PreCred_{ctx}^U\}$ to AA.

5. AA first Verifies the signatures on $Claim_{ctx}^U$ and, $PreCred_U^{ctx}$. Then, Verifies the Π_{link}.

6. AA extracts C_v from $PreCred_{ctx}^U$ and $C_{v'}$ from $Claim_{ctx}^U$. Compute $\beta = H(g, h, C_v, C_{v'}, c_1, c_2)$.

7. Randomly select $\alpha_1, \alpha_2 \in \mathcal{Z}_p$, $k = [\alpha_1 s_1 + \alpha_0 s_1, \alpha_0 s_2 + \alpha_1 s_3, -\alpha_0 \beta, -\alpha_1 \beta, -\alpha_0, -\alpha_1]$

8. Let $l = [g, h, P, Q, c_1, c_2]$. If $\prod_{i=0}^5 l_i^{k_i} = 1$ continue, otherwise, output fail.

9. AA extracts $Claims$ from $PreCred_{ctx}^U$ and use them together with $Claim_{ctx}^U$ to run CredGen to get $Cred_{ctx}^U$.

The $CFPGen$ operation is achieved through the execution of $CredVer$, $KeyGen$, and $DIDGen$ by CP. CP first runs the $CredVer$ to check the authenticity of the beneficiary B's $Cred_B$. It then executes a predefined smart contract to publish the crowdfunded project (CFP) for B. Smart contracts are collaboratively developed by multiple AAs to implement the operational procedures and regulations of contractual performance. The CP will execute the *KeyGen* and *DIDGen* to produce the DID_{CFP}, which will be utilized to request access to the CFP account. Notably, the CFP account is affiliated with an identical financial institution as that of B's account. Once the predetermined fundraising goal is achieved, the transfer of donations from DID_{CFP} to DID_B takes place.

Protocol 2 Transaction Record Generation

Input: Donation information=$\{DID_D, v\}$User U inputs $PreCred_{ctx}^U$, $Claim_{ctx}^U$.

Output: record=$\{\Pi_{Amt}, \Pi_{Ass}, \Pi_C, AT_D, AT_{CFP}, AT'_{CFP}, C_v\}$

1. Amount: $C_v = g^v h^r$ is a commitment to donation amount v with a witness r.

2. Asset: $C_{v'} = g^{v'} h^{r'}$ is a commitment to account assets $v' = \sum_{i=0}^n v_i$. where n represents the total number of donations so far, r' is a random number different from r.

3. Audit Token: $AT_D = pk_D^r$, $AT_{CFP} = pk_{CFP}^r$, $AT'_{CFP} = pk_{CFP}^{r'}$

4. Proof of Amount: This is a range proof which is used to ensure that the donation amount is not negative. $\Pi_{Amt} = ZOK(v, r : C_v = g^v h^r \wedge v \geq 0)$

5. Proof of Asset: This is a range proof which is used to ensure that the donation asset is not greater than the target value. $\Pi_{Ass} = ZOK(v', r' : C_{v'} = g^{v'} h^{r'} \wedge v' \leq target)$

6. Proof of Consistency: We use a non-interactive variant of the Chaum-Pedersen zero-knowledge

proofs $\Pi_C = ZOK_1(x_1^r, y_1^r \wedge x_1^{b_1}, y_1^{b_1}, c_1, s_1) \wedge ZOK_2(x_2^{sk}, y_2^{sk} \wedge x_1^{b_2}, y_1^{b_2}, c_2, s_2)$ to prove that the randomness used in Audit Token and commitment are the same, and AA does use asset v' to

generate asset proof instead of arbitrary values. where ZOK_1 and ZOK_2 are used to prove the knowledge of r and the knowledge of CFP's secret key sk respectively, x_1^r and x_2^{sk} are two generalized Schnorr proofs, b_1 and b_2 are two random numbers, $c_1 = H(AT_D)$, $c_1 = H(AT'_{CFP})$,

$s_1 = b_1 + rc_1$, $s_2 = b_2 + skc_2$, $x_1 = pk_D$, $y_1 = AT_D$,

$x_2 = (\prod_{i=0}^n C_{v_i})/C_{v'}$, $y_2 = (\prod_{i=0}^n AT_{CFP_i})/AT'_{CFP}$. Note that, $x_1^r = y_1, x_2^{sk} = y_2$.

After receiving the donor's donation, the AA to which DID_{CFP} will execute the *Donate* operation to generate a transaction record. Inspired by [21], Pederson commitments are utilized to hide transaction amounts and produce audit tokens for the donor to verify the committed transaction. NIZK proofs are utilized for enabling the validation and auditing of encrypted transactions. The details are shown in Protocol 2.

The *Audit* operation ensures the transaction's legitimacy by computing the commitments and verifying the zero-knowledge proofs in the record. The donor D can use his sk to check whether the transaction amount v in the record is correct by computing $AT_D \cdot g^{sk \cdot v} = (C_v)^{sk_D}$. He also need to check whether $x_1^{s_1} = (x_1^r)^{c_1} x_1^{b_1}$, $y_1^{s_1} = (y_1^r)^{c_1} y_1^{b_1}$, $x_2^{s_2} = (x_2^{sk_{CFP}})^{c_2} x_2^{b_2}$, $y_2^{s_1} = (y_2^{sk_{CFP}})^{c_1} y_2^{b_1}$. Range proofs Π_{Amt} and Π_{Ass} can also be verified by other users.

7 Security Analyse

Theorem 1. *If the DL problem holds in group G, then CFChain provides Sybil-resistance.*

Proof. Recall that IA will use table $MIssueT$ to record the correspondence between the user's DID_{main} and his unique identifier. Similarly, table $CIssueT$ ensures that each DID_{main} can get one DID_{ctx} in the same context. That is to say, when the adversary controls n users, it can get n $DIDs_{ctx}$ and the corresponding $Cred_{ctx}$s. The effectiveness of this scheme relies on $Cred$'s validity, which means users cannot create their $DIDs_{ctx}$ with someone else's DID_{main}. We exploit the soundness of Π_{link} to prove the validity of $Cred$, the proof is as follows. During DID_{ctx} generation, the challenger outputs the DID_{main}'s commitment values on the $Claim$ and $PreCred$, respectively, together with $\Pi_{Link} = (s_1, s_2, s_3, \beta)$. The simulator rewinds β and receives another proof $\Pi'_{Link} = (s'_1, s'_2, s'_3, \beta')$. Then we have $C_v^{\beta-\beta'} = g^{s_1-s'_1}h^{s_2-s'_2}$, $C_{v'}^{\beta-\beta'} = g^{s_1-s'_1}h^{s_3-s'_3}$. By setting $v = (s_1 - s'_1)/(\beta - \beta')$, $r = (s_2 - s'_2)/(\beta - \beta')$, $r' = (s_3 - s'_3)/(\beta - \beta')$, we have $C_v = g^v h^r$, $C'_v = g^v h^{r'}$, which are valid Pedersen commitments of value v. Since Pedersen commitment has computationally binding properties based on the discrete logarithm hypothesis [23], Π_{link} is sound. Thus, in G_{sybil}, $\mathcal{A}$ cannot generate more than n legitimate $Cred_{ctx}$s.

Theorem 2. *If the CDH problem holds in group G, then CFChain is unforgeable in the chosen-text model.*

Proof. In the user context-based identity authentication stage, the user needs to aggregate the BLS signatures on the required $Claim$s to prove the authenticity of the identity. According to [7], when an aggregate signature is an aggregation of signatures on distinct messages, it is secure in the chosen-text model. We add a timestamp when generating the $Claim$ to meet this requirement. Additionally, while verifying credentials, users need to sign the challenge with their private key since users' key never leaves their device, the unforgeability for this phase follows straightforwardly. So the adversary cannot win in G_{UF}.

Theorem 3. *CFChain provides unlikability in the random oracle model.*

Proof. In CFChain, users use different $DIDs_{ctx}$ in different contexts to prevent identities from being associated. The user's DID_{main} will be recorded in $Cred_{ctx}$ in the form of a commitment. Since the Pedersen commitment provide perfectly hiding property [23], the adversary cannot associate different $DIDs_{ctx}$ through the $Cred_{ctx}$. During the $Cred_{ctx}$ generation process, Π_{link} was used to prove the $Claim$'s validity. Π_{link} is zero-knowledge in the random oracle model. The proof is as follows. The simulator picks some random $\{C_v, C_{v'}, s_1, s_2, s_3\}$, and computes $c_1 = C_v^{-\beta}g^{s_1}h^{s_2}$, $c_2 = C_{v'}^{-\beta}g^{s_1}h^{s_3}$, the simulator set $\beta = H(g, h, C_v, C_{v'}, c_1, c_2)$ in the random oracle model. Hence, Π_{link} does not leak DID_{main}. So in G_{UL}, $\mathcal{A}$ cannot distinguish which DID_{main} is used through the generated $Cred_{ctx}$.

Theorem 4. *If the DDH problem holds in group G, then CFChain provides Transaction privacy in the random oracle model.*

Proof. If a PPT adversary $\mathcal{A}$ wins the game G_{TP} with non-negligible probability, we can utilize it to construct a PPT adversary $\mathcal{A}_{\mathcal{DDH}}$ who wins the DDH game with non-negligible advantage. In G_{TP}, $\mathcal{A}_{\mathcal{DDH}}$ plays the role of $\mathcal{C}$. We use $\mathcal{C}_{\mathcal{DDH}}$ to denote the challenger of $\mathcal{A}_{\mathcal{DDH}}$.

- $\mathcal{C}_{\mathcal{DDH}}$ sends the challenge (h, x, y, z) to $\mathcal{A}_{\mathcal{DDH}}$ where (x, y, z) are distributed either as (h^a, h^b, h^{ab}) or as (h^a, h^b, h^c) in the challenge.
- $\mathcal{A}_{\mathcal{DDH}}$ calls $\mathcal{A}$ on input (g, h, x) after sampling a random generator g, with x now acting as the crowdfunding project's public key.
- $\mathcal{A}$ sends transactions with transaction amounts of v1 and v2 to $\mathcal{C}$.
- $\mathcal{A}_{\mathcal{DDH}}$ picks a random bit $b \in \{0, 1\}$ and prepares $cm_k = g^{v_b}y$, $Token = z$ and sends them to $\mathcal{A}$.
- Finally, if $\mathcal{A}$'s guess for b is correct, $\mathcal{A}_{\mathcal{DDH}}$ responds a DDH quadruple (h, h^a, h^b, h^{ab}) otherwise it responds a random quadruple(h, h^a, h^b, h^c).

Its inputs are all appropriately constructed with respect to $sk = a$ and $r = b$. Therefore, if $\mathcal{A}$ wins the game G_{TP} with non-negligible advantage, so does $\mathcal{A}_{\mathcal{DDH}}$ in the DDH game. In addition, through Theorem B.1. and Theorem B.2. in [28], the zero-knowledge proofs Π_{amt} and Π_{ass} do not leak committed values' information-theoretic privacy. In conclusion, CFChain provides transaction privacy if the DDH problem holds in group G.

8   Performance Evaluation

8.1   Functionality Comparison

We compare our scheme with traditional international charity organizations (ICO), the non-anonymous cryptocurrency (NCC) schemes (Bitcoin [20], and Ether [30]), CharityCoin [12], and eDonation [6], which are shown in Table 2. Here, $\sqrt{}$ indicates the scheme has this feature, while × does not have this feature. Traditional ICOs are centralized, and they can know the real identity of users and the content of donations. Since attackers can deanonymize users of Bitcoin and Ether by reusing addresses [8], we classify them as NCC. Charity-Coin is based on Ethereum and defines seller and buyer authentication using smart contracts. eDonation builds a donation policy based on Zerocash, which uses zero-knowledge proof to protect user privacy and uses attribute-based signatures to identify users. Our scheme uses distributed identities to provide user privacy and unlinkability of identities. User revocability is achieved by revocation of verifiable credentials. The private audit function is exclusively implemented by the scheme in the study.

Table 2. Functionality Comparison with existing schemes.

Scheme	Distributed	Transaction privacy	Unlinkability	Authentication	Sybil resistance	revocability	Private Audit
ICO	×	×	×	√	√	√	×
NCC	√	×	×	×	×	×	×
CharityCoin	√	×	×	√	×	×	×
eDonation	√	√	√	√	√	√	×
Ours	√	√	√	√	√	√	√

8.2   Implementation

The key components of CFChain are the blockchain, user authentication scheme, and crowdfund scheme.

Blockchain is divided into public chain, consortium chain, and private chain. Private chains are governed by a singular entity, making them unsuitable for crowdfunding scenarios that require collaboration among multiple institutions. In consortium chains, transaction confirmation and processing are achieved through negotiation among participants, as opposed to the bidding mechanism employed in public chains. Consequently, it does not necessitate supplementary transaction expenses, such as gas fees on Ethereum [30]. Therefore, consortium chains can effectively decrease transaction costs and cater to practical business requirements in the context of crowdfunding, given the involvement of a significant number of small transactions. We deploy CFChain on Hyperledger Fabric [5], a consortium chain, where each organization owns two peer nodes and one certificate authority (CA) node. The default settings for the orderer node's block creation are a two-second batch timeout and ten transactions per block. Additionally, the experiment's node has eight cores.

The Pedersen commitment [23] and BLS signature [7] are utilized in the creation of verifiable credentials. Users can aggregate BLS signatures obtained from *Claim*s that have been issued by various agencies into a singular signature, thereby conserving storage capacity. The BLS signature is comparatively shorter than the Schnorr or ECDSA signatures as it comprises a single point on the elliptic curve instead of two [26]. Furthermore, it obviates the necessity for supplementary communication during key aggregation and the utilization of random number generators. CFChain uses the elliptic curve $secp256k1$ of the $btcec$ library to compute commitments, the $bls12\_381$ elliptic curve in the BLS signature, and SHA-256 as the default hash function.

Bulletproofs [10] are utilized in the creation of crowdfunding records, specifically for the purposes of zero-knowledge and range proof. Various types of zero-knowledge proofs have been implemented in practice, including zk-SNARK, zk-STARK, and Bulletproofs. Despite its efficiency, zk-SNARK necessitates a trusted setup phase wherein multiple parameters are produced and must be maintained confidential. zk-STARKs and Bulletproofs do not necessitate a trust assumption. However, zk-STARK exhibits a greater proof size, which can be a drawback in scenarios with storage or bandwidth constraints. Although Bulletproof's verification cost is higher than zk-SNARK, it is still much lower than zk-STARK. Furthermore, proof size in Bulletproofs has a logarithmic relationship to the number of commitments, which results in relatively small proofs. We take Barreto-Naehrig 256 (BN256) as the default elliptic curve in Bulletproofs.

8.3 Performance

The experimental environments were set up on a desktop equipped with a 3.60 GHz Intel(R) Core(TM) i7-12700K CPU and 32 GB of RAM. The CFChain was constructed on the top of Hyperledger Fabric, with specific versions of Fabric v2.2.5. The chaincode APIs are written in Go, and the client APIs are written in NodeJS. Supporting components are Node.js v12.18.2 and NPM v6.14.5. All evaluations are run on Ubuntu 18.04 VMs. We collect data from 50 runs for each experiment and take the average as the experimental result. The key components' performance of the CFChain system was evaluated as follows.

User Authentication: User authentication can be divided into two phases: user material preparation and user material verification. In the first phase, the user needs to generate *Claims* according to the context's requirements and get the signatures from corresponding AAs. The average time to create a *Claim* is about 3 ms, which hardly grows with the number of attributes. And the average time to generate a BLS signature for each *Claim* is 3.5 ms. Users also need to combine these *Claims* and generate Π_{link}, with an average time of 36 ms. In the second phase, AA verifies the Π_{link} and the signatures on the $Claim_{ctx}$ and generates the $Cred_{ctx}$ for the user. The average verification time of Π_{link} is 37 ms. When the number of aggregated *Claims* is 5, the average verification time is 97.7 ms. The time to generate the *Cred* and upload it to the blockchain is 22 ms.

The study conducted a comparative analysis of the efficiency of user identification schemes employed in CFChain and eDonation. The stage of user material preparation in CFChain can be considered analogous to the attribute signature generation stage in eDonation. Similarly, the user material verification stage in CFChain can be equated to the signature verification stage in eDonation. To facilitate comparison, a single attribute for the claim has been established. The comparative outcomes are depicted in Fig. 2 and Fig. 3. The findings indicate that the process of user authentication in CFChain is characterized by reduced time consumption. The approach employs BLS aggregate signatures, thereby preserving the signature size. Every time an attribute is added, the size of the *Cred* increases by 64B (commitment's size), which is almost negligible. With the presence of five attributes, the overall magnitude of *Cred* is approximately 0.7 KB, which is significantly lesser than the signature size of eDonation, which stands at 45 KB.

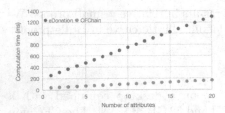
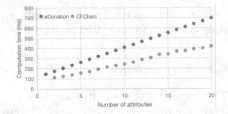

Fig. 2. User material preparation phase   **Fig. 3.** User material verification phase

Crowdfund: The procedure for producing a correct crowdfunding transaction can be bifurcated into the subsequent two steps. Initially, the crowdfunding account's managing entity will invoke the chaincode to generate a record after receiving the donation, which comprises two tokens, two range proofs, and a singular commitment. The aggregate duration of the process is about 0.81 s. Subsequently, the donor triggers the validation chaincode to authenticate the transaction record across the relevant organization associated with their account. The total time to run the chaincode to verify the record is approximately 0.19 s.

This study conducts a comparative analysis of the donation performance exhibited by CFChain in relation to CharityCoin [12] and eDonation [6]. Furthermore, an Ethereum private network was constructed on a local basis, and the $transfer$, $approve$, and $transferFrom$ functions were executed on the smart contract for CharityCoin to replicate the process of making a donation. Moreover, Hyperledger Caliper is utilized for the assessment of the donation process of the CFChain and Ethereum network. Additionally, a local simulation of [3] has been implemented to replicate the donation process in eDonation. Table 3 presents the comparison outcomes, wherein the Sender time denotes the duration expended by the donor, while the receiver time signifies the time taken for the recipient to receive a lawful transaction. The Zcash protocol execution in eDonation necessitates the operation of the protocol by both the donor and the beneficiary, resulting in a prolonged duration. It is noteworthy to mention that CharityCoin's transaction amount is not concealed, resulting in a shorter receive time. The CFChain network exhibits the highest throughput capacity.

Table 3. Performance comparison in the donation phase.

Scheme	eDonation	CharityCoin	CFChain
Send Time (s)	105.4	4.6	0.81
Receive Time (s)	107.2	0.02	0.19
Throughput (tps)	2.3	6.8	38

9   Conclusion

This paper proposes a blockchain-based charity crowdfunding platform called CFChain. The CFChain offers a high degree of transparency regarding the movement of funds through the utilization of blockchain technology, whereby crowdfunding records are uploaded onto the blockchain. Furthermore, CFChain employs distributed identities for the purpose of formulating a user authentication approach. This approach severs the connection between user identities across various contexts and is capable of withstanding sybil attacks. The approach involves assigning a distinct DID_{main} to each user as a means of thwarting Sybil attacks. Additionally, users have the ability to create multiple $DIDs_{ctx}$ for varying contexts, thereby achieving identity unlinkability. Pedersen commitments are utilized to conceal transaction amounts and create crowdfunding records with NIZK. This enables individuals to examine encrypted donation records and verify the accuracy of donations. Ultimately, the paper demonstrates the security of CFChain and executes a preliminary model on Hyperledger. The findings demonstrate the efficacy and feasibility of the subject matter.

Acknowledgements. This work is supported by JST SPRING (Grant No. JPMJSP2136), the Fundamental Research Funds for the Central Universities (No. 30106220482).

References

1. Yahoo.   https://finance.yahoo.com/news/crowdfunding-market-reach-42-93-150700017.html. Accessed 29 Jun 2022
2. Statista.   https://www.statista.com/statistics/1078273/global-crowdfunding-market-size/. Accessed 22 Aug 2022
3. Zcash "sapling" cryptography. https://github.com/zcash-hackworks/sapling-crypto. Accessed 28 Feb 2020
4. Weidentity: Digital identity for data sharing on open consortium chain (2022). Software available, https://fintech.webank.com/en/weidentity/
5. Androulaki, E., et al.: Hyperledger fabric: a distributed operating system for permissioned blockchains. In: Proceedings of the Thirteenth EuroSys Conference, pp. 1–15 (2018)
6. Biçer, O., Küpçü, A.: Anonymous, attribute based, decentralized, secure, and fair e-donation. Cryptology ePrint Archive (2020)
7. Boneh, D., Gentry, C., Lynn, B., Shacham, H.: Aggregate and verifiably encrypted signatures from bilinear maps. In: Biham, E. (ed.) EUROCRYPT 2003. LNCS, vol. 2656, pp. 416–432. Springer, Heidelberg (2003). https://doi.org/10.1007/3-540-39200-9_26
8. Béres, F., et al.: Blockchain is watching you: profiling and deanonymizing Ethereum users. In: 2021 IEEE International Conference on Decentralized Applications and Infrastructures (DAPPS), pp. 69–78 (2021)

9. Chaum, D., Pedersen, T.P.: Wallet databases with observers. In: Brickell, E.F. (ed.) CRYPTO 1992. LNCS, vol. 740, pp. 89–105. Springer, Heidelberg (1993). https://doi.org/10.1007/3-540-48071-4_7

10. Constantinides, K., et al.: BulletProof: a defect-tolerant CMP switch architecture, pp. 5–16. IEEE (2006)

11. Douceur, J.R.: The Sybil attack. In: Druschel, P., Kaashoek, F., Rowstron, A. (eds.) IPTPS 2002, Revised Paper. LNCS, vol. 2429, pp. 251–260. Springer, Heidelberg (2002). https://doi.org/10.1007/3-540-45748-8_24

12. Farooq, M.S., Khan, M., Abid, A.: A framework to make charity collection transparent and auditable using blockchain technology. Comput. Electr. Eng. **83**, 106588 (2020)

13. Fiege, U., et al.: Zero knowledge proofs of identity. In: Proceedings of the Nineteenth Annual ACM Symposium on Theory of Computing, pp. 210–217 (1987)

14. Hossain, M., Oparaocha, G.O.: Crowdfunding: motives, definitions, typology and ethical challenges. Entrep. Res. J. **7**, 1–14 (2017)

15. Kang, H., et al.: FabZK: supporting privacy-preserving, auditable smart contracts in hyperledger fabric. In: 2019 49th Annual IEEE/IFIP International Conference on Dependable Systems and Networks (DSN), pp. 543–555. IEEE (2019)

16. Lundkvist, C., et al.: uPort: a platform for self-sovereign identity (2017). https://whitepaper.uport.me/uPort_whitepaper_DRAFT20170221.pdf

17. Manda, V.K., Prasada Rao, S.S., Prasadarao, S.S.: Blockchain technology for the mutual fund industry. In: National Seminar on Paradigm Shifts in Commerce and Management, pp. 12–17 (2018)

18. Maram, D., et al.: CanDID: can-do decentralized identity with legacy compatibility, Sybil-resistance, and accountability. In: 2021 IEEE Symposium on Security and Privacy (SP), pp. 1348–1366. IEEE (2021)

19. Mehra, A., et al.: Vishrambh: trusted philanthropy with end-to-end transparency. In: HCI for Blockchain: a CHI 2018 Workshop on Studying, Critiquing, Designing and Envisioning Distributed Ledger Technologies, Montreal, QC, Canada (2018)

20. Nakamoto, S.: Bitcoin: a peer-to-peer electronic cash system. Decentralized business review, p. 21260 (2008)

21. Narula, N., et al.: zkLedger: privacy-preserving auditing for distributed ledgers. In: 15th {USENIX} Symposium on Networked Systems Design and Implementation ({NSDI} 18), pp. 65–80 (2018)

22. Patel, V.: New Jersey man gets 5 years in prison in GoFundMe fraud case. The Times (2022). https://www.nytimes.com/2022/08/07/nyregion/gofundme-scam-mark-damico-sentenced.html

23. Pedersen, T.P.: Non-interactive and information-theoretic secure verifiable secret sharing. In: Feigenbaum, J. (ed.) CRYPTO 1991. LNCS, vol. 576, pp. 129–140. Springer, Heidelberg (1992). https://doi.org/10.1007/3-540-46766-1_9

24. Picard, C.: Scammers hijack crowdfunding campaign for 6-year-old with leukemia (2016). https://www.goodhousekeeping.com/life/news/a39792/leukemia-patient-gofundme-hacked/

25. Renat, G., et al.: Karma-blockchain based charity foundation platform. In: 2021 IEEE International Conference on Blockchain and Cryptocurrency (ICBC), pp. 1–2. IEEE (2021)

26. Sahana, S.C., Bhuyan, B.: A provable secure short signature scheme based on bilinear pairing over elliptic curve. Int. J. Netw. Secur. **21**, 145–152 (2019)

27. Saleh, H., et al.: Platform for tracking donations of charitable foundations based on blockchain technology. In: 2019 Actual Problems of Systems and Software Engineering (APSSE), pp. 182–187 (2019)

28. Sasson, E.B., et al.: Zerocash: decentralized anonymous payments from bitcoin. In: 2014 IEEE Symposium on Security and Privacy, pp. 459–474. IEEE (2014)

29. Singh, A., et al.: Aid, charity and donation tracking system using blockchain. In: 2020 4th International Conference on Trends in Electronics and Informatics (ICOEI) (48184), pp. 457–462. IEEE (2020)

30. Wood, G., et al.: Ethereum: a secure decentralised generalised transaction ledger. Ethereum project yellow paper **151**(2014), 1–32 (2014)

31. Yin, J., et al.: SmartDID: a novel privacy-preserving identity based on blockchain for IoT. IEEE IoT J. **10**, 6718–6732 (2022)

32. Yuen, T.H.: PAChain: private, authenticated & auditable consortium blockchain and its implementation. Futur. Gener. Comput. Syst. **112**, 913–929 (2020)

33. Zhang, R., et al.: Security and privacy on blockchain. ACM Comput. Surv. (CSUR) **52**, 1–34 (2019)

TBAF: A Two-Stage Biometric-Assisted Authentication Framework in Edge-Integrated UAV Delivery System

Zheng Zhang[1], Huabin Wang[1], Aiting Yao[1], Xuejun Li[1], Frank Jiang[2], Jia Xu[1], and Xiao Liu[2(✉)]

[1] School of Computer Science and Technology, Anhui University, Hefei, China
zhengzhang@stu.ahu.edu.cn, {wanghuabin,xjli,xujia}@ahu.edu.cn
[2] School of Information Technology, Deakin University, Geelong, Australia
{frank.jiang,xiao.liu}@deakin.edu.au

Abstract. Edge-Integrated Unmanned Aerial Vehicles (UAVs) delivery systems have demonstrated the advantage of higher efficiency and lower latency in comparison with traditional intelligent delivery systems. But with its rapid development, a series of security and privacy issues have also emerged. For instance, it is of vital importance to maintain data safety due to UAVs exchanging sensitive data with servers through public channels, attackers can easily gain access to sensitive information by launching attacks including man-in-the-middle and impersonation attacks. Additionally, the requirements of frequent authentications between UAVs and edge servers can result in increased computation overhead, while UAVs are fast-moving and resource-constrained, and excessive computational overhead can degrade the user experience. To address these challenges, this paper proposes a Two-Stage Biometric-Assisted Authentication Framework (TBAF) that enhances security and efficiency. In TBAF, a novel secret sharing method is designed to distribute storage biometric templates with protection, ensuring the secret values which are biometric templates can only be accessed by authorized parties. Additionally, the two-stage authentication protocol reduces computation and communication overhead. Extensive formal and informal security analysis confirms the superior performance of the proposed protocol compared to existing solutions.

Keywords: Edge-Integrated · UAVs Delivery System · Authentication · Secret Sharing · Protocol

1 Introduction

In recent years, intelligent delivery systems have undergone rapid development due to their inherent advantages of cost-efficiency when compared to traditional delivery systems [1,2]. The UAV delivery system utilizes key technologies to address various challenges, such as 5G, IoT, cloud computing and edge computing. The challenges include data transmission, UAV flight control and user

Z. Tari et al. (Eds.): ICA3PP 2023, LNCS 14493, pp. 168–188, 2024.
https://doi.org/10.1007/978-981-97-0862-8_11

identity authentication [3]. Compared to traditional cloud computing environments, edge computing offers lower communication latency and higher computational efficiency advantages its closer proximity to user terminals. This can greatly assist UAV's flight and real-time communication with better efficiency and accuracy. Therefore, the edge-integrated UAVs delivery system presents a promising solution to overcome obstacles such as difficult terrain, traffic congestion, and high labor costs.

However, intelligent delivery system also poses potential risks of data privacy breaches and communication security threats while delivering its services. In an edge-integrated UAVs delivery system, the open and public flight environment of UAVs exposes them and their operators to potential hijacking attacks, cloning attacks, and physical attacks during mission operations [4]. These malicious acts by attackers can disrupt the seamless flow of delivery, such as changing the flight route to steal packages or obtaining the personal information of the package recipient to violate the recipient's privacy information.

Additionally, the frequent execution of authentication protocols by UAVs and multiple edge nodes brings high computational overhead [5]. During the UAV flight, UAVs need to frequently visit different edge nodes to obtain real-time airspace traffic flow, weather information and path planning information [6]. Meanwhile, the edge nodes also have to authenticate the UAV's identity information to ensure the safety of the UAV. Frequent authentication brings more computational overhead. While UAVs are fast-moving and resource-constrained, the high computational overhead will greatly reduce the flight duration of UAVs, resulting in a bad user experience [7].

To address the communication security and privacy issues, a Two-Stage Biometric-Assisted Authentication Framework (TBAF) has been proposed for the edge-integrated UAVs delivery system. The TBAF included a secure authentication protocol to ensure communication security between server and UAV-Operator (UO). Furthermore, a three-dimensional space-based secret share method is designed to enhance system security. In TBAF, a protected biometric template is regarded as a secret value and storage on multiple servers in a distributed manner. The secret value can only be recovered with the joint participation of the cloud, edge and UO. In addition, since the secret value is a protected biometric template, an attacker cannot obtain the user's private information even if the secret value is obtained. This approach offers double protection of user privacy.

For the sake of reducing the computation overhead caused by frequent authentication, a two-stage authentication protocol is proposed. In the first stage, UO end uses biometric information, Physical Unclonable Function (PUF) and password to complete mutual authentication with the cloud server. Then, the cloud server sends authentication results and assists information to the edge server. In the second stage, with the help of assist information, the edge server can realize authentication of the UO end while minimizing the number of complex operations. So, the two-stage method can degrade the computation overhead of edge server authentication UO end effectively. Moreover, considering edge server diversity and uncertainty, the second stage authentication protocol has adopted Zero Knowledge Proofs (ZKP) to strengthen the security of the system. The main contributions of this paper are as follows:

- A two-stage authentication protocol has been designed, the edge server can quickly complete the authentication process for UO end with the help of the first stage authentication results. Thus, TBAF reduces the computation overhead of the edge server when authenticating UO end.
- To enhance the security and privacy of UAVs delivery system, a Two-Stage Biometric-Assisted Authentication Framework has been proposed. In TBAF, we have utilized the multi-server environment and cancellable biometric protection method to develop a novel secret-sharing method. Secret information can only be recovered if all three parties are involved, which makes it more challenging for attackers to breach the system.
- The security and privacy of the TBAF have been tested and validated through formal and informal security analysis. The experimental results present the TBAF has superior performance in terms of computation and communication overhead.

2   Related Work

With the prevalence of UAVs, the research of UAVs and Internet of Drones (IoD) has attracted attention recently [8,9]. Lin et al. [8] analyzed the security challenges faced by IoD and proposed several possible solutions to improve the security of IoD, among which the security authentication of UAVs was considered an extremely important part.

To enable secure authentication, Tian et al. [10] proposed an authentication framework based on a digital signature mechanism in edge computing environment to realize secure and efficient authentication of UAVs. However, there are some scholars argued that the solution of the literature does not provide security against location threats and other physical attacks [10]. For example, Gope and Sikdar [11] proposed a double PUF-based authentication and secret key negotiation mechanism that uses PUFs to achieve the physical security of UAVs and the device does not need to store any secret key information. Shen et al. [5] proposed a two-stage authentication approach involving the cloud server, edge server and vehicle. The first stage authentication results were utilized to aid the second stage authentication, thus reducing the computational burden of performing multiple authentications with the edge server. However, their solution relied heavily on digital signatures and elliptic curve algorithms, resulting in significant computational overhead during the first stage authentication process. Alladi et al. [4] proposed a lightweight authentication protocol that utilizes PUF technology to achieve mutual authentication between the UAV-Ground station and UAV-UAV. Tian et al. [12] considered that UAVs need to perform mutual authentication with different ground stations in fast movement. They proposed a cross-domain mutual authentication mechanism for UAVs and ground stations.

Although these work achieve authentication of UAVs in different environments, they do not consider the data privacy of operators and UAVs Additionally, the same authentication protocol must be executed for each authentication, resulting in increased computational overhead [4, 12].

Biometric plays a crucial role in uniquely identifying users, and designing authentication protocols using biometric features has become a popular research topic. Kumar et al. [13] proposed a framework that combines biometric and Elliptic Curve Cryptography (ECC) assistance to complete mutual authentication of vehicle cloud servers in Vehicular Cloud Computing (VCC) environment. This framework is developed using cloud computing services, but due to the fast movement and high communication latency of vehicles, it puts a heavy computational strain on resource-constrained UAVs. Bera et al. [14] proposed a three-factor authentication protocol based on passwords, biometrics and mobile devices in smart city environments. However, the protocol introduces a trusted registry and an access point for relaying messages, and the trustworthiness issues of the registry and access point produced additional privacy anxiety. Bian et al. [15] proposed a user authentication and secret key negotiation mechanism based on PUF and fingerprint features, which achieves secure user authentication. That mechanism introduced excessive computational overhead due to the use of multiple fuzzy extractor operations. Zhang et al. [16] designed a biometric and PUF-based authentication protocol in a multi-server environment and a new secret-sharing technique for distributed storage of secret values. Whereas the secret sharing scheme is not suitable for the dynamic environment of edge computing and may result in authentication failure if some servers go offline.

In a nutshell, the majority of current UAV authentication protocols only focus on ensuring the identity security of the UAV, without considering the security of the operator controlling the UAV [4, 5, 10–12, 17] . This creates security vulnerability, as the UAV's flight process still heavily relies on the operator's control. Furthermore, most current authentication protocols based on biometrics has high computation overhead which is not acceptable for resource-constrained UAVs [13–16] .

3   Preliminaries

In this section, we describe the background knowledge involved in TBAF. Specifically, it includes PUF, secret sharing, zero knowledge proof, and cancellable biometric template protection.

3.1   Physical Unclonable Function

PUF is a hardware function that generates random values inside the chip due to differences in the Integrated Circuit (IC) fabrication process. PUF can be considered a biometric feature embedded in the device [18]. PUF can generate multiple Challenge-Response Pairs (CRPs), whose mathematical expression for the response is:

$$R = PUF(C) \tag{1}$$

where C is a set of possible challenge sets, and R is the output of the PUF also called the response value, the same challenge will generate the same response, and different challenges will generate different response values. Due to the function generate random differences arising during the fabrication of the intrinsic chip, so it is impossible for two different entities to have exactly the same PUF.

3.2 Blakley Secret Sharing

The secret sharing technique was initially proposed by Blakley et al. [19]. It refers to the encrypted distribution of secrets to multiple participants for joint storage to improve the security of secret storage. Suppose the secret S is transformed to a point on the t-dimensional space denoted $S = (S_1, S_2, ..., S_{(t)})$ and then randomly construct n planes passing through this point, the n planes do not overlap each other, where the plane equation is:

$$P_1 : t_1^1 X_1 + t_2^1 X_2 + \ldots + t_t^1 X_t = \gamma_1;$$
$$P_2 : t_1^2 X_1 + t_2^2 X_2 + \ldots + t_t^2 X_t = \gamma_2;$$
$$\ldots;$$
$$P_n : t_1^n X_1 + t_2^n X_2 + \ldots + t_t^n X_t = \gamma_n; \tag{2}$$

where any plane P_i has $t_1^1 S_1 + t_2^1 S_2 + .. + t_t^1 S_t = \gamma_1:$, send the parameter $t_1^i, t_2^i, ..., t_t^i, \gamma_i$ to the ith participant.

If any t of these n planes is known, the secret value S can be obtained by solving the following matrix. But if the information of at most $t - 1$ of the participants is known, it is also impossible to obtain any information about S.

$$\begin{bmatrix} t_1^1, t_2^1, \ldots, t_t^1 \\ t_1^2, t_2^2, \ldots, t_t^2 \\ \ldots \\ t_1^t, t_2^t, \ldots, t_t^t \end{bmatrix} \begin{bmatrix} X_1 \\ X_2 \\ \ldots \\ X_t \end{bmatrix} = \begin{bmatrix} \gamma_1 \\ \gamma_2 \\ \ldots \\ \gamma_t \end{bmatrix} \tag{3}$$

3.3 Zero Knowledge Proofs

ZKP is an encryption scheme to achieve data privacy protection. This statement describes a scenario where a prover can demonstrate to a verifier that they have a legitimate interest in something without disclosing their confidential information. ZKP have the following three properties [20]:

(1) Completeness. The prover's true statement must be able to convince the verifier.
(2) Soundness. A dishonest prover's statement must not pass the verifier's verification.
(3) Zero-knowledge. During the mutual communication between the prover and the verifier, the prover cannot disclose any knowledge content, and the verifier cannot extract any knowledge-related information from the message, i.e., the knowledge given to the outside world is "Zero".

3.4   Cancelable Biometric Template Protection

The cancelable biometric template protection method maps the original biometric to a protected biometric template through a one-way transformation function. Biometric is matched in the transformation domain. A generic cancelable biometric system is shown in Fig. 1. In the registration phase, the user inputs the biometric information to get the origin template, and the origin template is transformed to gain the protected biometric. When the user logs in, the same biometric information and transformation parameters are used to get the protected biometric and complete the biometric authentication in the transformation domain.

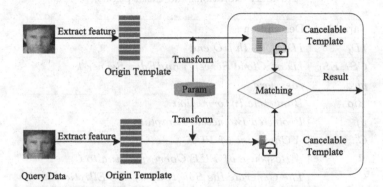

Fig. 1. A Block Diagram for General Cancelable Biometric System.

In this paper, to enable fast authentication and enhance the user experience, we adopted the classical Index-of-Maximum (IOM) Hashing [21] as our biometric template protection method. IOM uses a random matrix projection method to convert original biometric into protected biometric template and has been proven to be safe and efficient.

4   The Proposed Scheme

In this section, we describe the proposed scheme. Table 1 provides a list of notations used in this paper and their corresponding descriptions.

4.1   System Overview

In this subsection, we describe the involved entities of the TBAF with three main entities: Cloud Server, Edge Server, and UO end. Then, we elaborate on the five phases of the TBAF delivery process.

- The cloud server is responsible for maintaining the data of the distribution center, authenticating the identity of the UO that initiates the delivery request, and executing the route planning process for the authenticated UO.

- The edge server is an honest-but-cautious party that operates under the control of the cloud server. The edge server is honest in that it will perform the authentication process, provide airspace traffic flow conditions, and path navigation, and assist the UAV flight. However, it is also cautious, as there is a risk of the edge server maliciously stealing privacy information from the authentication process at the UO end.
- The UO end is responsible for collecting packages from the distribution center and completing the logistics distribution process with the assistance of the edge server.

Table 1. Notations and descriptions.

Variable	Description
ID_i	$ID\ of\ i-th\ UO\ end$
CS_j, ES_k	$ID\ of\ Cloud\ Server\ j\ and Edge\ Server\ k$
pwd	$password$
Bio	$Biometric\ Information$
pf	$Protected\ Biometric\ Template$
C_i	$A\ Challenge\ of\ PUF$
R_i	$A\ Response\ of\ PUF\ Corresponding\ to\ C_i$
G	$The\ Generate\ the\ Subgroup\ of\ the\ Elliptic\ Curve$
FA	$Auxiliary\ Parameters\ of\ Fuzzy\ Extracctor$
$PUF()$	$PUF\ Function$
$h()$	$Hash\ Function$
$FE.Gen()$	$Fuzzy\ Extroactor\ Generation\ Function$
$FE.Rec()$	$Fuzzy\ Extroactor\ Reproduction\ Function$
$SK_{i,j}$	$Session\ Key\ Between\ UO\ End\ and\ Cloud\ Server$
$SK_{j,k}$	$Session\ Key\ Between\ UO\ End\ and\ Edge\ Server$

As shown in Fig. 2, TBAF is divided into five phases.

Phase 1–2. The logistics center initiates a logistics delivery request, performs route planning and waits for an available UAV to respond to the demand. Then a registered UAV responds to the cloud server with a delivery request and requests an authentication.

Phase 3–4. After receiving the response from the UO, the cloud server must first verify the authenticity of the UO end. Once the identity of the UO is confirmed, the cloud server will send the authentication results and auxiliary authentication information to all subsequent edge nodes that UAV may encounter. Then the UAV has the right to get the package from the delivery station.

Phase 5. During the UAV flight, UAV may pass through an edge server controlled by the cloud server. The edge server has to authenticate the UO end using auxiliary information to ensure that the UO end is secure and provide airspace traffic flow, weather conditions, and auxiliary flight information. ZKP and cancellable biometric template protection methods are adopted to guarantee that edge servers do not maliciously steal the private information of UO end. After all edge nodes have authenticated the identity of UAV, the UAV arrives at destination and delivers the package.

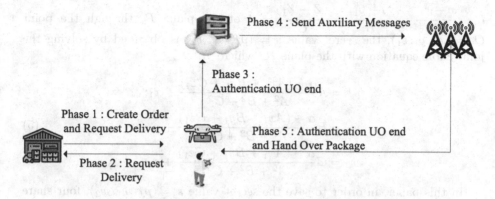

Fig. 2. TBAF Delivery Process.

4.2   Three-Dimensional Space-Based Secret Sharing

All edge servers are not always online simultaneously, the t-out-of-n edge servers are expected to recover via designated secrets. Furthermore, the true secret value can only be recovered by the cloud server, the UO end, and multiple edge servers. This prevents any collusion between parties to steal secret data. The details of the three-dimensional space-based secret sharing are as follows.

Secret Sharing Process: The protected biometric template pf and transformation parameter w are considered secret values, w is divided into two parts w_1 and w_2. Then we can calculate the secret value $s = (pf, w_1, w_2)$ to be protected. Finally, 3 secret pieces of information are shared to enhance the protection of biological data.

Step 1. Select a plane $P_s : AX + By + Cz = \alpha$ randomly through the points $s = (pf, w_1, w_2)$, which $Apf + Bw_1 + Cw_2 = \alpha$.

Step 2. Make a vertical line l perpendicular to the plane P_s through the point s, pick any point $Q = (x_1, y_1, z_1)$ from l which $Q \neq s$.

Step 3. Select N planes $P = \{P_1, P_1, \ldots P_N\}$ through point Q as share planes, which any plane $P_i : a_i x + b_i y + C_i z = \alpha_i$ has $a_i x_1 + b_i y_1 + c_i z_1 = \alpha_i$.

Secret Recovery Process: Three secrets are recovered by sharing values. The secret recovery process for biometric authentication is given below.

Step 1. Obtain the plane P_s where the secret s is located by sharing the value A, B, C, α.

Step 2. Take any t secret planes from the N secret planes where $t \geq 3$, the shared coordinates $Q = (x_1, y_1, z_1)$ can be obtained by solving the following system of equations:

$$P_1 : a_1x + b_1y + c_1z = \alpha_1;$$
$$P_2 : a_2x + b_2y + c_2z = \alpha_2;$$
$$\cdots; \tag{4}$$
$$P_t : a_tx + b_ty + c_tz = \alpha_t;$$

Step 3. After obtaining $Q = (x_1, y_1, z_1)$, and making the perpendicular $l : \dfrac{x - x_1}{A} = \dfrac{y - y_1}{B} = \dfrac{Z - Z_1}{C} = t$ of the plane P_s through the point $Q = (x_1, y_1, z_1)$, the secret value $s = (pf, w_1, w_2)$ is obtained by solving the joint cubic equation with the plane P_s, which:

$$pf = A\frac{\alpha - (Ax_1 + By_1 + Cz_1)}{A^2 + B^2 + C^2} + x_1;$$
$$w_1 = B\frac{\alpha - (Ax_1 + By_1 + Cz_1)}{A^2 + B^2 + C^2} + y_1; \tag{5}$$
$$w_2 = C\frac{\alpha - (Ax_1 + By_1 + Cz_1)}{A^2 + B^2 + C^2} + z_1;$$

In this paper, in order to save the secret value $s = (pf, w_1, w_2)$, four share values A, B, C, α and N shared planes are created as P, where A, B are stored in UO end, B, C are stored in the cloud server and the secret plane is stored in the edge server.

4.3 Registration

Each legitimate UO end should be registered with Cloud Sever CS_j before deployment. A UO end U_i has its own identity respectively. The operation of the UO end in the registration phase is shown how in Fig. 3. The specific registration steps are as follows.

In the registration phase, U_i input password pwd, biometric information Bio, generate a transformation parameter w and get a protected biometric template pf by IOM_Hashing. Then U_i interacts with the cloud server to get $\{PID, C_i, K_{uo}, A, B\}$ and computes $K_{uo}^*, FA^*, HV, HV_{FA}$. Finally, U_i storages $\{A, B, K_{uo}^*, C_i, HV, HV_{FA}\}$.

CS_j gets $\{pf, R_i, w\}$ from U_i, pf and w are regarded as secret using three-dimensional space-based secret sharing method to generate A, B, C, α and planes P. Then CS_j generate a privacy key K_{uo} and challenge C_i of U_i, computes HV_{cs}, storages $\{PID, K_{uo}, B, C, \alpha, C_i, HV_{cs}\}$ and send P to edge server.

4.4 Authentication and Key Negotiation

In this subsection, we describe the detailed design process of the two-stage authentication protocol and key negotiation, including the mutual authentication of cloud server and UO, the authentication process of edge server and UO end.

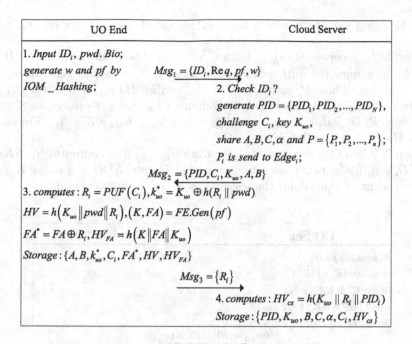

Fig. 3. Registration of UO End.

Stage 1: UO End and Cloud Server Mutual Authentication. As shown in the Fig. 4, the authentications phase procedures. First UO attempts to log in to the cloud server so that can get a package from the delivery station and start distribution.

Step 1: U_i inputs biometric information Bio and password pwd. U_i first computes R_i and K_{uo}, and verifies HV. Only real UO end have the right pwd and R_i. Then U_i computes HV_2 and sends $Msg^1_{Auth} = \{HV_2, PID_i, Req_{auth}\}$ to CS_j.

Step 2: Upon receiving Msg^1_{Auth} from U_i, CS_j checks PID_i to see whether PID_i in the database and verifies if HV_2 and $h(K_{uo} \parallel R_i \parallel PID_i)$ are equal, if not true or not in the database, CS_j will terminate the authentication. Otherwise, CS_j takes out share B and generates a random number r_1.Then CS_j computes $V_1 = h(CS_j \parallel K_{uo}) \oplus r_1$, $V_2 = B \oplus r$ and sends $Msg^2_{Auth} = \{V_1, V_2\}$ to U_i.

Step 3: On receiving Msg^2_{Auth}, computes r_1, B and verify that B is equal to B in the database. If not equal, U_i terminate the authentication, or else takeout share A and produce a random number r_2. Then computes $V_3 = K_{uo} \oplus r_2$, $V_4 = h(K_{uo} \parallel r_2 \parallel B) \oplus A$, $V_5 = h(K_{uo} \parallel A \parallel B \parallel r_2)$ and sends $Msg^3_{Auth} = \{V_3, V_4, V_5\}$ to CS_j.

Step 4: When receiving Msg^3_{Auth}, CS_j computes r_2 and verifies V_5. If not equal, CS_j terminate the authentication, or else takeout share planes from edge servers and recovery pf and w. Then, CS_j generates a random number r_3, computes $V_6 = K_{uo} \oplus w$, $V_7 = h(w \parallel K_{uo} \parallel r_3)$ and sends $Msg_{Auth}^4 = \{V_6, V_7, r_3\}$ to U_i.

Step 5: U_i receives Msg_{Auth}^4 from CS_j, computes w and verifies V_7. If not true, U_i terminate the authentication, otherwise, U_i uses Bio generate pf' by IOM_Hashing. Then, U_i computes FA, K, verifies HF_{FA}. If not true, U_i terminate authentication, otherwise, U_i computes $V_8 = K_{uo} \oplus K$, $V_9 = K \oplus R_i$, $SK_{ij} = h(PID_i \parallel R_i \parallel K \parallel r_3)$ and $V_{10} = h(SK_{ij} \parallel K_{uo} \parallel K \parallel R_i)$. Finally, U_i sends $Msg_{Auth}^5 = \{V8, V9, V10\}$ to CS_j.

Step 6: When CS_j receives Msg_{Auth}^5 from U_i, CS_j first computes R_i, $SK_{ij} = h(PID_i \parallel R_i \parallel K \parallel r_3)$. Then CS_j verifies that V_{10} and $h(SK_{ij} \parallel K_{uo} \parallel K \parallel R_i)$ are equivalent, if equivalent, the session key is successfully created.

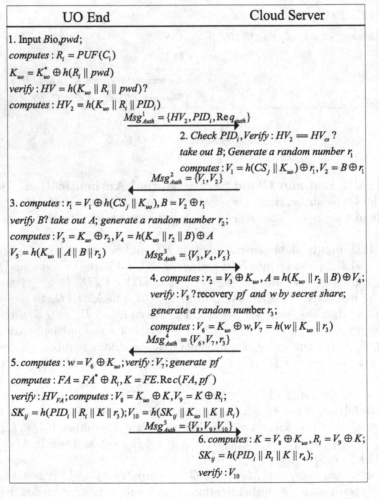

Fig. 4. UO End and Cloud Sever Mutual Authentication.

Stage 2: Edge Server Authentication UO End. In an edge-integrated UAV system, UAVs need to access edge servers more frequently and edge servers are less trusted than cloud servers. Therefore, in this paper, an authentication protocol is designed between UO and Edge Sever with ZKP. The specific authentication steps of the UO and edge server are as follows.

Initialization Phase: the cloud server CS_j selects a random point G on the elliptic curve EC, computes $\tau = h(R_i) \cdot G$ and sends $\{\tau, EC, C_i, G, pf\}$ to the edge server ES_k where the UAV may pass, which C_i and R_i are UAV's CRP-PUF and pf is the operator's protected biometric template.

Authentication Phase: Fig. 5 shows the authentication between U_i and ES_k

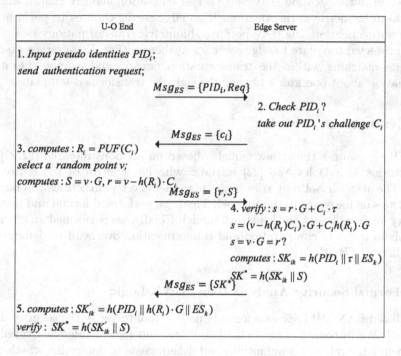

Fig. 5. Edge Sever Authentication UO End.

Step 1: U_i launches an authentication request with pseudo-identity PID_i and sends $MSG_{ES}^1 = \{PID_i, Req\}$ to ES_k.

Step 2: Upon receiving the message Msg_{ES}^1, first check if PID_i exists in the database. If not in the database, ES_k terminate the authentication, otherwise, takes out the challenge C_i and send C_i to U_i.

Step 3: After U_i receiving the C_i from ES_k, U_i selects a v random point and computes R_i, r and $S = v \cdot G$, which, as per elliptic curve algebra, is v times the addition of the elliptic curve point G to itself. U_i drafts the message $Msg_{ES}^3 = \{r, S\}$ and delivers it to ES_k.

Step 4: ES_k receives the Msg^3_{ES}, computes $S = r \cdot G + C_i \cdot \tau$ and compares it with the received r. If not valid, ES_k terminate the authentication, otherwise ES_k computes $SK_{ik} = h\left(PID_i \parallel \tau \parallel ES_k\right)$, $SK^* = h\left(SK_{ik} \parallel S\right)$ and send$Msg^4_{ES} = \{SK^*\}$ to U_i.

Step 5: After obtaining Msg^4_{ES}, U_i computes $SK'_{ik} = h\left(PID_i \parallel \tau \parallel ES_k\right)$ and verifies whether $SK^* = h\left(SK'_{ik} \parallel S\right)$ holds. If not valid, U_i terminate the authentication, otherwise U_i and ES_k successfully establish the session key SK_{ik}.

To protect the security of the operator's biometric information, this paper adopts IOM_Hashing cancelable biometric template protection method. After edge server authentication UAV, edge server can autonomously choose whether to perform biometric authentication. If an edge server chooses to perform biometric authentication, the UAV performs biometric transformations locally and sends protected template to edge sever for authentication. Edge server executes template matching within the transformation domain and cannot obtain any information about operator's privacy through the transformed template.

5   Evaluation

The TBAF assumes the attack model is based on previous research [12,22] and the attacker is a Dolev-Yao [23] intruder who has control of the entire network. The attacker not only tries to get past the server's authentication but also eavesdrops on messages in the channel. Then, we performed formal and informal security analyses based on this attack model. Finally, we performed an efficiency analysis in terms of computation and communication overhead to demonstrate the utility of TBAF.

5.1   Formal Security Analysis Using BAN Logic

We utilized BAN [24] logic as a formal analysis tool to verify the security of our protocol. We introduce some basic definitions and logical inference rules, and then define the relevant assumptions and design goals to determine whether our protocol can achieve the desired design goals through detailed logical reasoning. Based on the proposed scheme, we use U_i and CS_i as principals for formal analysis.

As shown in Table 2, we explain the basic symbols and definitions involved in BAN logic. The inference rules about BAN logic are shown below, and these are used to help us complete the subsequent theoretical proofs.

R1: Message-meaning rule: $\dfrac{P|\equiv P \xleftrightarrow{K} Q, P \triangleleft X_K}{P|\equiv Q|\sim X}$

R2: Nonce-verification rule: $\dfrac{P|\equiv \#(X), P|\equiv Q|\sim X}{P|\equiv Q|\equiv X}$

R3: Jurisdiction rule: $\dfrac{P|\equiv Q|\Rightarrow X, P|\equiv Q|\equiv X}{P|\equiv X}$

R4: Belief rule: $\dfrac{P|\equiv(X,Y)}{P|\equiv X}$, $\dfrac{P|\equiv X, P|\equiv Y}{P|\equiv(X,Y)}$ and $\dfrac{P|\equiv Q|\equiv(X,Y)}{P|\equiv Q|\equiv X}$

R5: Fresh rule: $\dfrac{P|\equiv \#(X)}{P|\equiv \#(X,Y)}$

Table 2. Basic Operation Time.

Notation	Meanings
$\#X$	Statement X is fresh
X_K	The statement X is encrypted by K
$P| \equiv X$	Principal P believes the statement X
$P \triangleleft X$	Principal P receives the statement X
$P| \sim x$	Principal P has send statement X
$P| \Rightarrow X$	Principal P has jurisdiction over the statement X
$P \overset{K}{\longleftrightarrow} Q$	K is the shared key between P and Q

Hypothetical and Goals: To complete the proof process using BAN, We make relevant assumptions $H_1 \sim H_6$ and preset the goals $G_1 \sim G_4$ between U_i and GS_j.

$$H_1 : CS_j |\equiv CS_j \overset{K_{u\infty}}{\longleftrightarrow} U_i \quad H_2 : U_i \equiv CS_j \overset{K_{u\infty}}{\longleftrightarrow} U_i$$

$$H_3 : CS_j |\equiv \#(r_2), CS_j| \equiv \#(R_i) \quad H_4 : U_i |\equiv \#(r_1), U_i| \equiv \#(r_3)$$

$$H_5 : CS_j |\equiv U_i \Rightarrow CS_j \overset{sK_{ij}}{\longrightarrow} U_i \quad H_6 : U_i \equiv CS_j \Rightarrow CS_j \overset{sK_{ij}}{\longrightarrow} U_i$$

$$G_1 : U_i |\equiv CS_j| \equiv CS_j \overset{SK_{ij}}{\longleftrightarrow} U_i \quad G_2 : CS_j |\equiv U_i| \equiv CS_j \overset{SK_{ij}}{\longleftrightarrow} U_i$$

$$G_3 : U_i |\equiv CS_j \overset{SK_{ij}}{\longleftrightarrow} U_i \quad G_4 : CS_j |\equiv U_i \overset{SK_{ij}}{\longleftrightarrow} U_i$$

Inference proof: Based on the aforementioned assumptions and goals, we complete the proof of the proposed scheme in the following messages. First of all, we summarize the messages conveyed during the authentication process, as follows:

Message 1: $U_i \rightarrow CS_j : \{HV_i, PID_i, Req_{Auth}\}$
Message 2: $CS_j \rightarrow U_i : \{V_1, V_2\}$
Message 3: $U_i \rightarrow CS_j : \{V_3, V_4, V_5\}$
Message 4: $CS_j \rightarrow U_i : \{V_6, V_7, r_3\}$
Message 5: $U_i \rightarrow CS_j : \{V_8, V_9, V_{10}\}$

We can reduce the above messages to the following two messages depending on the content of the transmission.

$$M_1 : U_i \rightarrow CS_j :< HV_2, r_2, A, R_i, K, SK_{ij} > \quad M_2 : CS_j \rightarrow U_i :< r_1, B, W, r_3 >$$

Convert M_1 and M_2 to BAN logic basic notation as:

$$M_1 : CS_j \triangleleft \{HV_2, r_2, A, R_i, K, CS_j \overset{SK_{ij}}{\longrightarrow} U_i\}_{K_{uo}} M_2 : U_i \triangleleft \{w, r_3, B, CS_j \overset{SK_{ij}}{\longleftrightarrow} U_i\}_{K_{uo}}$$

Based on the above work, the logical reasoning process is as follows. According to the message M_1 and hypothesis H_1 and H_3, applying the rule R_1, R_2 and R_5 we can get:

$S_1 : CS_j \mid\equiv U_i \equiv \{HV_2, r_2, A, R_i, K, CS_j \overset{SK_{ij}}{\longleftrightarrow} U_i\}$

Apply rule R_4 according to formula S_1, we can get:

$S_2 : CS_j \mid\equiv U_i \mid\equiv \{CS_j \overset{SK_{ij}}{\longrightarrow} U_i\}$ **(G2)**

According to S_2, hypothesis H_5 and inference rule R_3, we get:

$S_3 : CS_j \mid\equiv \{CS_j \overset{SK_{ij}}{\longrightarrow} U_i\}$ **(G4)**

Applying message M_2, assuming H_2 and H_4, rule R_1, R_5 and R_2 we get:

$S_4 : U_i \mid\equiv CS_j \mid\equiv \{w, r_3, B, CS_j \overset{SK_{ij}}{\longrightarrow} U_i\}$

Apply rule R_4 according to formula S_4, we can get:

$S_5 : U_i \mid\equiv CS_j \mid\equiv \{CS_j \overset{SK_{ij}}{\longrightarrow} U_i\}$ **(G1)**

According to S_5, hypothesis H_6 and inference rule R_3, we get:

$S_6 : U_i \mid\equiv \{CS_j \overset{SK_{ij}}{\longrightarrow} U_i\}$ **(G3)**

Through the above derivation process, we can obtain the derivation results S_5, S_2, S_3, S_6, which correspond to our objectives G1-G4, from which can be seen that the protocol we designed satisfies the BAN logic proof.

5.2   Informal Security Analysis

In this subsection, we conduct a detailed non-formal analysis of our proposed scheme assuming that the public channel is controlled by an attacker. We aim to demonstrate more scientifically that the proposed scheme can effectively achieve mutual authentication, key negotiation, and privacy protection.

User Anonymity. Firstly, during registering, we generate a random set of pseudo-random identities for UO. The UO selects a pseudo-identity each time to complete the authentication process. Additionally, we use the CRO PUF to generate shared secret keys for resource-constrained UAV devices. The personalized hardware configuration structure ensure that different devices have different CRO response pairs, achieve anonymity of UAVs from a hardware perspective.

Freshness of Session Key. Our proposed scheme combines registration information with random numbers to calculate the session secret key SK_{ij} and SK_{ik}. The presence of random numbers makes the session secret key different each time it is created. This property guarantees the freshness of the session secret key of the proposed scheme.

Database Attack. We propose a combination of using cancelable biometric template protection method and a secret sharing scheme to generate multiple shares to be stored by multiple parties. There are no single or two parties can recover private information about the user. Moreover, due to the presence of cancelable biometric template protection method, even if the attacker recovers the secret information, they cannot obtain the valid biometric information through the user's biometric vector. On the other hand, our proposed protocol, the UO, cloud server, and edge server do not directly store the necessary parameters regarding operator privacy and mutual authentication. Therefore, it is completely impossible for an attacker to try to pass authentication or obtain privacy information by attacking the database.

Impersonation Attacks

Cloud Server Impersonation Attack. In our proposed scheme, UO verifies the authenticity of the cloud server identity. Only the real cloud server can know the shared information in the secret sharing, UO will verify whether the cloud server has the real shared information. In addition, only the real cloud server can get the shared plane from the multi-party edge server and finally recover the secret value. Therefore, for a fake cloud server to successfully disguise and complete the authentication, it not only needs to obtain the secret information but also to gain the trust of the UO and the edge server. This is very difficult to achieve.

Edge Server Impersonation Attack. The edge server is under the jurisdiction of the cloud server. To successfully launch an impersonation attack, the trust of the cloud server. Additionally, the edge server does not store any data information about the privacy of the UAV. Authentication of the UAV is completed through the way of ZKP, making it impossible for the attacker to obtain any valuable information even if the attack is successful.

UO Impersonation Attack. The attacker may attempt to gain access to the cloud server by disguising themselves as legitimate operator or UAV. In our proposed scheme, the UAV uses CRO PUF to complete authentication, which is a hardware facility of the UAV device and difficult for an attacker to obtain. In addition, for the operator, their biometric information needs to be used, and each authentication needs to be collected in real-time, making it difficult for an attacker to achieve.

Man-in-the-Middle Attacks (MIMA). During the execution of mutual authentication of UO and cloud server and authentication of UO by edge server. The transmission of messages $Msg^1_{Auth} \sim Msg^5_{Auth}$ and $Msg^1_{ES} \sim Msg^4_{ES}$ is involved, and the attacker will try to intercept and tamper with them. Firstly, A and B are messages known only to the UO and the cloud server. An attacker cannot obtain any private information through the public channel. Furthermore, UO, cloud servers and edge servers verify the authenticity of messages each time they are received. Only messages that pass the verification are considered as reliable transmissions. Therefore, launching a man-in-the-middle attack is not possible to achieve the purpose of gaining access to the server.

Cross-Matching Attacks. When operators access different servers using the same biometric traits, each server generates a pseudo-identity for the operator and uses the unlinkability of cancelable biometric protection to assign unique transformation parameters to the operator to get the corresponding pf, which is different for different servers. Thus, the application of these two techniques ensures that our authentication scheme is resistant to Cross-Matching Attacks.

5.3 Efficiency Analysis

The computation time and communication overhead of TBAF are considered in our experiments.

Table 3. Basic Operation Time.

Operation	Description	Time(ms)
T_h	Hash Operation	0.0026 ms
$T_p uf$	PUF Operation	0.13 ms
T_{Gen}	Fuzzy Extractor Gen()	2.67 ms
T_{Rec}	Fuzzy Extractor Rec()	3.35 ms
$T_s m$	Scalar Multiplication	0.86 ms

Experimental Environment Settings. A Raspberry Pi 3B is adopted as the emulation environment for the UAV. It was capable of running common mathematical and cryptographic operations such as XOR, Pseudo-Random Number Generation (PRNG), Scalar Multiplication, Hash (SHA-1) and concatenations. We used Python 3.8 as our programming language for these operations on a 3.61 GHZ Intel Core i7-12700k CPU with 64 GB of RAM, and running on Ubuntu-20.04.

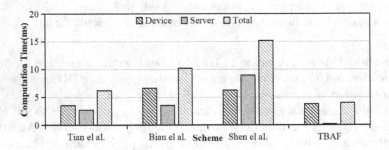

Fig. 6. Computation Overhead of One Authentication.

Computation Overhead. The computation overhead is measured by the computation time of each device during the authentication process. To evaluate the computation overhead, we summarize the execution time required for the main operations, including Hash, PUF, Fuzzy Extractor, and Scalar Multiplication, as shown in Table 3.

We compare the computation overhead incurred for the first authentication (UO and cloud server). As shown in Fig. 6, the labels in the figure represent the computation time of the device, server and total in performing the authentication and key negotiation process, respectively. In case of one authentication, the protocol designed by Tian et al. [12] generated a computation time of 3.493 ms at the device and 2.683 ms at the server, for a total computation time of 6.176 ms. The protocol designed by Bian et al. [15] generated a computation time of 6.6436 ms at the device. The computation time is 3.532 ms at the server, and the total computation time is 10.1756 ms. The two-stage authentication protocol has been designed by Shen et al. [5]. In the first authentication, the computation time at the device and server are 6.244 ms and 8.92 ms. While the proposed protocol mostly uses less time-consuming computations such as XOR and Hash operations. The computation time of the proposed protocol is 3.766 ms at the end device, 0.182 ms at the server, and 3.948 ms in total, which is much lower than the computation overhead of the other methods.

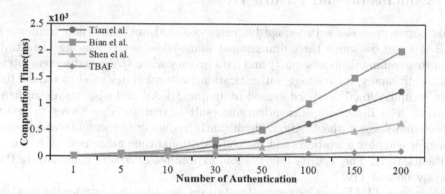

Fig. 7. Computational overhead of multiple authentication.

During the flight, multiple authentication protocols need to be executed because multiple edge servers will be passed in the UAV. We compare the computation overhead variation in the case of multiple authentications. As shown in Fig. 7, our authentication protocol has better performance in terms of computation overhead as the number of authentications increases. Because we use a two-stage authentication scheme and minimize the number of scalar multiplication used while ensuring security.

Table 4. Communication Overhead (bytes).

Scheme	Tian et al.	Bian et al.	Shen et al.	TBAF
UO-Cloud	3020	1080	8784	2080
UO-Edge			3036	846

Communicational Overhead. The communicational overhead is measured by the number of bytes in every communication message sent. In TBAF, we set each hash processing, identity ID, random number, and scalar point multiplication to the standard size of 160 bytes per message and challenge value, and timestamp to 32 bytes. The communication overhead of our designed authentication protocol for the initial authentication is 2080 bytes, while for the second authentication, the communication overhead from the cloud server to the edge is 192 bytes, and the communication overhead between the edge server and the UAV authentication is 672 bytes. Resulting in a total communication overhead of 864 bytes in the second phase. Compared to other methods, TBAF has significantly lower communication overhead in the second phase. Table 4 shows a comparison of the communication overhead of TBAF and other methods.

6   Conclusion and Future Work

This paper proposes a two-stage biometric-assisted authentication framework (TBAF) and designs a three-dimensional space-based secret-sharing algorithm to ensure communication security and data privacy when UAVs and servers interact. Additionally, a two-stage authentication protocol is designed to reduce the high computational overhead caused by frequent UAV and edge server authentication. The first-stage authentication result is sent to edge server to assist the second-stage authentication, significantly reducing the number of calculations for complex operations and the computational time an caused by frequent authentication. The security analysis and experimental results demonstrate the effectiveness of TBAF.

However, TBAF does have some limitations. In order to minimize the computational overhead, TBAF assumes that the edge server will not provide malicious information. As a result, the UO does not authenticate the edge server, which introduces certain security risks. Therefore, implementing mutual authentication between the UO end and edge servers will be a part of our future work. Additionally, mutual authentication between UAVs is also an important aspect that is not considered in this article. Incorporating mutual authentication between UAVs into TBAF will be included in our future work.

Acknowledgements. This work was supported by the National Natural Science Foundation of China Project (No. 61972001).

References

1. Lee, S., Kang, Y., Prabhu, V.: Smart logistics: distributed control of green crowdsourced parcel services. Int. J. Prod. Res. **54**(23), 6956–6968 (2016)
2. Suma, S., Mehmood, R., Albugami, N., Katib, I., Albeshri, A.: Enabling next generation logistics and planning for smarter societies. Procedia Comput. Sci. **109**, 1122–1127 (2017)

3. Zhang, J., Cui, J., Zhong, H., Bolodurina, I., Lu, L.: Intelligent drone-assisted anonymous authentication and key agreement for 5G/B5G vehicular ad-hoc networks. IEEE Trans. Netw. Sci. Eng. **8**(5), 2982–2994 (2020)
4. Alladi, T., Bansal, G., Chamola, V., Guizani, M.: SecAuthUAV: a novel authentication scheme for UAV-ground station and UAV-UAV communication. IEEE Trans. Veh. Technol. **69**(12), 15068–15077 (2020)
5. Shen, M., Lu, H., Wang, F., Liu, H., Zhu, L.: Secure and efficient blockchain-assisted authentication for edge-integrated internet-of-vehicles. IEEE Trans. Veh. Technol. **71**(11), 12250–12263 (2022)
6. Xu, J., Liu, X., Li, X., Zhang, L., Yang, Y.: EXPRESS: an energy-efficient and secure framework for mobile edge computing and blockchain based smart systems. In 35th IEEE/ACM International Conference on Automated Software Engineering, pp. 1283–1286 (2020)
7. Alzahrani, B., Barnawi, A., Chaudhry, S.: A resource-friendly authentication protocol for UAV-based massive crowd management systems. Secur. Commun. Netw. **2021**, 1–12 (2021)
8. Lin, C., He, D., Kumar, N., Choo, K., Vinel, A., Huang, X.: Security and privacy for the internet of drones: challenges and solutions. IEEE Commun. Mag. **56**(1), 64–69 (2018)
9. Abualigah, L., Diabat, A., Sumari, P., Gandomi, A.: Applications, deployments, and integration of internet of drones (IoD): a review. IEEE Sens. J. **21**(22), 25532–25546 (2021)
10. Tian, Y., Yuan, J., Song, H.: Efficient privacy-preserving authentication framework for edge-assisted Internet of Drones. J. Inf. Secur. Appl. **48**, 1–11 (2019)
11. Gope, P., Sikdar, B.: An efficient privacy-preserving authenticated key agreement scheme for edge-assisted internet of drones. IEEE Trans. Veh. Technol. **69**(11), 13621–13630 (2020)
12. Tian, C., Jiang, Q., Li, T., Zhang, J., Xi, N., Ma, J.: Reliable PUF-based mutual authentication protocol for UAVs towards multi-domain environment. Comput. Netw. **218**, 1–13 (2022)
13. Kumar, V., Ahmad, M., Kumari, A., Kumari, S., Khan, M.: SEBAP: a secure and efficient biometric-assisted authentication protocol using ECC for vehicular cloud computing. Int. J. Commun. Syst. **34**(2), 1–21 (2021)
14. Bera, B., Das, A., Balzano, W., Medaglia, C.: On the design of biometric-based user authentication protocol in smart city environment. Pattern Recogn. Lett. **138**, 439–446 (2020)
15. Bian, W., Gope, P., Cheng, Y., Li, Q.: Bio-AKA: an efficient fingerprint based two factor user authentication and key agreement scheme. Futur. Gener. Comput. Syst. **109**, 45–55 (2020)
16. Zhang, H., Bian, W., Jie, B., Xu, D., Zhao, J.: A complete user authentication and key agreement scheme using cancelable biometrics and PUF in multi-server environment. IEEE Trans. Inf. Forensics Secur. **16**, 5413–5428 (2021)
17. Alladi, T., Chamola, V., Kumar, N.: PARTH: a two-stage lightweight mutual authentication protocol for UAV surveillance networks. Comput. Commun. **160**, 81–90 (2020)
18. Kim, B., Yoon, S., Kang, Y., Choi, D.: PUF based IoT device authentication scheme. In: 2019 International Conference on Information and Communication Technology Convergence (ICTC), pp. 1460–1462. IEEE (2019)
19. Blakley, G.: Safeguarding cryptographic keys. In: Managing Requirements Knowledge, International Workshop, pp. 313–313. IEEE Computer Society (1979)

20. Gaba, G., Hedabou, M., Kumar, P., Braeken, A., Liyanage, M., Alazab, M.: Zero knowledge proofs based authenticated key agreement protocol for sustainable healthcare. Sustain. Urban Areas **80**, 1–12 (2022)
21. Jin, Z., Hwang, Y.J., Lai, Y., Kim, S., Teoh, A.: Ranking-based locality sensitive hashing-enabled cancelable biometrics: index-of-max hashing. EEE Trans. Inf. Forensics Secur. **13**(2), 393–407 (2017)
22. Bansal, G., Sikdar, B.: S-MAPS: scalable mutual authentication protocol for dynamic UAV swarms. IEEE Trans. Veh. Technol. **70**(11), 12088–12100 (2021)
23. Dolev, D., Yao, A.: On the security of public key protocols. IEEE Trans. Inf. Theory **29**(2), 198–208 (1983)
24. Burrows, M., Abadi, M., Needham, R.: A logic of authentication. ACM Trans. Comput. Syst. (TOCS) **8**(1), 18–36 (1990)

Attention Enhanced Package Pick-Up Time Prediction via Heterogeneous Behavior Modeling

Baoshen Guo[1], Weijian Zuo[1,3], Shuai Wang[1(✉)], Xiaolei Zhou[2(✉)],
and Tian He[3]

[1] Southeast University, Nanjing, China
{guobaoshen,shuaiwang}@seu.edu.cn
[2] The Sixty-Third Research Institute, National University of Defense Technology,
Nanjing, China
zhouxiaolei@nudt.edu.cn
[3] JD Logistics, Beijing, China
{zuoweijian1,tim.he}@jd.com

Abstract. The logistics industry has developed rapidly with the popularity of online-to-offline businesses in recent years. First-mile package pick-up is one of the most critical and expensive parts of the whole logistics service chain, which is finished by couriers in practice. Accurate prediction of package pick-up time at the customers' addresses is essential to improve customers' experience and increase platforms' profits. For some logistics service providers, couriers conduct heterogeneous tasks (i.e., first-mile pick-up and last-mile delivery) simultaneously in a certain area to improve efficiency. However, existing works neglect the impact of the package delivery process, which produces inaccurate prediction results due to the coupling of the pick-up and delivery process. Considering the delivery process in pick-up time prediction introduces two additional challenges: (i) *Limited pickup requests.* In practice, couriers have a limited number of package delivery tasks in a delivery trip, which hinders the direct application of existing deep learning models for the prediction. (ii) *Dynamic package pickup requests.* Package pick-up requests are generated dynamically, which affects the courier's route. In this paper, we propose HTAPT, a heterogeneous tasks aware package pick-up time prediction framework, which consists of two modules: (i) *Pre-trained stay time prediction module* to learn the embedding of the courier's stay time. (ii) *Attention enhanced pick-up time and route prediction module* to predict the delivery route and pick-up arriving time of the courier under the pick-up influence. The evaluation results with real-world order data from JD Logistics, which is one of the largest logistics companies in China show HTAPT improves the prediction accuracy by up to 10% compared with the state-of-the-art methods.

Keywords: Last-mile delivery · Pick-up time prediction · Machine learning

Z. Tari et al. (Eds.): ICA3PP 2023, LNCS 14493, pp. 189–208, 2024.
https://doi.org/10.1007/978-981-97-0862-8_12

1   Introduction

With the rapid development of the mobile Internet, the online-to-offline business has injected huge vitality into the development of the logistics industry. So far, there are tens of millions of employees in the logistics industry, and the number of parcels delivered every day has reached hundreds of millions.

To keep up with the advancements of the era, the traditional logistics industry has also incorporated various personalized services to enhance the overall user experience. Like the door-to-door pick-up service, the user will not leave the house. Just make a request on the app, and then the courier will pick up the packages at the door. Due to the real-time nature of the pick-up task and the uncertainty of the location, the couriers cannot plan the path in advance. In addition to the pick-up task, the couriers also have nearly 200 delivery tasks to complete every day. Therefore, in order to ensure that the delivery task can be completed on time, the couriers have difficulty immediately responding to the pick-up task. In addition, due to business process requirements, users need to wait for the courier check packages and confirm the user identity information. It is essential to estimate the package pick-up time accurately to reduce the waiting anxiety of users and promote the user experience.

Existing studies on package pick-up time prediction problems include estimated time of arrival (ETA) in transportation systems [4,7,9,14,20–22] and mobility prediction applications in logistics [5,8,10,17,24–26]. But these works are unsuitable for the pick-up time prediction problem because (i) Pick-up time in last-mile logistics is affected by two heterogeneous tasks (i.e., delivery task and pick-up task) due to the timeliness and uncertainty, while existing ETA studies in transportation focus on one task; (ii) Pick-up time prediction should consider both staying time in residence for delivery packages [13] and outdoor routing time on the road while existing studies on logistics focus on the route prediction and routing time on the road.

In this paper, by conducting an in-depth data-driven analysis using real-world data collected from a large logistics company, we find the following opportunities to achieve accurate package pick-up time prediction. (i) *Couriers patterned mobility behaviors:* In last-mile logistics, each courier is responsible for a small and relatively fixed delivery zone, resulting in patterned routine delivery and pick-up behaviors, which help us model the static residence features (e.g., number of floors) and learn the dynamic relationship between stay time and package characteristics (e.g., the weight of the package). (ii) *Spatial-constrained prediction difficulty reduction:* It is complex to predict the delivery sequence at a single order level because couriers usually deliver more than one hundred orders one day. Through analyzing couriers' historical delivery behaviors, we find that couriers usually send all orders to one building at one time before going to the next building. So we only need to predict which buildings the couriers will go to next, which brings the possibility of problem-solving.

Leveraging these opportunities, we still find that there are the following challenges in practical application: (i) *Limited pick-up requests:* Compared with the delivery orders of 100 orders in a day, the number of pick-up tasks is less than

10. It is challenging to model environmental features and predict pick-up time with limited historical pick-up data. Moreover, due to the work mobility of the courier, the data drift problem is caused; (ii) *Correlations among heterogeneous tasks:* Dynamic pick-up requests have strong correlations with delivery tasks. Due to the timeliness and uncertain spatial-temporal distribution of pick-up orders, couriers may change the existing delivery route to complete the pick-up task within the time constraints.

To address these challenges, we propose HTAPT, a Heterogeneous Tasks Aware Package pick-up Time prediction framework. HTAPT consists of (i) a pre-trained staying time prediction module, which makes use of a large number of delivery data to address the challenges of limited pick-up data and improve the staying time prediction accuracy; (ii) an attention-enhanced pickup time and route estimation module to capture the mutual impacts among heterogeneous delivery and pick-up tasks. The contributions of this paper are summarized as follows:

- To the best of our knowledge, we are the first to conduct package pick-up time prediction taking both historical couriers' pick-up behaviors and delivery behaviors into consideration. Based on the in-depth behavior analysis with large enterprise logistics data, we utilize patterned couriers' heterogeneous behaviors and spatial constraints to make our prediction framework more practical and applicable to a real-world environment.
- We design a heterogeneous task-aware package pick-up time prediction framework named HTAPT. To address the challenge of limited pick-up data, we first designed a pre-trained staying time prediction module to obtain the AOI representation in the pick-up process leveraging the large number of delivery data. Then, to capture the impact of the pickup task on the delivery, we design an attention-enhanced pick-up time and route prediction module to estimate the uncertain route sequence with both delivery and pickup tasks, which enables the accurate and parallel pickup time prediction with multiple on-the-fly pickup requests.
- We conduct extensive experiments based on half-year real-world logistics data, including 100 thousand orders. The evaluation results show that HTAPT improves the prediction accuracy significantly compared with some state-of-the-art baselines. Furthermore, we implement and deploy HTAPT at JD Logistics [11], which is one of the largest logistics companies in China. After deployment, the online results show that the number of users' complaints declined, which underscores the system's effectiveness in enhancing user satisfaction and pick-up efficiency.

2   Overview

In this section, we first explain some concepts of last-mile logistics. Then, we give the formulation of the heterogeneous task-aware pick-up time prediction.

2.1 Preliminaries

Definition 1 (Delivery Station and Delivery Zone). In logistics scenarios, considering the delivery ability of the courier and the efficiency of package transfer, the city is divided into irregular areas (i.e., delivery stations) according to road networks, administrative divisions, and other geographic information. Each delivery station has a service point to store and transfer goods and has a certain number of couriers to finish delivery and transfer. The delivery station is further divided into delivery regions with finer granularity according to certain division rules, which are called Delivery Zone.

Definition 2 (AOI). AOI is the abbreviation of the area of interest. In a geographic information system, a plane enclosed by the boundary of buildings is called AOI. In the road Zone shown in Fig. 1, we regard a plane enclosed by the boundary of each building as the AOI of the building, denoted by A_i. A road zone $R_1 = \{A_1, A_2, ...A_n\}$ is actually a collection of AOI.

Definition 3 (Courier track). The routine work of one courier is traveling between AOI to deliver orders and pick up orders within AOI. We use the access order of AOI to represent the delivery trajectory of the courier, denoted as $route_i = \{A_i, A_{i+1}, ..., A_n | A_n \in R_1\}$. As shown in Fig. 1, the arrow direction of the lines indicates the AOI access sequence of the courier C_1, then the corresponding $route_1 = \{A_1, A_2, A_3, A_6, A_4, A_5, A_9, A_8, A_7 | A_n \in R_1\}$. The daily trajectory of the courier at the AOI level is relatively stable because the delivery scope of each courier is relatively stable.

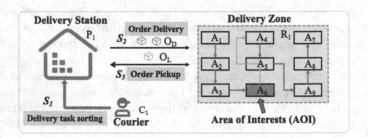

Fig. 1. Workflow of Last-mile Logistics

Definition 4 (Staying Time in AOI). We use $O_L^i = \{o_{l1}, o_{l2}, ...o_{ln}\}$ to represent all orders belonging to one AOI A_i. When the courier delivers orders in A_i, the courier keeps staying in A_i during this period until all orders in the AOI are finished. We define the stay time of A_i as $S_i = F(A_i) = \{max(o_{li}^t - o_{lj}^t) | o_{li}, o_{lj} \in O_L^i\}$.

Definition 5 (Workflow of Couriers). Figure 1 shows the workflow of a courier in last-mile delivery, where C_1 represents one courier, R_1 represents the delivery zone of the courier, which is composed of AOIs. $O_D = \{o_{d1}, o_{d2}, ...o_{dn}\}$

is the delivery tasks including multiple delivery orders, $O_L = \{o_{l1}, o_{l2}, ...o_{ln}\}$ represents the order set of pick-up requests, P_1 represents the location of the delivery station, which is the service point for express package sorting ($\mathbf{s_1}$), the starting point of the last-mile delivery ($\mathbf{s_2}$), and the end point of first-mile pick-up ($\mathbf{s_3}$).

2.2 Problem Formulation

In this paper, we aim to predict the package pick-up time when a new pick-up request occurs and is assigned to the courier. Given the courier's current location A_p, the spatial-temporal factors of the pick-up request in AOI A_q, and the existing route plan of unfinished delivery tasks $A^w = \{A_{p+1}, ..., A_n\}$, we estimate the pick-up time with the following function:

$$T_i = S_{(A_p, A_q)} + M_{(A_p, A_q)} \tag{1}$$

We define AOI set $A^r = \{A_i | A_i \in A^w\}$ as AOIs that couriers need to pass in courier's route plan from current location A_p to the AOI A_q of pick-up requests. $S_{(A_p, A_q)}$ represents the sum of the courier's stay time of AOIs in A^r. And $M_{(A_p, A_q)}$ represents the travel time in the road from A_p to A_q according to the courier's route plan, where the route is denoted as $route_{(A_p, A_q)} = \{A_p, A_{p+1}, ..., A_q | A_i \in R_1\}$.

3 Model Design

In this section, we first overview the framework of HTAPT. Then we give detailed descriptions of the pre-trained time prediction module and attention-enhanced route estimation module. Lastly, we introduce the training and inference of the proposed package pick-up time prediction framework.

3.1 Overview

Figure 2 elaborates the overview of HTAPT, which consists of three modules:

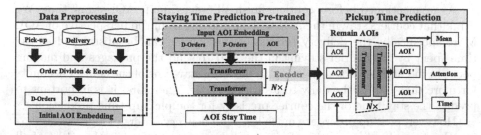

Fig. 2. Overview of HTAPT. In HTAPT, we first perform data preprocessing to integrate pickup order features, delivery order features, and AOI static features into initial AOI embeddings. Then, we propose a transformer-based encoder to pre-train the AOI embedding based on stay-time prediction. Lastly, we perform attention-enhanced pickup time and route prediction.

- *Order-AOI Matching Construction:* Given the sequence of delivery orders $\{o_{l1}, o_{l2}, ...o_{ln}\}$ of one courier, we first extract the binding relationship between orders and AOIs. Based on the geocoding service, we convert the text address information of orders into coordinates (i.e., latitude, longitude). Then, we establish the mapping between orders and AOIs by matching orders' coordinates to the boundary fence information of AOIs. However, due to the certain deviation of the geocoding service and the small fence area of AOI, some orders cannot match a specific AOI, resulting in the absence of AOI features, which eventually leads to an inaccurate prediction of staying time in residence.
- *Pre-trained Stay Time Prediction:* The stay time of the courier in each AOI is affected by the number of delivery tasks in the AOI, the static attributes of the AOI (e.g., number of floors), and the spatial-temporal factors of pick-up requests. As shown in Fig. 2, we obtain the embedding of AOIs using massive historical delivery orders based on a pre-trained mechanism, which embeds the influence of static features of the AOI and order quantity on the staying time. With the latent representation of each AOI extracted by the pre-trained AOI embedding module, we further enhance the accuracy of staying time prediction with pick-up orders in the fine-tuning process.
- *Attention Enhanced Pickup Time Prediction:* The main objective of this module is to predict the route plan of the courier and to infer the final pick-up time. Due to the random appearance of the collection request and the strong performance time constraint, it will affect the couriers' decision of which AOI to deliver next. Different delivery routes will consume different delivery time and the pick-up time of couriers will be different. Therefore it is essential to predict couriers' route considering the uncertainty of pick-up requests, which also help us to estimate the package pick-up time accurately.

3.2   Pre-trained Stay Time Prediction

Motivation for Stay Time Prediction. Through the data-driven analysis, we find that couriers usually deliver all orders of one AOI at one time and update the status information of each order when all orders of the AOI are finished, which helps us to aggregate delivery and pick-up tasks in the same AOI and enhance the matching efficiency between orders and AOIs.

As we mentioned before, couriers finish all orders in one AOI at one time. We think that couriers stay in the AOI without moving from the perspective of AOI, More specifically, the delivery process of the courier can be abstracted as the time spent by the courier staying in the AOI to send the packages and moving between the AOI. As is shown in Fig. 3, the daily stay time of the courier accounts for more than half of the daily total working time. Therefore, it is important to predict the stay time of the courier precisely for the pick-up time prediction.

However, pick-up orders are much less than delivery orders in logistics scenarios. For example, the quantity of pick-up orders accounts for less than 10% of all orders in Beijing. Therefore, according to the method of the traditional training model, a large number of orders need to be discarded because these orders are

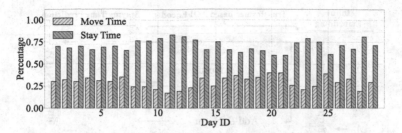

Fig. 3. Daily Stay Time and Moving Time of Couriers

not affected by pick-up requests. However, the pick-up requests mainly affect the route of the courier, and the impact on the stay time is limited. So we design a pre-trained module to leverage delivery orders that are not affected by pick-up to obtain the staying time-related embedding of each AOI.

Pre-trained Model for Stay Time Prediction. As is shown in Fig. 4, we divide the stay time into two parts, where A_i represents the AOI static attribute and O_i represents the order features. Specifically, static attributes of AOI consist of the height distribution (e.g., number of floors), area (e.g., delivery scope) of AOI, whether AOI is equipped with elevators, and whether AOI can enter. Contextual features such as weather conditions, traffic conditions, holidays, and day of the week are also considered AOI features. These attributes affect the delivery efficiency of residents of the AOI and thereby affecting their staying time in the AOI.

As for orders' features O_i, we consider both individual features and aggregate features of orders. The individual features of each order include the weight, volume, floor information, and order types (i.e., delivery orders or pick-up orders). Only considering the individual impact on the staying time of each order is not enough because of mutual influence among a batch of orders. Therefore we design the attention module to capture the mutual impact between orders. Specifically, we use $\mathbf{O} = \{O_1, O_2, \cdots, O_N | O_i \in A_i\}$ to represent orders belonging to the A_i and use $\vec{h}_i$ to represent the characteristics of order i. For A_i, we use $\mathbf{h} = \{\vec{h}_1, \vec{h}_2, ..., \vec{h}_N\}, \vec{h}_i \in \mathbb{R}^F$ to represent the characteristic sequence of nodes, where N is the number of nodes and F is the dimension of features in each node.

Transformer-Based AOI Encoding. To capture the relationship between different orders and enhance the feature expression ability, we design an attention-based AOI encode model with N transformer blocks. Each block consists of a multi-head attention layer and a feed-forward layer.

Firstly, we feed initial AOI embeddings into the multi-head attention layers. The multi-head attention mechanism has the superior ability to capture effects

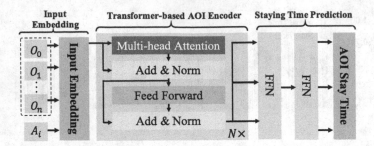

Fig. 4. Pre-trained Module for Stay Time Prediction

between orders from different perspectives. Each attention head is denoted as

$$
\begin{aligned}
head_i &= Attention(W^Q h_i, W^K h_i, W^V h_i) \\
&= softmax(\frac{W^Q h_i (W^K h_i)^T}{\sqrt{d_k}}) W^V h_i
\end{aligned}
\tag{2}
$$

where W^Q, W^K, and W^V are the parameters for the query, key, and value, respectively. d_k is the dimension of the key K and query vector Q. Then, the multi-head attention aggregation is calculated as

$$
MHA_i(h_1, h_2, ..., h_n) = \sum_{n=1}^{N} head_n W^O
\tag{3}
$$

where $head_n$ is the single-head attention. W^O is the parameter matrix. After the multi-head attention layer, we have a fully connected feed-forward network with ReLu activation functions in the feed-forward layer, which is defined as

$$
FF(\hat{h}_i) = W_{ff}^1 \times ReLu(W_{ff}^0 \times \hat{h}_i + b_{ff}^0) + b_{ff}^1
\tag{4}
$$

where W_{ff}^0, W_{ff}^1, b_{ff}^0, b_{ff}^1, W_{bn} and b_{bn} are parameters. $ReLu$ is the ReLu activation. After the N multi-head attention layers and feed-forward layers, we obtain the aggregated embedding of AOI representation h^{agg}. The AOI embeddings are pretrained through the staying time prediction task.

Pretrained Staying Time Prediction. We divide the daily delivery sequence of the courier in the pre-trained stage with fixed time intervals (i.e., one hour) because in one pick-up requests need to be finished in one hour in package pickup scenarios. With the multiple transformer encoder, we encode the AOI static features and different orders' features into a hidden representation h^{agg}.

Then, we leverage two feed-forward network layers to further transform the hidden representation of multiple features and to infer the final staying time of package pick-up requests. Lastly, based on MSE loss between the predicted staying time and the predicted staying time, the proposed pre-trained staying time prediction module captures the features of both AOI and orders and outputs the hidden representation for the pick-up time prediction task.

3.3   Attention Enhanced Route Estimation

Influence of Pick-Up Behaviors on Delivery Process. The biggest difference between pick-up and delivery tasks is that there are fulfillment time constraints (e.g., 1 h) on the pick-up process. Failing to finish the pick-up process within time constraints may lead to the cancellation of the pick-up request by the user and harm the user experience.

To complete the pick-up task within the specified time, the courier needs to adjust the route of existing delivery tasks. We utilize the editing distance [16] to represent the similarity of the daily delivery sequence of the courier and measure the stability of the couriers' daily delivery route. Noted that the peak period of pick-up requests starts at 10 o'clock, to capture the impacts of pick-up requests on delivery routes, we divide couriers' routes into two intervals (i.e., before 10 o'clock and after 10 o'clock) and calculate their edit distance, respectively.

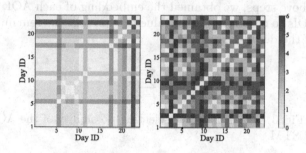

Fig. 5. Editing Distance of Courier Delivery Sequence

In Fig. 5, the left figure shows the editing distance of the courier's daily route sequence before 10 o'clock, with an average edit distance of 1.49. The right figure shows the editing distance after 10 o'clock, offering an average value of 3.44. To ensure fairness, the AOIs delivered by the courier are similar every day, and a smaller edit distance means the route sequences are more stable. Comparing the edit distance before 10 o'clock (without pick-up requests) and after 10 o'clock (with pick-up), we find that pick-up requests have a strong impact on couriers' delivery route and increase the probability of route change.

Attention Enhanced Pickup Time Prediction. We designed an attention [19] enhanced route prediction module to conduct route estimation in heterogeneous task-aware logistics. In Fig. 6, a courier stays in A_0 delivery packages and accepts a pick-up request in A_2. At this time, the AOI set that the courier needs to deliver is $D = \{A_1, A_2, A_3, A_4, A_5\}$ and the pick-up request of A_3 has already appeared. The aim is that we need to predict A_{i+1} that the courier will visit. We first feed the embedding of the current AOI sets of the courier into multiple transformer layers to capture the relationships between these AOIs.

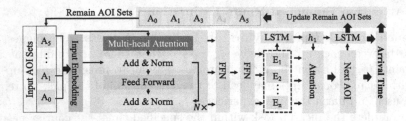

Fig. 6. Illustration of *Attention Enhanced Pickup Time Prediction*

$$\hat{A}_i = \mathbf{BN}^l(A_i^{l-1} + \text{MHA}_i^l(head_1^{l-1}, ..., head_{n_{head}}^{l-1}))$$
$$E_i = \mathbf{BN}^l(\hat{A}_i + FFN(\hat{A}_i)) \tag{5}$$

Through the above steps, we obtained the embedding of each AOI, and then we averaged all AOIs to obtain the mean value to represent the remaining aois, and as the input of the decoder layer.

$$\bar{E} = \frac{1}{n}\sum_{i=1}^{n} E_i \tag{6}$$

As is shown in Fig. 6, we calculate the selecting possibility of the AOI in D using the output of LSTM.

$$u_{(j)i} == \begin{cases} \alpha^T \cdot tanh(\mathbf{W_1}E_i \oplus \mathbf{W_2}h_1) & \text{if } i \neq \pi_{t'} \quad \forall t' < t \\ -\infty & \text{otherwise.} \end{cases} \tag{7}$$

where $\mathbf{W_1}$ and $\mathbf{W_1}$ and α^T are learn-able parameters. $u_{(j)i}$ is the attention score, which means the compatibility of couriers' arrival at this AOI. We finally use softmax to output the probability.

$$p_{(j)i} = p_\theta(\pi_t = i|s, \pi_{1:t-1}) = \frac{u_{(j)i}}{\sum_m u_{(j)m}} \tag{8}$$

At each step, our model predicts the next AOI to be visited based on selecting the possibility and gives the arrival time of the selected AOI.

$$t_{arrival} = \text{MLP}(E_i \oplus h_2) \tag{9}$$

Based on the next AOI selection and arriving time results, we update the remaining AOI set of the courier as well as the features of these AOI. The updated AOI set is fed into the proposed prediction model literately until the courier arrives at the pick-up AOI.

3.4 Training and Prediction

In the pre-trained stage, to keep the scale consistent between pick-up time pre-diction and the real-world pick-up time scale, we divide couriers' entire delivery process with a one-hour time slice. In the pre-trained stage, we use MSE as the loss function to predict staying time, which is defined as follows:

$$\ell(x,y) = L = \{l_1,\ldots,l_n\}^\top, \quad l_n = (x_n - y_n)^2 \tag{10}$$

Apart from the staying time, in the pick-up route prediction and pick-up time prediction stage, we utilize cross entropy as the loss function of route prediction, which is defined as follows:

$$L = \frac{1}{N}\sum_i L_i = -\frac{1}{N}\sum_i\sum_{c=1}^{M} y_{ic}\log(p_{ic}) \tag{11}$$

Lastly, we obtain the final package pick-up time prediction results with the attention-enhanced staying time prediction and pick-up route estimation.

4 Evaluation

In this section, we first show the data utilized in this work, followed by experi-mental settings and metrics. After that, the evaluation results are presented.

4.1 Dataset Description

To evaluate the performance of our work, we conduct experiments based on a real-world package pick-up and delivery dataset collected from one of the largest logistics companies in China. The dataset is collected from Oct. 1, 2021, to Apr. 1, 2022, and involves 100 thousand orders. Table 1 shows the key fields and examples of our datasets. The details are as follows:

- **Pick-up and delivery order data:** For the pick-up data, we record the order ID, courier id, order creation time, pick-up address, and the promised pick-up time. The main fields for historical delivery order data consist of order id, station id, courier id, and destination address.
- **Couriers' reporting data:** Couriers working for the platform are required to report the status of the orders in the last-mile delivery process, e.g., order collection and delivery time, with the Personal Digital Assistants (PDAs).
- **Spatial-temporal contexts:** We also obtain the AOI information as our spatial contexts, which consists of the AOI id and the geographical bound-aries. The temporal contextual information includes weather, day of the week, holiday, and traffic conditions.

Table 1. A sample of package pick-up and delivery Dataset

Pick-up Order	Order ID	Create time	Destination	Promised Time	Weight	Volume
	JDVA***21	2020/9/1 11:25	BeijingXXX	1 h	0.7 kg	10 cm * 20 cm * 10 cm
Delivery Order	Order ID	Courier ID	Destination	Delivered Time	Weight	Volume
	JDVA***21	226**21	BeijingXXX	2020/9/1 12:14	0.3 kg	10 cm * 20 cm * 40 cm
Reporting Data	Order ID	Courier ID	Timestamp	Status	longitude	latitude
	JDVA***21	226**21	2020/9/1 10:30	Pick-up	116.4460	39.9343
Contexts	AOI	Weather	Holiday	Traffic	AOI type	transportation
	POLYGON(.)	Rainy	National Day	busy	residence	wheelbarrow

4.2 Experimental Settings

Dataset Split: We used six months of data as the pre-trained part of the model, and then added the pick-up data as input for subsequent fine-tuning of the model. Use one month's data as a test for the model. The validation set used t-1 data to verify model accuracy. *Implementation:* We implement the model and baselines with Pytorch 1.10.2 and Python 3.8 environment and train these in an edge server with Intel(R) Xeon(R) CPU E5-2680 v4 @ 2.40 GHz (CPU) and one NVIDIA Tesla P40 (GPU).

4.3 Metrics

Route Prediction Metrics. We utilize edit distance [16], accuracy [27], and Kendall τ [12,25] as metrics to evaluate the delivery courier's route prediction.

- **Edit distance** is used to measure the similarity of two route sequences by counting the minimum number of operations, (i.e., insertion, deletion, and substitution for transferring one sequence into the other one).
- **Accuracy** is the difference between the actual and the predicted route and is calculated as $accuracy = \frac{\sum_{i=1}^{N} Diff(l_i^a, l_i^p)}{N}$, where N is the number of locations in the delivery courier's route. l_i^a and l_i^p are values in the i-th position of the delivery courier's actual and predicted route, respectively. $Diff(x,y) = 1$ if $x = y$, otherwise $Diff(x,y) = 0$.
- **Kendall rank correlation coefficient** is used to quantify the ordinal association between two sequences. Let $s = ((l_1, \hat{y}_1, y_1), (l_2, \hat{y}_2, y_2), ..., (l_n, \hat{y}_n, y_n))$ as a route, $\hat{y}_i$ and y_i are the positions of location l_i in the predicted and actual route, respectively. For arbitrary two different locations l_i and l_j, if $\hat{y}_i > \hat{y}_j$, and $y_i > y_j$, or if $\hat{y}_i < \hat{y}_j$, and $y_i < y_j$, then the two locations are concordant. Otherwise, they are discordant. The Kendall τ is defined as: $\tau = \frac{N_c - N_d}{N_c + N_d}$, where N_c and N_d are the numbers of concordant and discordant pairs, respectively.

Pick-Up Time Prediction Metrics. We utilize **RMSE** and **MAPE** to measure the performance of the package pick-up time prediction, which are widely used in estimating the time of arrival and mobility prediction tasks.

- **RMSE**: which computes root mean square error. If $\hat{y}_i$ is the predicted value of the ith sample and y_i is the corresponding true value. Then the root mean squared error (RMSE) estimated over n_{samples} is defined as

$$\text{RMSE}(y, \hat{y}) = \sqrt{\frac{1}{n_{\text{samples}}} \sum_{i=0}^{n_{\text{samples}}-1} (y_i - \hat{y}_i)^2}$$

- **MAPE**: which is the mean absolute percentage error. The idea of this metric is to be sensitive to relative errors. It is for example not changed by a global scaling of the target variable.

$$\text{MAPE}(y, \hat{y}) = \frac{1}{n_{\text{samples}}} \sum_{i=0}^{n_{\text{samples}}-1} \frac{|y_i - \hat{y}_i|}{\max(\epsilon, |y_i|)}$$

where ϵ is a positive number to avoid undefined results when y is zero.

4.4   Baselines

Baselines for Route Prediction: To evaluate the effectiveness of the route prediction module, we utilize the following three types of baselines: (i) Traditional ranking-based methods: In *Greedy-distance*, the courier selects the nearest AOI as the next stop each time. *XGBoost* [3] use the XBG Rank model to predict the next stop locations of the route. (ii) Optimization-based methods: In this category, we choose Google Ortools [15] to generate couriers' routes. (iii) Attention-based deep learning methods: FDNET [6] predicts the probability of each feasible location the driver will visit next based on RNN and attention modules. DeepRoute [23] predict couriers' future package pick-up routes according to the couriers' decision experience and preference.

Baselines for Pickup Time Prediction: We select three types of baselines: (i) four widely-used machine learning-based time prediction baselines, including LR, MLP, XGB [3], RF [2]. (ii) Three state-of-the-art learning-based baselines, including MLP, DeepTTE [20], and DeepMove [4]. (iii) Attention-based pickup time prediction methods, including FDNET [6] and DeepRoute [23].

4.5   Main Performance

Route Prediction Evaluation. We first give the route prediction performance comparisons between the proposed model and baselines. As shown in Table 2, we compare the route prediction performance of our proposed approach to other baselines under different settings (i.e., the number of Aois in each route). We found that HTAPT outperforms other baselines in both three metrics, offering the highest Accuracy, the highest Kendall, and the lowest Edit Distance, which demonstrates the effectiveness of the proposed method on accurate route prediction.

Table 2. Performance comparisons of route prediction

Metric	aois ∈ (0,4]			aois ∈ (4,8]			aois ∈ (8,12]		
	Edit Distance	Accuracy	τ	Edit Distance	Accuracy	τ	Edit Distance	Accuracy	τ
Greedy-distance	4.31	0.23	0.61	7.93	0.27	0.57	9.9	0.25	0.56
XGBoost Ranking	2.83	0.59	0.89	4.11	0.57	0.84	5.56	0.53	0.82
Google Ortools	2.61	0.63	0.88	3.83	0.61	0.85	4.9	0.55	0.81
FDNET	1.63	0.83	0.91	2.69	0.77	0.88	4.12	0.72	0.84
DeepRoute	1.57	0.87	0.93	2.47	0.79	0.91	3.78	0.75	0.88
HTAPT	**1.21**	**0.91**	**0.98**	**2.16**	**0.88**	**0.97**	**3.24**	**0.85**	**0.97**

Table 3. Performance comparisons of pick-up time prediction

datasets	Station1		Station2		Station3	
method	RMSE	MAPE	RMSE	MAPE	RMSE	MAPE
LR	10.1 ± 0.6	30.1 ± 0.8	12.4 ± 0.7	35.3 ± 0.9	11.7 ±1.5	30.5 ± 0.8
XGB	10.7 ± 0.4	27.6 ± 0.9	10.5 ± 0.7	25.9 ± 1.3	9.1 ± 0.5	24.5 ± 0.7
RF	11.3 ± 0.3	32.1 ± 1.8	11.9 ± 0.6	33.4 ± 1.5	10.3 ± 0.8	30.4 ± 1.3
MLP	14.8 ± 0.2	24.4 ± 1.1	12.8 ± 0.3	25.8 ± 0.7	15.6 ± 0.3	25.5 ± 1.6
DeepTTE	10.2 ± 0.5	25.3 ± 2.6	12.2 ± 0.7	27.2 ± 2.1	11.3 ± 0.7	26.3 ± 2.3
DeepMove	9.2 ± 0.6	21.7 ± 3.0	9.7 ± 0.7	22.7 ± 2.6	9.1± 0.7	20.7 ± 2.5
DeepRoute	7.5± 0.3	17.6 ± 1.1	7.1± 0.6	16.7 ± 1.2	7.3± 0.5	17.3 ± 1.2
FDNET	7.2± 0.5	17.1 ± 1.3	7.4± 0.7	17.6 ± 1.6	7.1± 0.3	16.7 ± 0.9
HTAPT	**6.3±0.3**	**13.4 ± 0.8**	**6.2 ± 0.2**	**13.6 ± 0.6**	**6.1 ± 0.4**	**12.7 ± 0.5**

Pick up Time Prediction Evaluation. As shown in Table 3, we evaluate the proposed method and other baselines in three representation stations with RMSE and MAPE metrics. We find that HTAPT achieves the lowest RMSE and the lowest MAPE in three stations. Specifically, compared with the FDNET model, HTAPT reduces around 12.5% of RMSE and reduces 21.6% of MAPE among these three stations.

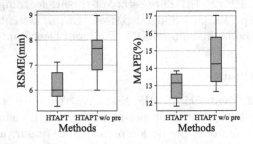

Fig. 7. Ablation study of the pre-trained module

4.6   Ablation Studies of the Pre-trained Module

To verify the effectiveness of the proposed Pre-trained Module, we compared HTAPT with its variant HTAPT w/o pre, which is the HTAPT without pre-trained module. As shown in Fig. 7, for our model, if the pre-trained process is removed, not only will the RMSE and MAPE decline, but also the stability of the model will decline. Compared with HTAPT w/o pre, HTAPT reduces 12.7% of RMSE and reduces 15.1% of MAPE.

4.7   Real-World Deployment

System Interface: We implement and deploy the HTAPT at the JD Logistics [11]. Figure 8 shows the interface of HTAPT plugged into the platform, which real-time displays the predicted pickup arrival time. When the user makes a pick-up request, the system will obtain the user's shipping address, the current location information of the courier, and the unfinished delivery order. Then according to the context information such as the traffic situation at that time, the door-to-door time of the courier will be calculated and pushed to the user to facilitate their schedule.

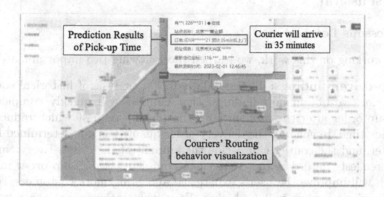

Fig. 8. System interface of HTAPT

Real-World Field Study: As is shown in Fig. 9, we provide a case study to show the effectiveness of HTAPT, which is performed in one representative delivery station in Beijing.

From Fig. 9, we find that HTAPT achieves accurate pick-up route prediction and offers a relatively low pick-up time prediction error. Through informing users of the specific pick-up time of the courier, users' waiting anxiety can be alleviated to a certain extent. After deployment, it is found that the number of users complaint has a certain degree of decline. In the courier order delivery recommendation system, the door-to-door pick-up time output by HTAPT helps the system to get a better delivery path.

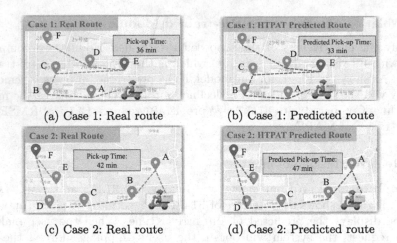

(a) Case 1: Real route   (b) Case 1: Predicted route

(c) Case 2: Real route   (d) Case 2: Predicted route

Fig. 9. Real-world case studies. We show two cases comparisons between ground-truth routes and predicted routes.

5 Discussion

5.1 Lessons Learned

During working on this work, we learned the following two important insights:

- **Data-driven findings:** Through data-driven analysis of historical couriers' behavior, we obtain new findings, that couriers will quickly complete the pick-up order after the user initiates the pick-up request which reduces the latency time. In general, the delivery order quantity is not determined by the courier, and the courier can decide the amount of the order. Every time the user requests an order, the courier can complete the pick-up order in time, which will improve the user's experience and bring more customers, so that the courier's income will be higher. From the station perspective, stations with higher average income generally have larger average K-values.
- **Couriers' behavior findings:** Due to the high mobility of couriers, the same road area may have different couriers in history. In the case of the same site manager and distribution range, the income of couriers varies greatly, and the analysis of the reason is mainly due to the change in the collection amount of couriers.

5.2 Limitations and Future Works

In this work, we focus on the courier pickup time prediction problem. However, there are two main limitations in this work:

- We have evaluated HTAPT with the data collected from delivery stations in Beijing, which may have a bias in other delivery stations due to different

densities of population and consumption ability. However, we believe HTAPT has the potential to perform well with attention considering the courier habit.
– We currently only predict courier pickup times in this paper. In the future, we think that based on the predicted pick-up time of couriers, we can output some decision-making methods to help the system schedule the appropriate couriers or help couriers arrange the appropriate delivery order, improve the work efficiency of couriers and the user experience at the same time.

6   Related Works

In this section, we introduce existing works related to our heterogeneous tasks aware package pick-up time prediction problem, including estimate time of arrival and route prediction problem in logistics and transportation systems.

6.1   Estimate Time of Arrival Applications

Wu et al. [26] propose a novel, spatial-temporal sequential neural network model to predict package delivery time and take full advantage of the sequential features of the latest delivery route and historical frequent and relative delivery patterns. Wang et al. [20] design an end-to-end multi-task Deep learning framework for Travel Time Estimation (called DeepTTE) to estimate the travel time of the whole path directly by taking both geographic spatial correlations and the temporal dependencies into consideration. Jun et al. [18] propose a KNN regressor to predict staying time in last-mile logistics, but it does not take into account the fine-grained floor information in waybills. Wang et al. [21] propose a Wide-Deep-Recurrent (WDR) learning model to accurately predict the travel time along a given route at a given departure time in ride-sharing services. Hong et al. [9] propose HetETA in ETA task taking structural spatial-temporal graph transportation data into account. Ruan et al. [17] a meta-learning-based neural network model to predict the service time of couriers in each AOI, and it is helpful for some downstream tasks (e.g., task assignment, route planning) in last-mile logistics. Arthur et al. [1] utilize an origin-destination (OD) formulation for the prediction of parcel delivery time.

6.2   Route Prediction in Transportation and Logistics Systems

Gao et al. [5] propose a deep network named FDNET to predict the probability of each feasible location the driver will visit next and predict the delivery time based on the routing probability through mining a large amount of food delivery data. Wen et al. [24] propose a novel attention-based encoder and decoder model to predict couriers' future package routes according to couriers' decision experience learned from their historical spatial-temporal behaviors. Jie et al. [4] design an attentional recurrent network for mobility prediction from lengthy and sparse trajectories leveraging a multi-modal sequential embedding recurrent neural network and a historical attention model to capture the multi-level periodicity. Zhu et al. [29] present the order fulfillment cycle time prediction model that incorporates representations of couriers, restaurants, and delivery destinations to enhance prediction efficacy. Zhou et al. [28] focus on fully leveraging multi-source data to improve the accuracy of route prediction and propose a multi-source data fusion framework. Zhang et al. [27] design multiple features to model the decision-making of individual couriers in instant delivery and predict couriers' routes with an Xgboost-ranking-based learning algorithm.

6.3   Uniqueness of Our Work

The heterogeneous tasks aware package pick-up time prediction problem is essentially different from existing time prediction works and estimated time of arrival applications in other transportation systems because (i) we conduct pick-up time prediction by taking *heterogeneous tasks' uncertain route* into consideration and disentangling the correlations between the pick-up process and delivery process. (ii) we design an attention-enhanced module to *model couriers' delivery behaviors and pick-up behaviors simultaneously* and leverage massive delivery data to address the challenges of shortage of pick-up data.

7   Conclusion

In this paper, we propose a `HTAPT`, a heterogeneous tasks aware package pick-up time prediction framework to estimate pick-up time considering couriers' delivery process via heterogeneous couriers' behaviors modeling. Specifically, `HTAPT` first designed a pre-trained stay time estimation module to predict couriers' stay duration within each AOI, which utilizes a pre-train mechanism leveraging massive delivery records to address the challenge of limited pick-up data. Then, `HTAPT` adopts an attention-enhanced network to conduct route prediction and output the final package pick-up time by integrating the above two modules. We evaluate our framework with a real-world logistics dataset, and the experimental results show that `HTAPT` outperforms state-of-the-art baselines.

Acknowledgments. This work was supported in part by Science and Technology Innovation 2030 - Major Project 2021ZD0114202, National Natural Science Foundation of China under Grant No. 62272098.

References

1. de Araujo, A.C., Etemad, A.: End-to-end prediction of parcel delivery time with deep learning for smart-city applications. IEEE Internet Things J. 8(23), 17043–17056 (2021)
2. Breiman, L.: Random forests. Mach. Learn. 45(1), 5–32 (2001)
3. Chen, T., Guestrin, C.: XGBoost: a scalable tree boosting system. In: Proceedings of the 22nd ACM SIGKDD International Conference on Knowledge Discovery and Data Mining, KDD 2016, pp. 785–794. Association for Computing Machinery, New York (2016)

4. Feng, J., et al.: DeepMove: predicting human mobility with attentional recurrent networks. In: Proceedings of the 2018 World Wide Web Conference, WWW 2018, pp. 1459–1468. International World Wide Web Conferences Steering Committee, Republic and Canton of Geneva, CHE (2018)
5. Gao, C., et al.: A deep learning method for route and time prediction in food delivery service. In: Zhu, F., Ooi, B.C., Miao, C. (eds.) The 27th ACM SIGKDD Conference on Knowledge Discovery and Data Mining, KDD 2021, Virtual Event, Singapore, 14–18 August 2021, pp. 2879–2889. ACM (2021)
6. Gao, C., et al.: A deep learning method for route and time prediction in food delivery service. In: Proceedings of the 27th ACM SIGKDD Conference on Knowledge Discovery & Data Mining, pp. 2879–2889 (2021)
7. Gao, C., et al.: Applying deep learning based probabilistic forecasting to food preparation time for on-demand delivery service. In: Zhang, A., Rangwala, H. (eds.) The 28th ACM SIGKDD Conference on Knowledge Discovery and Data Mining, KDD 2022, Washington, DC, USA, 14–18 August 2022, pp. 2924–2934. ACM (2022)
8. Guo, B., et al.: Towards equitable assignment: Data-driven delivery zone partition at last-mile logistics. In: Proceedings of the 29th ACM SIGKDD Conference on Knowledge Discovery and Data Mining, pp. 4078–4088 (2023)
9. Hong, H., et al.: HetETA: heterogeneous information network embedding for estimating time of arrival. In: Proceedings of the 26th ACM SIGKDD International Conference on Knowledge Discovery and Data Mining, KDD 2020 pp. 2444–2454. Association for Computing Machinery, New York (2020)
10. Hong, Z., et al.: CoMiner: nationwide behavior-driven unsupervised spatial coordinate mining from uncertain delivery events. In: Proceedings of the 30th International Conference on Advances in Geographic Information Systems, pp. 1–10 (2022)
11. JD Logistics: JD logistics (2022). https://www.jdl.com/
12. Kendall, M.G.: A new measure of rank correlation. Biometrika **30**(1/2), 81–93 (1938)
13. Li, Q., Zheng, Y., Xie, X., Chen, Y., Liu, W., Ma, W.Y.: Mining user similarity based on location history. In: Proceedings of the 16th ACM SIGSPATIAL International Conference on Advances in Geographic Information Systems, GIS 2008. Association for Computing Machinery, New York (2008)
14. Mesa, J.P., Montoya, A., Toro, M., et al.: A two-stage data-driven metaheuristic to predict last-mile delivery route sequences. Eng. Appl. Artif. Intell. **125**, 106653 (2023)
15. Perron, L., Furnon, V.: OR-Tools. https://developers.google.com/optimization/
16. Ristad, E.S., Yianilos, P.N.: Learning string-edit distance. IEEE Trans. Pattern Anal. Mach. Intell. **20**(5), 522–532 (1998)
17. Ruan, S., et al.: Service time prediction for delivery tasks via spatial meta-learning. In: Zhang, A., Rangwala, H. (eds.) The 28th ACM SIGKDD Conference on Knowledge Discovery and Data Mining, KDD 2022, Washington, DC, USA, 14–18 August 2022, pp. 3829–3837. ACM (2022)
18. Song, J., Wen, R., Xu, C., Tay, J.W.E.: Service time prediction for last-yard delivery. In: 2019 IEEE International Conference on Big Data (Big Data), pp. 3933–3938 (2019). https://doi.org/10.1109/BigData47090.2019.9005585
19. Vaswani, A., et al.: Attention is all you need. In: Advances in Neural Information Processing Systems, vol. 30 (2017)

20. Wang, D., Zhang, J., Cao, W., Li, J., Zheng, Y.: When will you arrive? Estimating travel time based on deep neural networks. In: McIlraith, S.A., Weinberger, K.Q. (eds.) Proceedings of the Thirty-Second AAAI Conference on Artificial Intelligence, (AAAI-18), The 30th Innovative Applications of Artificial Intelligence (IAAI-18), and The 8th AAAI Symposium on Educational Advances in Artificial Intelligence (EAAI-18), New Orleans, Louisiana, USA, 2–7 February 2018, pp. 2500–2507. AAAI Press (2018)

21. Wang, Z., Fu, K., Ye, J.: Learning to estimate the travel time. In: Proceedings of the 24th ACM SIGKDD International Conference on Knowledge Discovery and Data Mining, KDD 2018, pp. 858–866. Association for Computing Machinery, New York (2018)

22. Wen, H., et al.: Graph2Route: a dynamic spatial-temporal graph neural network for pick-up and delivery route prediction. In: Proceedings of the 28th ACM SIGKDD Conference on Knowledge Discovery and Data Mining, KDD 2022, pp. 4143–4152. Association for Computing Machinery, New York (2022)

23. Wen, H., et al.: DeepRoute+: modeling couriers' spatial-temporal behaviors and decision preferences for package pick-up route prediction. ACM Trans. Intell. Syst. Technol. 13(2), 1–23 (2022)

24. Wen, H., et al.: Package pick-up route prediction via modeling couriers' spatial-temporal behaviors. In: 37th IEEE International Conference on Data Engineering, ICDE 2021, Chania, Greece, 19–22 April 2021, pp. 2141–2146. IEEE (2021)

25. Wen, H., et al.: Package pick-up route prediction via modeling couriers' spatial-temporal behaviors. In: 2021 IEEE 37th International Conference on Data Engineering (ICDE), pp. 2141–2146. IEEE (2021)

26. Wu, F., Wu, L.: DeepETA: a spatial-temporal sequential neural network model for estimating time of arrival in package delivery system. In: The Thirty-Third AAAI Conference on Artificial Intelligence, AAAI 2019, The Thirty-First Innovative Applications of Artificial Intelligence Conference, IAAI 2019, The Ninth AAAI Symposium on Educational Advances in Artificial Intelligence, EAAI 2019, Honolulu, Hawaii, USA, 27 January–1 February 2019, pp. 774–781. AAAI Press (2019)

27. Zhang, Y., et al.: Route prediction for instant delivery. In: Proceedings of the ACM on Interactive, Mobile, Wearable and Ubiquitous Technologies, vol. 3, no. 3, pp. 1–25 (2019)

28. Zhou, Z., Zhou, X., Lu, Y., Yan, H., Guo, B., Wang, S.: Multi-source data-driven route prediction for instant delivery. In: 2021 17th International Conference on Mobility, Sensing and Networking (MSN), pp. 374–381 (2021). https://doi.org/10.1109/MSN53354.2021.00064

29. Zhu, L., et al.: Order fulfillment cycle time estimation for on-demand food delivery. In: Gupta, R., Liu, Y., Tang, J., Prakash, B.A. (eds.) The 26th ACM SIGKDD Conference on Knowledge Discovery and Data Mining, KDD 2020, Virtual Event, CA, USA, 23–27 August 2020, pp. 2571–2580. ACM (2020)

Optimizing Pointwise Convolutions
on Multi-core DSPs

Yang Wang[1,2,3], Qinglin Wang[1,2](✉) (iD), Xiangdong Pei[1,2], Songzhu Mei[1],
and Jie Liu[1,2]

[1] National Key Laboratory of Parallel and Distributed Computing, National
University of Defense Technology, Changsha 410073, China
[2] Laboratory of Digitizing Software for Frontier Equipment, National University of
Defence Technology, Changsha 410073, China
wangqinglin.thu@gmail.com
[3] Beijing Institute of Astronautical Systems Engineering, Beijing 100076, China

Abstract. Pointwise convolutions are widely used in various convolutional neural networks, due to low computation complexity and parameter requirements. However, pointwise convolutions are still time-consuming like regular convolutions. As a result of increasing power consumption, low-power embedded processors have been brought into high-performance computing field, such as multi-core digital signal processors (DSPs). In this paper, we propose a high-performance multi-level parallel direct implementation of pointwise convolutions on multi-core DSPs in FT-M7032, a CPU-DSP heterogeneous prototype processor. The main optimizations include on-chip memory blocking, loop ordering, vectorization, register blocking, and multi-core parallelization. The experimental results show that the proposed direct implementation achieves much better performance than GEMM-based ones on FT-M7032, and a speedup of up to 79.26 times is achieved.

Keywords: CNNs · Pointwise Convolution · Direct Convolution · DSPs · Parallel algorithm

1 Introduction and Related Work

Convolutional neural networks (CNNs) are extensively used in diverse fields such as computer vision and scientific computing [3,4,15,28]. As CNNs develop, more convolutional layers with small filters are applied in the models, such as pointwise convolutions in which the filter size is only 1×1. And this type of convolutional layer is commonly utilized in mainstream backbone networks, such as ResNet [6] and GoogleNet [21], and lightweight networks, such as MobileNetV1 [8] and MobileNetV2 [20]. Thus, it is very important to implement high-performance pointwise convolutions on targeted platforms.

supported by the National Natural Science Foundation of China under grant nos. 62002365 and 62025208.

The dominant methods for implementing convolutions are matrix multiplication-based, Winograd-based, Fast Fourier Transform (FFT)-based, and direct algorithms [2,5,7,9,10,12,22,23,26]. For the matrix multiplication-based method, the convolutions are converted into matrix multiplication operations in an explicit or implicit way. For example, Wang et al. [26] implemented two-dimensional convolutions using implicit matrix multiplication. Thus, the performance of convolutions largely relies on the performance of matrix multiplication on hardware platforms in this method. The fast methods including Winograd-based and FFT-based ones can effectively decrease the computational complexity of convolutions, while they are only applicable to convolutions with large filters. Since the direct method has no extra memory overhead and can gain high performance, numerous direct implementations for various types of convolutions have been proposed on different platforms, such as regular convolutions on Intel CPUs [7] and ARM Mali GPUs [16]. Lu et al. proposed two novel optimization techniques to improve the performance of pointwise convolutions by enhancing data reuse in row and column directions on NVIDIA mobile graphics processing units (GPUs) [13,14]. Wang et al. proposed a parallel direct algorithm for pointwise convolutions on ARMv8 multi-core CPUs [24]. However, there is little work on the direct implementation of pointwise convolutions on multi-core DSPs.

Multi-core digital signal processors (DSPs) have been brought into the high-performance computing field due to the low-power characteristic [11]. To diminish power consumption, DSPs usually adopt Very Long Instruction Word (VLIW) architecture, software-controlled on-chip memories, and Direct Memory Access (DMA) engines for data moving, which are unique and different from the architectures of modern CPUs and GPUs. There have been many parallel implementations of algorithms and applications on multi-core DSPs, such as matrix multiplications [19,27], matrix transpose [18], and GEMM-based convolutions [25], but the parallel direct optimization of pointwise convolutions targeting multi-core DSPs has not been found.

FT-M7032 is a CPU-DSP heterogeneous prototype processor which consists of one 16-core ARMv8 CPU for process management and four 8-core DSPs for offering major peak performance [27]. To improve the performance of pointwise convolutions on FT-M7032, this paper proposes a high-performance parallel direct implementation for pointwise convolutions targeting multi-core DSPs. In parallelization, many common optimization techniques are carried out, such as vectorization, register blocking, and multi-core parallelization. The experimental results demonstrate that the direct implementation gets the computation efficiency of 11.42% - 58.61% and outperforms the GEMM-based one with speedups of $1.43\times-79.26\times$ on multi-core DSPs in FT-M7032. Compared with the implementations in Pytorch [17] and ARM Computer Library [1] running on the ARMv8 CPU in FT-M7032, the proposed direct implementation gets a speedup of up to 35.84 times. To the best of our knowledge, this is the first work about the direct parallelization of pointwise convolutions on multi-core DSPs.

The structure of this paper is as follows. Section 2 outlines the definition of pointwise convolutions and the architecture of FT-M7032 processors. Section 3 describes our parallel direct implementation of pointwise convolutions on multi-core DSPs in FT-M7032 processors in detail. Section 4 shows the analyses of the performance results. Last, the conclusion and future work are given in Sect. 5.

2   Backgound

2.1   Pointwise Convolution

For the forward propagation pass, pointwise convolutions work on input feature maps tensor I with filters tensor F to produce output feature maps tensor O. The backward propagation and weight gradient update passes obtain the input feature maps gradient tensor dI and filter gradient tensor dF based on output feature maps gradient tensor dO, respectively. With blocked data layout which is very beneficial to vectorization, the three passes above of pointwise convolutions are figured by Eqs. 1, 2, and 3.

$$O_{n,k_d,h_o,w_o,k_l} \mathrel{+}= I_{n,c_d,h_o \times S,w_o \times S,c_l} \times F_{c_d \times L + c_l, k_d, 0, 0, k_l}, \tag{1}$$

$$dI_{n,c_d,h_o \times S,w_o \times S,c_l} \mathrel{+}= dO_{n,k_d,h_o,w_o,k_l} \times F_{c_d \times L + c_l, k_d, 0, 0, k_l}, \tag{2}$$

$$dF_{c_d \times L + c_l, k_d, 0, 0, k_l} \mathrel{+}= dO_{n,k_d,h_o,w_o,k_l} \times I_{n,c_d,h_o \times S,w_o \times S,c_l}, \tag{3}$$

where $n \in [0, N)$, $k_d \in [0, K_d)$, $h_o \in [0, H_o)$, $w_o \in [0, W_o)$, $k_l \in [0, L)$, $c_d \in [0, C_d)$, $c_l \in [0, L)$, N is the mini-batch size, C and K are the number of input and output channels, C_d and K_d represent the number of blocks in C and K dimensions, $C = C_d \times L$, $K = K_d \times L$, L is the number of lanes in vector units of DSPs, $H_{i/o}$ and $W_{i/o}$ denotes the spatial dimensions of different tensors, and S is the stride size. In this paper, only the unit-stride pointwise convolutions are involved so the stride size is 1 in the following.

2.2   Architecture of FT-M7032 Heterogeneous Processors

An FT-M7032 heterogeneous processor consists of a 16-core ARMv8 CPU and four GDPSP clusters, shown in Fig. 1. The 16-core CPU where the Linux operating system runs is mainly for process management and multi-node communication, and its single-precision peak performance is 281.6 GFlops with 2.2 GHz working frequency. Each GPDSP cluster, also called a multi-core DSP, includes eight DSP cores and global shared memory (GSM), which is connected by an on-chip crossbar network. Each core can offer 345.6 GFlops single-precision peak performance with 1.8 GHz working frequency so that the total peak performance of each GPDSP cluster can achieve up to 2764.8 GFlops. The 16-core CPU and

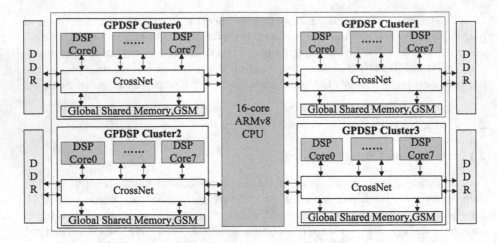

Fig. 1. Architecture of FT-M7032 Processors

four GPDSP clusters share the same memory space. Specifically, the CPU can access the whole main memory space in FT-M7032, while each GPDSP cluster can only access a specific part with 42.6 GBytes/s bandwidth. Therefore, four GPDSP clusters can communicate with each other via the CPU, and be mainly utilized by process-level parallelization.

The micro-architecture of each DSP core is shown in Fig. 2. Each core primarily includes a scalar processing unit (SPU), a vector processing unit (VPU), an instruction dispatch unit (IFU), and a DMA engine. SPU is used to support parallel execution of five scalar instructions, where the size of scalar memory (SM) is 64 KB. VPU is applied to carry out vector instructions, and the capacity of array memory (AM) is 768 KB. There are three 64-bit float-point fused multiply-add (FMAC) units in each of 16 vector processing elements (VPEs), so VPU can perform three vector 32-bit FMAC (VFMAC) operations with 32 lanes per cycle. VPU also has two parallel vector load-store units (VLoad/VStore), each of which can convey data of up to 2048 bytes per cycle between AM and vector registers. There are 64 1024-bit vector registers in total. SPU can directly transfer data to VPU through broadcast operations and shared registers. These DSP cores adopt VLIW architecture, and IFU can issue up to 11 instructions per cycle, including at most five scalar instructions and six vector instructions. The DMA engine is in charge of fast data transmission between different memories.

3   Parallel Direct Implementation

3.1   Overview of Our Implementation

Pointwise convolutions are computationally equivalent to matrix multiplication. Therefore, when directly mapping pointwise convolutions on multi-core CPUs and GPUs, the optimization methods for matrix multiplication are carried out

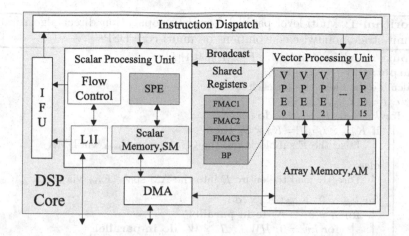

Fig. 2. Micro-architecture of each DSP core in FT-M7032 Processors

for high efficiency. This paper also follows this rule above and incorporates the architectural features of the GPDSP cluster in the FT-M7032 and the relatively small number of parameters in pointwise convolution for targeted algorithm design and optimization.

3.2 Multi-level Parallel Forward Propagation Algorithm

When the stride size is 1, the spatial dimensions H and W of feature maps can be merged into a single dimension denoted as $H \times W$. In this section, we propose a multi-level parallel direct algorithm named directConv1x1Fwd() for computing the forward propagation pass of pointwise convolutions in convolutional neural networks, shown in Algorithm 1. The implementation of the Conv1x1FwdAsm kernel function within directConv1x1Fwd() is presented in Algorithm 2. Since the storage cost of the filter tensor F in pointwise convolutions is typically low, directConv1x1Fwd() prioritizes loading F into the on-chip AM space or GSM space. To accommodate the unit-strided convolutions, directConv1x1Fwd() merges the dimensions H and W directly into one (Line 10). In the following, we primarily employ directConv1x1Fwd() as an exemplar to elucidate the meticulous design of a multi-level parallel forward propagation algorithm for realizing high-performance pointwise convolution.

On-Chip Memory Blocking and Loop Ordering. GPDSP clusters are equipped with on-chip storage spaces, namely SM, AM, and GSM spaces. In order to achieve high-performance computing objectives, algorithms commonly load relevant tensor data into these spaces in blocking format prior to performing calculations using the on-chip data. Furthermore, loop ordering is necessary to optimize the locality of the on-chip data within the storage space and reduce the overhead of accessing off-chip DDR storage.

Algorithm 1: Multi-level parallel forward propagation direct algorithm for unit-stride pointwise convolutions on multi-core DSPs

---

Input : $I[N][C_d][H_i][W_i][L]$, $F[C][K_d][1][1][L]$
Output: $O[N][K_d][H_o][W_o][L]$

1 Calculate the block size for each level
2 **for** $c_{gd} = 0 \colon C_{dgb} \colon C_d$ **do**
3   **for** $k_{dg} = 0 \colon K_{dgb} \colon K_d$ **do**
4   **if** $K_{dab} \times C_{dab} \mathrel{!=} K_d \times C_d$ **then**
5   Load the F subblock into the GSM space F_{gsm} via DMA
6   **else**
7   Directly load the entire F into the AM space F_{am} via DMA
8   **for** $k_{da} = 0 \colon K_{dab} \colon K_{dgb}$ **do**
9   **for** $n = 0 \colon 1 \colon N$ **do in parallel**
10   **for** $hw = 0 \colon HW_{ob} \colon H_o \times W_o$ **do in parallel**
11   **if** $c_{gd} \mathrel{!=} 0$ **then**
12   Load the O subblock into the AM space O_{am} via DMA
13   **for** $c_{da} = 0 \colon C_{dab} \colon C_{dgb}$ **do**
14   **if** $K_{dab} \times C_{dab} \mathrel{!=} K_d \times C_d$ **then**
15   Load the F_{gsm} subblock into the AM space F_{am} via DMA
16   **for** $c_{ds} = 0 \colon C_{dsb} \colon C_{dab}$ **do**
17   Load the I subblock into the SM space I_{sm} via DMA
18   Call Conv1x1FwdAsm()
19   Store the O_{am} subblock in the DDR space O via DMA

Within the design of directConv1x1Fwd(), the SM space stores the blocking data of the input feature tensor I, while the AM space accommodates the blocking data of both the output feature tensor O and the filter tensor F. By prioritizing the loading of the filter tensor F as a whole, this design utilizes the GSM space to buffer the blocking data of F. In this section, the subscripts sm, am, and gsm indicate the on-chip storage space positions of the tensors corresponding to the SM, AM, and GSM spaces, respectively. To load the relevant subblocks into their respective on-chip storage spaces, the corresponding dimensions of the filter tensor, input feature tensor, and output feature tensor must be divided for the GSM, SM, and AM spaces, labeled with the subscripts gb, sb, and ab, respectively.

In the directConv1x1Fwd() algorithm, the blocking data of F is stored in the GSM space, while the blocking data of tensors I and O need to be loaded from DDR to the SM space and AM space, respectively, during internal iterative calculations. To prevent simultaneous reading of both tensors from DDR during calculation, this section establishes the conditions $HW_{ab} = HW_{sb} = HW_{ob}$ and

$C_{dsb} \leqslant C_{dab}$ to balance the block parameters of the SM space and the AM space. In total, we derive the on-chip storage blocking limit conditions as presented in Eq. 4.

$$
\begin{aligned}
sizeof(\boldsymbol{F_{gsm}}) &\leqslant sizeof(\text{GSM}) \\
sizeof(\boldsymbol{I_{sm}}) &\leqslant sizeof(\text{SM}) \\
sizeof(\text{AM}) &\geqslant sizeof(\boldsymbol{O_{am}}) + sizeof(\boldsymbol{F_{am}}) \\
sizeof(\boldsymbol{F_{gsm}}) &= C_{dgb} \times K_{dgb} \times L \times L \\
sizeof(\boldsymbol{I_{sm}}) &= C_{dsb} \times HW_{ob} \times L \\
sizeof(\boldsymbol{O_{am}}) &= K_{dab} \times HW_{ob} \times L \\
sizeof(\boldsymbol{F_{am}}) &= K_{dab} \times C_{dab} \times L \times L \\
C_{dsb} &\leqslant C_{dab} \leqslant C_{dgb} \leqslant C_d \\
K_{dab} &\leqslant K_{dgb} \leqslant K_d \\
HW_{ob} &\leqslant H_o \times W_o
\end{aligned}
\tag{4}
$$

To optimize the data locality in on-chip memories, the loop order in the original direct implementation of pointwise convolution was rearranged to achieve the loop order in directConv1x1Fwd(). The outermost two loops, c_{gd} and k_{dg}, are utilized to load the largest subblock of $\boldsymbol{F}$ into on-chip storage at once. If the size of $\boldsymbol{F}$ does not match the size of the allocated AM space for $\boldsymbol{F_{am}}$, i.e., $C_{kad} \times K_{dab} \neq K_d \times C_d$, then the $\boldsymbol{F}$ subblock will be cached in the GSM space $\boldsymbol{F_{gsm}}$ using DMA (Line 5). Otherwise, $\boldsymbol{F}$ will be directly loaded into the AM space $\boldsymbol{F_{am}}$ (Line 7). The three loops, k_{da}, n, and hw, are employed to load and store the output feature map tensor $\boldsymbol{O}$, followed by the loop c_{da} to determine the subblock of $\boldsymbol{F_{gsm}}$ that needs to be loaded into the AM space $\boldsymbol{F_{am}}$. The innermost loop, c_{ds}, is used to identify the subblock of the input feature map tensor $\boldsymbol{I}$ that must be loaded from DDR space to the SM space $\boldsymbol{I_{sm}}$. Within the c_{ds} loop, a subblock of $\boldsymbol{I}$ is loaded into the SM space using DMA, and then Conv1x1FwdAsm() is called once with the loaded data to perform the calculation.

Vectorization and Register Blocking. The employed second optimization technique is vectorization and register blocking. Once the relevant subblocks of tensors are loaded into the SM and AM spaces, effectively utilizing the execution units within a single DSP core to reduce computational costs becomes a critical concern. The objective of this approach is to minimize the runtime of Conv1x1FwdAsm() by maximizing the computational capacity of each DSP core. Specifically, it utilizes vectorization to harness the power of the 16 parallel VPEs in the VPU of each DSP core. Furthermore, register blocking techniques are utilized to conceal the pipeline latency of the VPU's execution units and take advantage of multiple vector floating-point multiply-add fusion units (VFMAC) within the VPU.

Vectorization is applied along the K dimension where the calculation associated with each element is independent, and there are L consecutive elements when accessing the K dimension of the related tensors (O and F). To enhance data locality in registers, this method employs register blocking in the K_{dab}, HW_{ob}, and C dimensions, as described in Algorithm 2, and fully unrolls the cc, j, and i loops (Lines 8, 9, and 11) to conceal pipeline latency. The implementation of register blocking is subject to limitations imposed by the number of registers and the pipeline latency of the relevant functional units, as specified in Eq. 5, where $\text{Latency}_{\text{VFMAC}}$ and $\text{Latency}_{\text{FP32Bcast}}$ represent the latency time of the VFMAC units and FP32 Broadcasting, and $\text{Num}_{\text{VFMAC}}$ represents the number of the VFMAC units in VPUs of DSP cores.

Algorithm 2: Vectorized algorithm for the forward propagation of point-wise convolutions based on SPU and VPU in each DSP core

---

Input : $I_{sm}[C_{dsb}][HW_b][L]$, $F_{am}[C_{dsb} \times L][K_{dab}][L]$
Output: $O_{am}[K_{dab}][HW_b][L]$

1 for $k_d = 0: K_{drb}: K_{dab}$ do
2   for $hw = 0: HW_{rb}: HW_{ob}$ do
    // Load the $O_{am,k_d,hw}$ subblock into the vector register
3   for $j = 0: 1: K_{drb}$ do
4   for $i = 0: 1: HW_{rb}$ do
5          $\text{VR}_{j \times HW_{rb}+i} = \text{VLoad}(O_{am,k_d+j,hw+i})$

6   for $cb = 0: 1: C_{dsb}$ do
7   for $cl = 0: C_{lrb}: L$ do
        // The following loop will be fully unrolled in the assembly implementation
8   for $cc = cl: 1: cl + C_{lrb}$ do
9   for $j = 0: 1: K_{drb}$ do
10             $\text{VR}_f = \text{VLoad}(F_{am,cb \times L+cc,k_d+i})$
11   for $i = 0: 1: HW_{rb}$ do
12               $\text{VR}_s = \text{SVBcast}(\text{FEXT}(\text{SLoad}((I_{sm,cb,hw+j,cc}))))$
13               $\text{VR}_{j \times HW_{rb}+i} = \text{VFMAC}(\text{VR}_f, \text{VR}_s, \text{VR}_{j \times HW_{rb}+i})$

    // Store the data in the vector register to $O_{am,k_d,hw}$
14   for $j = 0: 1: K_{drb}$ do
15   for $i = 0: 1: HW_{rb}$ do
16         $\text{VStore}(\text{VR}_{j \times HW_{rb}+i}, O_{am,k_d+j,hw+i})$

$$K_{drb} \times HW_{rb} \geqslant \text{Latency}_{\text{VFMAC}} \times \text{Num}_{\text{VFMAC}}$$
$$K_{drb} \times HW_{rb} \times C_{lrb} \geqslant \text{Latency}_{\text{FP32Bcast}} \times \text{Num}_{\text{VFMAC}} \tag{5}$$

Multi-core Parallelization and Blocking Size Calculation. The third optimization method involves distributing tasks on multiple DSP cores and

determining the appropriate block sizes for computation. In the algorithm for
multi-level parallel implementation of pointwise convolution forward propagation, the calculation tasks are partitioned based on two loops: n and hw. A task
pool is created, where each DSP core independently handles a task from the
pool. The tasks from the task pool are processed in parallel by eight DSP cores
until all tasks are completed.

In the previous parts, we have discussed the constraints that govern the
blocking sizes of on-chip and register storage in this study. However, determining the appropriate block sizes remains an unresolved issue. The selected
blocking sizes not only affect the efficiency of tensor access but also influence
the overall data communication between off-chip and on-chip memories in the
directConv1x1Fwd() algorithm. In the deep neural network library for the FT-
M7032 heterogeneous general-purpose multi-core DSP, tensors are stored in the
row-major format. After applying blocking, tensors require cross-stride reading.
Larger blocking sizes in the tensor's inner dimensions facilitate more efficient
access when using cross-stride reading. The Eq. 6 presents the calculation of the
total amount of data transferred between off-chip and on-chip storage in direct-
Conv1x1Fwd(), where $sizeof(\boldsymbol{F})$, $sizeof(\boldsymbol{I})$, and $sizeof(\boldsymbol{O})$ denote the sizes of
tensors $\boldsymbol{F}$, $\boldsymbol{I}$, and $\boldsymbol{O}$, respectively. Therefore, we calculate the blocking size in
directConv1x1Fwd() while satisfying the conditions specified in Eqs. 4 and 5,
guided by the following three principles. First, ensure that the larger blocking
parameter is an integer multiple of the smaller blocking sizes (e.g., HW_{ob} must be
an integer multiple of HW_{rb}). Second, minimize the value of $Total_{conv1x1FwdS1}$
as much as possible. Third, maximize the blocking size of the tensor's inner
dimensions.

$$\text{Total}_{\text{conv1x1FwdS1}} = sizeof(\boldsymbol{F}) + \frac{K_d}{K_{dab}} \times sizeof(\boldsymbol{I}) + sizeof(\frac{C_d}{C_{dgb}}) \times \boldsymbol{O} \quad (6)$$

3.3   Multi-level Parallel Algorithms for Backward Propagation and Weight Gradient Update Propagation

The backward propagation pass of pointwise convolution involves taking the
output feature map gradient $\boldsymbol{dO}$ and the convolution kernel $\boldsymbol{F}$ as input tensors
and generating the input feature map gradient $\boldsymbol{dI}$ as the output tensor, shown
in Eq. 2. The filter gradient $\boldsymbol{dF}$ is computed from the output feature maps
gradient $\boldsymbol{dO}$ and the input feature maps $\boldsymbol{I}$ in the weight gradient update pass,
shown in Eq. 3. The computational mode of the two passes above also is the
matrix multiplication. Compared to the forward propagation pass, the main
difference is that the matrix multiplications involve the matrix transposition in
these two passes. Therefore, we get the multi-level parallel direct algorithms for
the left two passes of pointwise convolutions, based on the parallel optimization
approaches described in Sect. 3.2 and the vectorization matrix transpose kernel
trnKernel-32 on multi-core DSPs proposed in [18].

4 Performance Evaluation

This section gives the test results of our direct implementation on multi-core DSPs and compares it with other implementations of pointwise convolutions on FT-M7032.

4.1 Experiment Setup

We chose ResNet50 [6] and MobileNetV1 [8] as representatives of widely-used backbone networks and lightweight networks, respectively. The performance of the pointwise convolution implementation is evaluated by employing the point-wise convolution layers from these models. The specific configurations are presented in Table 1. For the pointwise convolution tests, a batch size N of 64 is used for all tested network layers.

This subsection introduces three metrics, namely computing time T_{conv}, computing performance P_{conv}, and computing efficiency E_{conv}, to evaluate the performance of convolution implementations. The relation among these metrics is outlined in Eq. 7. P_{peak} represents the peak performance of a given hardware platform, such as a single GPDSP cluster and a 16-core ARMv8 CPU. Additionally, $TotalOp_{conv}$ is the total floating-point operations involved in the convolution computation. For pointwise convolutions, the formula for $TotalOp_{conv}$ is given by $2 \times N \times K \times H_o \times W_o \times C \times 1 \times 1$.

$$P_{conv} = \frac{TotalOp_{conv}}{T_{conv}},$$
$$E_{conv} = \frac{P_{conv}}{P_{peak}}. \tag{7}$$

4.2 Performance

This section compares the direct implementation of pointwise convolutions with two GEMM-based implementations on FT-M7032. The first is a GEMM-based implementation method optimized for multi-core DSPs [25], in which matrix multiplication and all tensor transformations run on multi-core DSPs. The second is the GEMM-based implementation in Pytorch [17], which runs solely on the 16-core ARMv8 CPU of FT-M7032. These two GEMM-based implementations are referred to as ftmEconv and Pytorch-conv, respectively. Furthermore, we compare the performance of the forward propagation pass with ARM Computer Library (ACL), which does not implement the left two passes. The absolute performance of three passes in different implementations of pointwise convolutions running on the FT-M7032 processor is presented in Figs. 3, 4, and 5. We can find that our direct implementation outperforms all the other implementations on FT-M7032. In addition, ftmDconv-Dlt avoids all additional memory overhead in ftmEconv and Pytorch-Conv.

Table 1. The parameter configuration of the pointwise convolutional layers

Layer ID	Model	$C \times H_i \times W_i$	$H_f \times W_f$	S	P
1	Resnet50 [6]	$64 \times 56 \times 56$	$64 \times 1 \times 1$	1	0
2		$64 \times 56 \times 56$	$256 \times 1 \times 1$	1	0
3		$256 \times 56 \times 56$	$64 \times 1 \times 1$	1	0
4		$256 \times 56 \times 56$	$128 \times 1 \times 1$	1	0
5		$128 \times 28 \times 28$	$512 \times 1 \times 1$	1	0
6		$512 \times 28 \times 28$	$128 \times 1 \times 1$	1	0
7		$512 \times 28 \times 28$	$256 \times 1 \times 1$	1	0
8		$256 \times 14 \times 14$	$1024 \times 1 \times 1$	1	0
9		$1024 \times 14 \times 14$	$256 \times 1 \times 1$	1	0
10		$1024 \times 14 \times 14$	$512 \times 1 \times 1$	1	0
11		$512 \times 7 \times 7$	$2048 \times 1 \times 1$	1	0
12		$2048 \times 7 \times 7$	$512 \times 1 \times 1$	1	0
13	Mobilenetv1 [8]	$32 \times 112 \times 112$	$64 \times 1 \times 1$	1	0
14		$64 \times 56 \times 56$	$128 \times 1 \times 1$	1	0
15		$128 \times 56 \times 56$	$128 \times 1 \times 1$	1	0
16		$128 \times 28 \times 28$	$256 \times 1 \times 1$	1	0
17		$256 \times 28 \times 28$	$256 \times 1 \times 1$	1	0
18		$256 \times 14 \times 14$	$512 \times 1 \times 1$	1	0
19		$512 \times 14 \times 14$	$512 \times 1 \times 1$	1	0
20		$512 \times 7 \times 7$	$1024 \times 1 \times 1$	1	0
21		$1024 \times 7 \times 7$	$1024 \times 1 \times 1$	1	0

Figure 3 shows the computational performance of the forward propagation pass in four implementations, where the horizontal axis denotes the layer ID of different pointwise convolutional layers and the vertical axis represents the computational performance P_{conv} obtained by each implementation. The results indicate that ftmDconv-Pt achieves performance ranging from 336.57 GFlops to 1593.51 GFlops, resulting in a computational efficiency of 12.17% to 57.64%. Notably, ftmDconv-Pt has a significant speedup of 5.93 times to 35.84 times and 3.76 times to 24.07 times when compared with Pytorch-Conv and ACL algorithms, respectively. In the comparison with ftmEconv, the speedup is in the range of 1.55 times to 5.57 times, and the main reason for the observed performance speedup is that the direct implementation has no additional memory overhead and shows much better on-chip data locality.

For the backward propagation pass, we also compare the computational performance P_{conv} of ftmDconv-Pt with that of ftmEconv and Pytorch-Conv on all the tested network layers, as shown in Fig. 4. The ftmDconv-Pt implementation achieves performance ranging from 315.76 GFlops to 1620.33 GFlops, resulting in a computational efficiency of 11.42% to 58.61%. When compared with Pytorch-

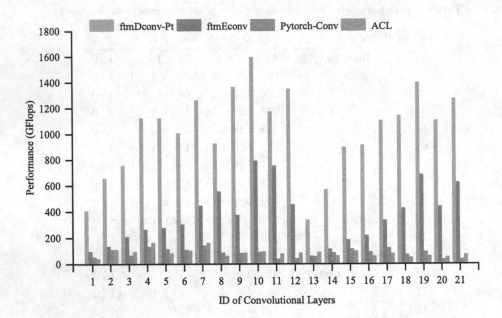

Fig. 3. Performance of various forward propagation algorithms for pointwise convolutions on FT-M7032 processors

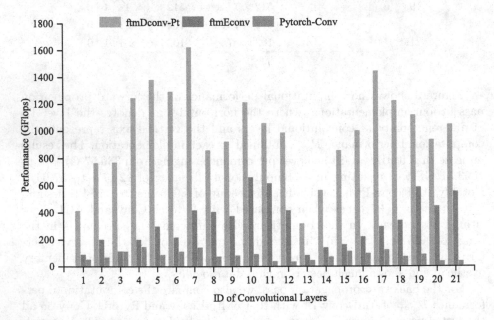

Fig. 4. Performance of various backward propagation algorithms for pointwise convolutions on FT-M7032 processors

Conv, ftmDconv-Pt achieves a significant speedup of 6.90 times to 29.14 times. In the comparison with ftmEconv, the maximum speedup is 6.80 times.

Figure 5 compares the computational performance P_{conv} of the direct implementation of the weight gradient update pass with that of ftmEconv and Pytorch-Conv on all the tested network layers. For all the tested network layers, ftmDconv-Pt achieves the performance of 366.216 GFlops - 1582.35 GFlops, resulting in a computational efficiency of 13.24% - 57.23%. When compared with Pytorch-Conv, ftmDconv-Pt obtains a speedup of 2.66 times to 13.27 times. In the comparison with ftmEconv, the maximum speedup is 79.26 times.

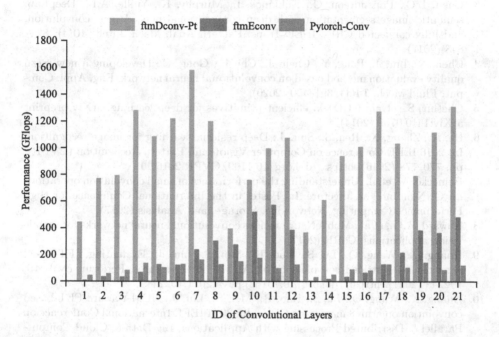

Fig. 5. Performance of various weight gradient update algorithms for pointwise convolutions on FT-M7032 processors

5 Conclusions and Future Work

This paper presents a high-performance parallel algorithm for the direct implementation of pointwise convolutions on multi-core DSPs in FT-M7032 heterogeneous processors. The parallel implementation can take full advantage of the parallel functional units and multi-level on-chip memories in multi-core DSPs. The primary optimizations involve multi-level memory blocking, loop ordering, vectorization, and multi-core parallelization. The experimental results on pointwise convolutional layers of popular networks show the proposed direct implementation outperforms other implementations on FT-M7032 heterogeneous processors, and get the maximum speedup of up to 79.26 times.

In the future, we will focus on the direct implementations for other types of convolutions on multi-core DSPs.

References

1. Arm Corporation: Arm computer library: A software library for machine learning. https://www.arm.com/technologies/compute-library (2023). Accessed 3 Jan 2023
2. Chaudhary, N., et al.: Efficient and generic 1d dilated convolution layer for deep learning. arXiv preprint arXiv:2104.08002 (2021)
3. Chen, L.C., Papandreou, G., Kokkinos, I., Murphy, K., Yuille, A.L.: DeepLab: semantic image segmentation with deep convolutional nets, Atrous convolution, and fully connected CRFs. IEEE Trans. Pattern Anal. Mach. Intell. **40**(4), 834–848 (2017)
4. Chen, X., Liu, J., Pang, Y., Chen, J., Chi, L., Gong, C.: Developing a new mesh quality evaluation method based on convolutional neural network. Eng. Appl. Comput. Fluid Mech. **14**(1), 391–400 (2020)
5. Chetlur, S., et al.: CUDNN: efficient primitives for deep learning. arXiv preprint arXiv:1410.0759 (2014)
6. He, K., Zhang, X., Ren, S., Sun, J.: Deep residual learning for image recognition. In: 2016 IEEE Conference on Computer Vision and Pattern Recognition (CVPR), pp. 770–778 (2016). https://doi.org/10.1109/CVPR.2016.90
7. Heinecke, A., et al.: Understanding the performance of small convolution operations for CNN on intel architecture. In: Poster in the International Conference for High Performance Computing, Networking, Storage, and Analysis (2017)
8. Howard, A.G., et al.: MobileNets: efficient convolutional neural networks for mobile vision applications. CoRR (2017)
9. Huang, X., Wang, Q., Lu, S., Hao, R., Mei, S., Liu, J.: Evaluating FFT-based algorithms for strided convolutions on ARMv8 architectures. Perform. Eval. **49**, 102248 (2021). https://doi.org/10.1016/j.peva.2021.102248
10. Huang, X., Wang, Q., Lu, S., Hao, R., Mei, S., Liu, J.: NUMA-aware FFT-based convolution on armv8 many-core CPUs. In: 2021 IEEE International Conference on Parallel & Distributed Processing with Applications, Big Data & Cloud Computing, Sustainable Computing & Communications, Social Computing & Networking (ISPA/BDCloud/SocialCom/SustainCom), pp. 1019–1026 (2021). https://doi.org/10.1109/ISPA-BDCloud-SocialCom-SustainCom52081.2021.00142
11. Igual, F.D., Ali, M., Friedmann, A., Stotzer, E., Wentz, T., van de Geijn, R.A.: Unleashing the high-performance and low-power of multi-core DSPs for general-purpose HPC. In: SC 2012: Proceedings of the International Conference on High Performance Computing, Networking, Storage and Analysis, pp. 1–11. IEEE (2012)
12. Kim, M., Park, C., Kim, S., Hong, T., Ro, W.W.: Efficient dilated-winograd convolutional neural networks. In: 2019 IEEE International Conference on Image Processing (ICIP), pp. 2711–2715. IEEE (2019)
13. Lu, G., Zhang, W., Wang, Z.: Optimizing GPU memory transactions for convolution operations. In: 2020 IEEE International Conference on Cluster Computing (CLUSTER), pp. 399–403. IEEE (2020)
14. Lu, G., Zhang, W., Wang, Z.: Optimizing Depthwise separable convolution operations on GPUs. IEEE Trans. Parallel Distrib. Syst. **33**(1), 70–87 (2021)

15. Mehta, S., Rastegari, M., Caspi, A., Shapiro, L., Hajishirzi, H.: ESPNet: efficient spatial pyramid of dilated convolutions for semantic segmentation. In: Ferrari, V., Hebert, M., Sminchisescu, C., Weiss, Y. (eds.) ECCV 2018. LNCS, vol. 11214, pp. 561–580. Springer, Cham (2018). https://doi.org/10.1007/978-3-030-01249-6_34

16. Mogers, N., Radu, V., Li, L., Turner, J., O'Boyle, M., Dubach, C.: Automatic generation of specialized direct convolutions for mobile GPUs. In: Proceedings of the 13th Annual Workshop on General Purpose Processing using Graphics Processing Unit, pp. 41–50 (2020)

17. Paszke, A., et al.: Pytorch: an imperative style, high-performance deep learning library. In: Advances in Neural Information Processing Systems, vol. 32 (2019)

18. Pei, X., et al.: Optimizing parallel matrix transpose algorithm on multi-core digital signal processors (in Chinese). J. Natl. Univ. Defense Technol. **45**(1), 57–66 (2023)

19. Safonov, I., Kornilov, A., Makienko, D.: An approach for matrix multiplication of 32-bit fixed point numbers by means of 16-bit SIMD instructions on DSP. Electronics **12**, 78 (2022)

20. Sandler, M., Howard, A., Zhu, M., Zhmoginov, A., Chen, L.C.: Mobilenetv 2: Inverted residuals and linear bottlenecks. In: Proceedings of the IEEE Conference on Computer Vision and Pattern Recognition, pp. 4510–4520 (2018)

21. Szegedy, C., et al.: Going deeper with convolutions. In: Proceedings of the IEEE Conference on Computer Vision and Pattern Recognition, pp. 1–9 (2015)

22. Wang, Q., Li, D., Huang, X., Shen, S., Mei, S., Liu, J.: Optimizing FFT-based convolution on ARMv8 multi-core CPUs. In: Malawski, M., Rzadca, K. (eds.) Euro-Par 2020. LNCS, vol. 12247, pp. 248–262. Springer, Cham (2020). https://doi.org/10.1007/978-3-030-57675-2_16

23. Wang, Q., Li, D., Mei, S., Lai, Z., Dou, Y.: Optimizing Winograd-based fast convolution algorithm on Pythium multi-core CPUs (in Chinese). J. Comput. Res. Dev. **57**(6), 1140–1151 (2020). https://doi.org/10.7544/issn1000-1239.2020.20200107

24. Wang, Q., Li, D., Mei, S., Shen, S., Huang, X.: Optimizing one by one direct convolution on ARMV8 multi-core CPUs. In: 2020 IEEE International Conference on Joint Cloud Computing, pp. 43–47. IEEE (2020). https://doi.org/10.1109/JCC49151.2020.00016

25. Wang, Q., et al.: Evaluating matrix multiplication-based convolution algorithm on multi-core digital signal processors (in Chinese). J. Natl. Univ. Defense Technol. **45**(1), 86–94 (2023). https://doi.org/10.11887/j.cn.202301009

26. Wang, Q., Songzhu, M., Liu, J., Gong, C.: Parallel convolution algorithm using implicit matrix multiplication on multi-core CPUs. In: 2019 International Joint Conference on Neural Networks (IJCNN), pp. 1–7 (2019). https://doi.org/10.1109/IJCNN.2019.8852012

27. Yin, S., Wang, Q., Hao, R., Zhou, T., Mei, S., Liu, J.: Optimizing irregular-shaped matrix-matrix multiplication on multi-core DSPs. In: 2022 IEEE International Conference on Cluster Computing (CLUSTER), pp. 451–461 (2022). https://doi.org/10.1109/CLUSTER51413.2022.00055

28. Yu, F., Koltun, V.: Multi-scale context aggregation by dilated convolutions. arXiv preprint arXiv:1511.07122 (2015)

Detecting SDCs in GPGPUs Through Efficient Partial Thread Redundancy

Xiaohui Wei⬤, Yan Wu⬤, Nan Jiang(✉)⬤, and Hengshan Yue(✉)⬤

College of Computer Science and Technology, Jilin University, Changchun, China
{weixh,yuehs}@jlu.edu.cn, {yanwu21,jiangnan22}@mails.jlu.edu.cn

Abstract. As General-Purpose Graphics Processing Units (GPGPUs) are widely employed in various precision-sensitive and safety-critical domains, guaranteeing the execution reliability of such applications under the impact of soft errors becomes a critical issue. Redundant Multi-Threading (RMT) provides a potentially low-cost mechanism for improving GPGPU reliability, but full protection comes with high time and resource costs. In this paper, we propose a partial thread protection mechanism for efficient Silent Data Corruption (SDC) detection in GPGPU programs. Firstly, we establish an accurate and efficient model for assessing the thread SDC vulnerability by capturing intra-thread error propagation and inter-thread error propagation. Then, based on the analysis results, we selectively replicate the SDC vulnerable threads. Experimental results indicate that our proposed thread SDC vulnerability assessment model closely aligns with the fault injection results, while introducing much lower execution overhead. Our partial thread redundancy mechanism provides a better trade-off between reliability and overhead compared with full RMT.

Keywords: GPGPUs · Soft Error · Silent Data Corruptions (SDCs) · Partial Thread Protection

1 Introduction

With remarkably high concurrency and improved programmability [7], Graphics Processing Units (GPUs) have been widely used in various general-purpose fields, such as scientific computing, financial analysis, automatic driving and other safety-critical systems [7,12], which is also known as general-purpose computing on GPUs (GPGPUs). Unlike the inherent error tolerance of graphics applications, GPGPUs applications typically have more stringent requirements for reliability [6]. However, with the scaling of transistor sizes and operating voltages, GPGPUs become more vulnerable to high energy particle strikes [10], which can cause bit flipping during execution, also called soft errors. Soft errors can potentially cause silent data corruptions (SDCs) in application outputs or

This work is supported by the National Natural Science Foundation of China (NSFC) (Grants No. 62272190, No. 62302190 and No. U19A2061).

directly result in application/system hang-ups or crashes. SDC is regarded as the most critical error type because there is no visible indication of program corruption, thereby significantly impacting the quality of application outputs.

Most modern GPGPU architectures typically employ Error Correcting Codes (ECCs) and parity to protect storage structures [21]. However, these mechanisms are unable to detect errors occurring in the execution units [5]. To ensure reliable execution, some fine-grained protection mechanisms, such as Selective Instruction Duplication (SelDup) and Redundant Multi-Threading (RMT), are employed in GPUs to detect SDCs at the software level [10]. The existing SelDup typically preferentially duplicate instructions with high SDC vulnerability, and compare their execution results to detect SDCs [16]. Although this partial protection provides some meaningful performance savings, it has two inherent limitations as shown in Fig. 1. When redundant instructions appear (1) in a non-branch structure, all threads will execute the redundant instructions sequentially, increasing the execution time (① in Fig. 1). (2) in a branch structure, it is possible that only a subset of threads need to execute the redundant instructions. However, due to the Single Instruction Multiple Threads (SIMT) scheme followed within a warp, the remaining threads in the warp have to wait, resulting in incurred time overhead and a decrease in resource utilization (② in Fig. 1). Redundant Multi-Threading (RMT) creates two copies of a thread, a leading thread and a trailing thread, that use the same input to execute in parallel and compare the results while the program is running [15]. With sufficient resources, the leading thread and the trailing thread can be executed in parallel without incurring thread wait overhead.

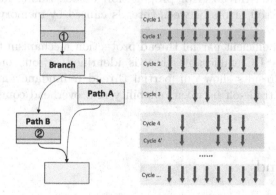

Fig. 1. The instruction duplication execution mode. Redundant instructions are highlighted in red. Cycle 1' and Cycle 4' denote the additional performance overhead when executing redundant instructions in non-branching(①) and branching structures(②), respectively. (Color figure online)

In this work, we propose an efficient partial thread redundancy method that only replicates SDC vulnerable threads for reliability overhead saving. To this end, the first essential task is to establish a model that can accurately assess the SDC vulnerable thread in GPGPUs. Our key insight is that the resilience of threads varies based on their unique execution instruction sequences. Furthermore, the SDC vulnerability of a thread may be influenced by other threads due to inter-thread memory access dependencies. We build an error propagation model to efficiently predict thread SDC vulnerability without any fault injection. Firstly, we establish an intra-thread error propagation model using data flow analysis, while accounting for the error masking and program Crash during error propagation. In addition, we also consider the effects of inter-thread error propagation due to thread memory access dependencies. Then, based on the model prediction results, we selectively protect SDC vulnerable threads within a block.

Although several studies have utilized RMT techniques to improve the reliability of GPU applications [8,15], these works primarily focus on optimizing the duplication technique without incorporating selective replication, which can lead to overprotection. Although there have been efforts proposing selective thread protection [18], coarse-grained thread resiliency analysis results in lower soft error coverage.

In summary, our main contributions in this work are as follows:

- We propose an effective error propagation model to estimate thread SDC vulnerability without any fault injection experiments.
- We build an intra-thread error propagation model and identify the situations that cause error propagation masking and program crash to improve the prediction accuracy of thread SDC probability.
- We build an inter-thread error propagation model, taking into account the propagation of soft errors between threads caused by memory access dependencies.
- We propose an efficient partial thread protection mechanism that selectively replicates the SDC vulnerable threads identified by our proposed model. Experimental results show our partial thread redundancy mechanism provides a better trade-off between reliability and overhead compared with full RMT.

2   Background

In this section, we provide a concise overview of the execution patterns of GPGPU applications. Following this, we introduce our fault model and establish the definitions of the key terms.

2.1   GPGPUs Architecture and Programming Model

A GPGPU application is composed of a control program launched by host CPU and one or more computing functions, also called kernels, executed on the GPU.

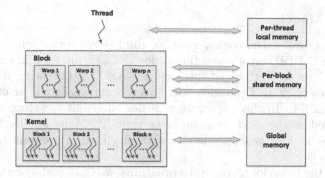

Fig. 2. Thread-Memory access hierarchy during GPGPU execution.

Each kernel is divided into multiple blocks, which are also known as Cooperative Thread-Arrays (CTAs). A CTA can guarantee synchronization among component threads by executing a local barrier instruction, whereas threads across different CTAs require additional explicit synchronization. Each CTA is sub-divided into 32 individual threads, referred to as warp, which serve as the fundamental instruction scheduling unit. The threads within a warp execute the same instruction accessing different data in a lock step. Such an execution mode is usually called Single Instruction Multiple Threads(SIMT) execution, which enables the GPU to achieve high throughput. Both CPU and GPU device maintain their own local memory, and there are significant differences between them. The memory hierarchy of modern GPUs can be categorized into the following levels: (1) global, (2) texture, (3) constant, (4) shared, and (5) thread-local memory. Constant and texture memory can be read-only and can be accessed by all threads, which are designed for special purposes. Figure 2 illustrates the thread-memory access hierarchy during GPGPU execution. Each thread has a private local memory. Thread-local memory is usually stored in the fast register file, which provides fast access and high-bandwidth data transfer for the threads. Each block has a shared memory that is accessible by all of its threads and lasts as long as the block itself. The shared memory space can be managed by software and is much faster than accessing global memory. Additionally, all threads in the kernel have access to the same global memory, which can be used for inter-thread communication and data sharing across multiple blocks. Thus, different threads may depend on each other because the load and store instructions access the same memory, which is referred to as inter-thread memory dependency. In case bit flips occur during the execution of an instruction in a thread, it may transmit the effect of the transient hardware faults to other threads through inter-thread memory dependencies, leading to potential errors. A thread can be affected by soft errors in two ways: (1) Intra-thread error propagation, where a transient hardware failure occurs during thread execution and propagates to memory, and (2) Inter-thread error propagation, where a thread reads corrupted data written by another thread and propagates the error to memory.

2.2   Fault Model

Many architectural technologies, such as single-error-correction double-error-detection (SEC-DED) error correction codes (ECCs), are used in existing commercial GPU to protect shared memory, DRAM, L1/L2 cache, register files. Therefore, this work only considers soft errors occurring in the functional units (eg., ALUs, LSUs). In this paper, we consider single-bit flip since as it is typically regarded as the most common error type [2]. Many studies have shown that multiple-bit flips generally have a similar effect on program resiliency as single-bit flip [3].

Based on the behavior of corrupted programs, we classify the effects of corruption into three categories: (1) Masked, the erroneous output is indistinguishable from the correct output of the program. (2) Silent Data Corruptions (SDCs), the program successfully completes its execution, but the corrupted output differs from the expected output. (3) Detected Unrecoverable Errors (DUEs), system errors manifest obvious symptoms, such as program crashes or hangs.

3   Thread-Level SDC Proneness Analysis

To accurately capture the SDC proneness of each thread, Fault injection is an intuitive method. But thousands of threads in GPGPU applications results in a tremendously large fault site space, leading to significant overhead. To mitigate this issue, we employ graph propagation modeling to analyze thread SDC proneness. Our work attribute the origin of a thread's vulnerability to two distinct factors: intra-thread error propagation and inter-thread error propagation. In this section, we first propose heuristics to explore the error-masking and DUE-incurring instructions in the propagation chain. Secondly, we consider the effect of error propagation within threads on thread SDC proneness. Finally, considering that errors may propagate between threads through memory, we develop an inter-thread error propagation model to further enhance the accuracy of error resilience estimation.

3.1   Vulnerability Identification

In order to more accurately identify the SDC proneness of each thread so as to ensure that threads with high vulnerability are protected in the subsequent design, we explore certain error-masking and DUE-incurring instructions in the program that aid us in estimating the probability of error propagation.

Propagation Masking. For a soft error, it may not affect the application output. In this context, we analyze several factors that contribute to error masking.

Logical Masking. Due to computational logic or conditional control, the effects of soft errors may be masked, which we call logical masking. For example, in the 4th instruction of Fig. 4a, the high 9 bits of register r5 are masked, having no impact on the final result. Consequently, the propagation probability is reduced to 23/32. We employ instruction analysis to identify such cases and assign appropriate weights to the corresponding propagation probabilities.

Dead Store. A soft error propagated to a store instruction may not affect thread resiliency. Dead store occurs when a value is written to memory by a store instruction but is never read or utilized, rendering it irrelevant to the final result [20]. We consider dead storage as two store instructions accessing the same memory, with the value of the previous store instruction overwritten by the value of the later instruction before it is used. We identify dead store instructions and exclude the SDC probability caused by them from the thread SDC vulnerability estimation.

Propagation Crash. The primary factor causing program crashes due to soft errors is out-of-bound memory access [1]. We consider a crash occurring as a result of load/store instruction operations accessing out-of-bounds memory addresses based on the acquired memory range. The crash probability is estimated by examining the memory size dynamically allocated to the kernel function, which will also not be included in the final thread SDC proneness.

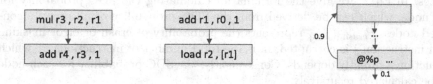

(a) Value dependency.   (b) Address dependency.   (c) Control-flow dependency.

Fig. 3. Instruction dependency.

3.2 Intra-thread Error Propagation Analysis

In this step, we utilize the Dependency Graph (DG) to capture the dependencies among instructions executed by an individual thread. In the DG, each node corresponds to an instruction, while edges represent dependencies between instructions, imposing constraints on the order of instruction execution[1]. Based on the characteristics of instructions, we classify instruction dependency into three distinct types: (1) Value dependency: the result of instruction x is used as the source operand for instruction y. (2) Address dependency, instruction y uses the result of instruction x as a memory address. (3) Control-Flow dependency, it occurs when the execution of instruction x depends on the outcome of a branch instruction, denoted as y. Figure 3 exemplifies these three dependencies, one for each case. The number of cycles of branches in the thread execution sequence is counted to configure the branch probability. For example, in Fig. 3c, the branch probability is 0.9, which means that during the execution of the thread, the branch has a 90% probability of being executed.

We define the impact of intra-thread error propagation for a thread as the probability that, a bit flip occurring during the execution of instruction i by a thread, and the error is ultimately propagate to the store instruction. The probability of SDC corruption for a store instruction can be expressed by Eq. 1:

$$P_{ts} = 1 - \prod_{k=1}^{n} (1 - v_k P_{C_k}(1 - P_{\text{crash}_k})) \qquad (1)$$

$$P_{C_k} = 1 - \prod_{j=1}^{m} (1 - v_j P_{C_j}(1 - P_{\text{crash}_j|\text{mask}_j})) \qquad (2)$$

In the equation, P_{ts} represents the SDC probability of a store instruction, n refers to the total number of instructions on which the stored instruction directly depends, P_{C_k} denotes the probability that the error propagates to instruction i on which the store instruction depends, and P_{crash_k} is the probability of crash, which is caused by address dependency between instruction k and store instruction. The branching probability, v, is assumed to be 1 for non-branching dependencies. In Eq. 2, we give the formula for calculating the SDC probability for each node, which can be derived from the SDC probability of all directly connected nodes. $P_{\text{crash}_j|\text{mask}_j}$ represents the probability of crash or error masking when instruction j is corrupted, m is the total number of instructions on which instruction k directly depends. Consequently, the SDC probability for each node can be calculated recursively.

We use the example in Fig. 4 to show how to use the above formula to calculate. When a bit flip occurs during 5th instruction execution by a thread, the probability of SDC in instruction 5 is 1, denoted as $P_{C_5} = 1$. The 6th instruction is the load instruction, which needs to consider the probability of crash caused by soft errors. So the SDC probability of the 6th instruction can be computed as $P_{C_6} = 1 - (1 - 1 * P_{C_5} * (1 - P_{crash})) = 1 - (1 - 1 * 1 * (1 - 0.62)) = 0.38$ (where 0.62 represents the crash probability used in the 2mm example). Similarly, the SDC probability of the store instruction is $P_{store} = 1 - [(1 - 1 * 1 * 0.38) * (1 - 1 * 0.38 * 1) * (1 - 0.01 * 0.38 * 1)] = 0.617$.

Due to the large number of threads in a GPU, constructing a DG for each individual thread is time-consuming. In order to minimize the number of threads requiring analysis, we group threads based on their execution instruction sequences, which provides a more efficient approach compared to considering the number of instructions executed by individual threads. Since threads in a group execute the same sequence, the error will propagate along the same path from where the bit flip occurs to the store instruction, thus merging them will not reduce evaluation accuracy. We first extract the PTX instruction sequences executed by the threads and group the threads that execute the same instructions. From each group, we select a representative thread and model its execution sequence.

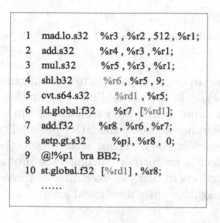

```
1   mad.lo.s32    %r3 , %r2 , 512 , %r1;
2   add.s32       %r4 , %r3 , %r1;
3   mul.s32       %r5 , %r3 , %r1;
4   shl.b32       %r6 , %r5 , 9;
5   cvt.s64.s32   %rd1 , %r5;
6   ld.global.f32  %r7 , [%rd1];
7   add.f32       %r8 , %r6 , %r7;
8   setp.gt.s32   %p1, %r8 , 0;
9   @!%p1  bra BB2;
10  st.global.f32  [%rd1] , %r8;
......
```

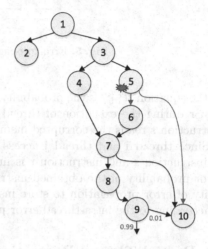

(a) A code segment example.   (b) Corresponding DG.

Fig. 4. A thread running example. (a) Shows a sequence of instructions executed by a thread, blue and red indicate conditions that cause error masking and crash and (b) shows the sequence of instructions represented by Dependency Graph. The red edges indicate address dependencies. (Color figure online)

3.3 Inter-thread Error Propagation Analysis

Faults can result in the corruption of memory values and continue to propagate through memory operations. If the erroneous stored value is accessed by load instructions executed by other threads, the impact of the soft error is further amplified. Figure 5 illustrates an instance where a thread stores results via the store instruction, which are subsequently accessed by other threads. During the execution of thread 1, a soft error occurs and propagates to memory via the store instruction, the erroneous value will be accessed by thread 2 and thread 3, as they access the same memory location, which can potentially lead to incorrect execution results in thread 2 and thread 3.

We utilize Eq. 3 to show how corrupted data written to memory by thread i affects the SDC probability of thread j reading corrupted data.

$$P_{\mathrm{ld_{j,s}}} = P_{\mathrm{td_{i,r}}} \qquad (3)$$

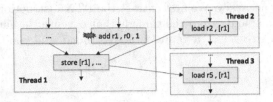

Fig. 5. Error propagation among threads.

In the equation, $P_{\mathrm{td}_{i,r}}$ is the probability that an error propagates to store instruction r during the execution of thread i. $P_{\mathrm{ld}_{j,s}}$ is the probability that the load instruction s reads the corrupted memory value during the execution of thread j. Since thread i and thread j access the same address, the SDC probability of instruction s and instruction r is numerically equal. Then, according to the error probability of the obtained instruction s, we further calculate the probability of error propagation to store instructions during the execution of thread j, according to the intra-thread error propagation rules mentioned above.

4   Partial Thread Protection Framework

To effectively enhance reliability, leveraging the model presented in Sect. 3, we first profile thread SDC vulnerability. Then we propose an efficient thread redundancy approach to prioritize the protection of SDC vulnerable threads.

4.1   Thread SDC Vulnerability Profiling

Our thread SDC vulnerability profiling phase consists of two stages: (1) Profiling and (2) Modeling. In the profiling stage, we execute the GPGPU programs and dynamically collect information about the execution sequences of threads. Once all the information is gathered, we conduct the static analysis phase of thread resilience. In the modeling stage, we first merge threads with identical execution sequences to reduce the number of threads that need to be analyzed. This is because the same thread execution sequence has the same instruction dependencies, leading to the same error propagation paths. Then, we utilize both the intra-thread error propagation model and inter-thread error propagation model to calculate the SDC Vulnerability of individual thread. Additionally, by considering the SDC probabilities of individual thread, we can estimate the overall SDC probability of the program.

4.2   Partial Thread Redundancy

The fundamental idea of thread redundancy is to use additional threads to execute the same instruction sequence with the same data as the original thread,

---

Algorithm 1: The Core Algorithm of Thread Resiliency Evaluation.

---

    input : Instructions set executed by a thread: I;
    output: The SDC probability of the thread: P_{SDC}

1   $intruNum \leftarrow$ get_Instructions_Number(I) ;
2   $P_{SDC} \leftarrow 0$;
3   $storeNum \leftarrow$ get_non-dead_Store_Intructions_Number(I);
4   **for** *all non-dead store instructions S* **do**
       // apply intra-thread error propagation model to calculate the
       SDC probability of store instruction S
5     $P_{intra} \leftarrow$ Intra_SDC_Probability(S);
6     $P_{SDC} \leftarrow P_{SDC} + \frac{1}{intruNum} \times P_{intra}$;
       // read corrupt values written by another thread
7   **if** *S corrupt due to inter-thread dependencies* **then**
         // apply inter-thread error propagation model
8       $P_{inter} \leftarrow$ Inter_SDC_Probability(S);
9       $P_{SDC} \leftarrow P_{SDC} + P_{inter}$;

10   $P_{SDC} \leftarrow \frac{1}{storeNum} \times P_{SDC}$;
11   **return** P_{SDC}

---

with the only difference being their thread ID numbers. Comparisons and notifications are inserted at certain points, often before storing data. We selectively protect vulnerable threads in a block, as thread synchronization between blocks requires explicit synchronization [8,15].

Our replication strategy comprises three steps. First, we consider a block as the unit of thread replication and identify vulnerable threads within the block, as shown in Fig. 6a, where SDC vulnerable threads are marked in red. Second, we duplicate these vulnerable threads within the block. We allocate a buffer in the local memory for thread communication. Before thread communication, the original thread and redundant thread execute in parallel. Third, before the store instruction is finished, the original thread reads the stored address and value operand from the buffer, which were previously stored by the redundant thread, and compares the values with its own execution result to verify whether an error has occurred as can be seen in Fig. 6b.

5 Experiment Methodology

We select from 7 benchmarks including 12 kernels from Polybench [14] and Rodinia [4], to evaluate our approach, which are listed in Table 1. These benchmarks encompass diverse domains including data mining, linear algebra, and deep learning, which have been extensively utilized in prior researches on GPGPU application resiliency and protection [9].

We perform reliability experiments on GPGPU-Sim [2], a widely used GPU architecture simulator. We obtain the set of instructions executed by the thread,

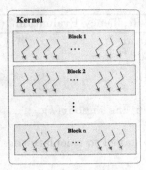

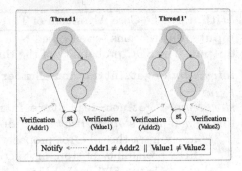

(a) Identify SDC vulnerable threads.

(b) Thread replication.

Fig. 6. Thread redundancy methodology.

and subsequently randomly select an instruction destination register to inject one fault per run. We obtain the ground truth of individual thread SDC vulnerability through an exhaustive FI experiments. We also use smaller inputs specific to each benchmark in our FI experiments to ensure accurate evaluation and maintain experimental control, following the approach of previous studies [1,17,19].

Table 1. Characteristics and modeling overhead of the benchmark used.

Suit	Application	Kernel name	Kernel ID	Uses shared memory	Total threads	FI Overhead (in hours)	Ours Overhead (in minutes)
Polybench	2MM	mm2_kernel1	K1	No	65536	32.931	18.466
		mm2_kernel2	K2	No	65536		
	2DCONV	Convolution2D_kernel	K1	No	16384	0.733	0.237
	BICG	bicg_kernel1	K1	No	512	1.162	1.438
		bicg_kernel2	K2	No	512		
	MVT	mvt_kernel1	K1	No	2048	8.552	11.789
		mvt_kernel2	K2	No	2048		
	COVAR	mean_kernel	K1	No	128	5.005	11.579
		reduce_kernel	K2	No	4096		
		covar_kernel	K3	No	128		
	CORR	corr_kernel	K1	No	256	27.339	50.662
Rodinia	HotSpot	calculate_temp	K1	Yes	9216	1.001	0.843

6 Evaluation

In this section, we first evaluate the accuracy of predicting the SDC probability for individual threads and the SDC probability for each benchmark. Subsequently, we discuss the overhead associated with partial thread protection.

6.1   SDC Probability Prediction Accuracy

Thread SDC Probability Accuracy. We evaluate the predicted SDC probability for individual thread in each kernel and compared it with the thread SDC probability obtained by random FI. We select threads appropriately from the thread group based on thread number proportion and perform multiple FI experiments on individual thread, totaling 5000 FI experiments for a kernel as previous work did [1].

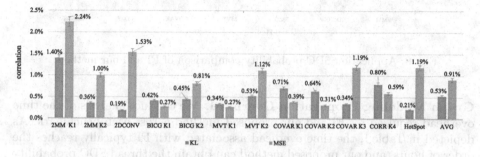

Fig. 7. Thread SDC probabilistic correlation comparison of FI and our method.

We use Kullback-Leibler (KL) divergence to measure the differences in thread SDC probability distributions within a kernel between FI and our method. As shown in Fig. 7, our method fits well with the distribution of FI results, and achieves the KL divergence difference of 0.53% on average.

We also used the Mean Square Error (MSE) to assess thread SDC vulnerability similarity between the FI and our method. The average MSE for thread SDC probabilities across all kernels is 0.91%, indicating a small difference between the FI results and our predicted results.

Benchmark SDC Probability Accuracy. The overall SDC probability of the program is the accumulation of SDC probabilities across all threads. Therefore, in order to further evaluate the accuracy of our method in estimating thread SDC vulnerability, we compare the overall SDC probability of the program obtained through our method and FI in this section .

As shown in Fig. 8, our method shows a small difference compared to the FI results, with a mean absolute error of 3.8%. Furthermore, we conducted a comparative analysis of Pearson correlation coefficients. The correlation between our method and the FI results yielded a coefficient of 0.944, implying a substantial consistency between our method and the results obtained by the FI experiments.

6.2   Overhead

In this section, we evaluate the overhead of our proposed approach from the two phases of graph modeling propagation and thread replication.

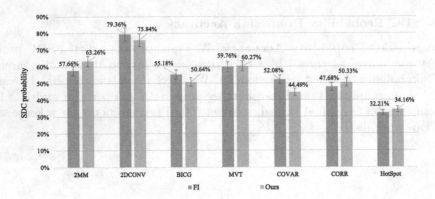

Fig. 8. Application SDC probability comparison of FI and our method.

Graph Modeling Propagation Overhead. We conduct analysis of the time overhead required for FI and our proposed method across each benchmark. As depicted in Table 1, the time overhead associated with FI typically reaches the order of hours, and our proposed method can obtain the thread SDC probability in less than one hour with the average execution time being reduced by one order of magnitude. We observe that thread execution instructions traced by GPGPU-sim account for over 95% of time overhead. It is worth noting that we also observe a relatively longer runtime for the CORR K3 benchmark. This can be attributed to the presence of numerous loop branches in CORR K3, which results in a diverse sequence of instructions executed by threads. Consequently, more dependency graphs need to be built to analyze different kinds of representative thread resilience.

Thread Redundancy Overhead. We also compare the performance saving of our partial thread redundancy technique with full RMT. We set the thread SDC probability threshold to 10%, considering any thread with an SDC probability below 10% as reliable. Consequently, we apply redundancy to those threads deemed unreliable. Figure 9 shows the number of additional threads required for redundant execution at different thread SDC thresholds for our method and full RMT. We observe that, at a thread SDC threshold of 10%, the number of redundant threads for 2Dconv and Hotspot decreased by 4% and 33.7%, and SDC coverage is 99.56% and 82.33%, respectively. However, for the remaining cases, the reduction in the number of threads is not significant. We note that in these cases, the thread SDC probabilities were concentrated within a certain range of $(0.35, 0.8)$, indicating that the impact of these thread execution results on the application outcomes can not be ignored. As the target output quality acceptable threshold decreases, the number of additional threads for redundant execution will be further reduced, resulting in more performance gains.

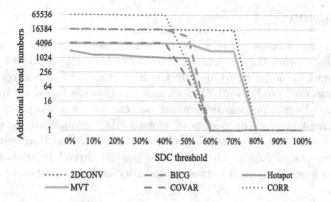

Fig. 9. The number of additional threads required for replication decreases as the acceptable thread SDC threshold increases.

7 Related Work

Error Propagation Modeling. Error propagation modeling analysis is widely used to evaluate application reliability because it does not require thousands of fault injection experiments. Lu et al. [13] proposed the SDCTune model based on the static and dynamic characteristics of the program, which can identify the variables in the program that are relatively prone to SDC based on the heuristic method developed. Khudia et al. [11] propose a pure-software application analysis scheme, which employs control flow, memory dependency, and value profiling to understand the common-case behavior of applications. However, none of this work is targeted towards GPU applications. GPU-Trident [1] proposes an accurate and scalable technique for modeling GPU program error propagation. However, it is unable to handle situations where there is control flow within a basic block, and it does not quantify the SDC probability of individual thread.

Redundancy Mechanism. Several software-level efforts have been proposed to improve the reliability of GPGPU applications. Öz I et al. [22] evaluated vulnerabilities in kernel functions using fault injection experiments to quantify the severity of data corruption by considering metrics other than SDC rates, providing guidance for selective redundant execution of GPGPU applications. Wei et al. [16] introduced a partial protection scheme at the instruction level, which leverages machine learning techniques to intelligently identify all SDC vulnerable instructions based on observed heuristic features. By duplicating the instructions associated with high SDC proneness, the scheme aims to reduce the reliability overhead. Yang et al. [18] proposed a partial thread protection mechanism that involves changing the mapping of threads to warps, increasing the proportion of reliable warps and Only replicate the unreliable warps. However, the use of coarse-grained analysis based on the thread dynamic instruction count as a criterion for distinguishing thread resilience may lead to less accurate analysis of thread SDC proneness.

8   Conclusion

In this paper, we propose an effective partial thread redundancy method that can effectively improve the execution reliability of GPGPU applications. We use graph propagation modeling to analyze the SDC tendency of a single thread in detail. Our method predicts all SDC vulnerable threads based on intra-thread and inter-thread error propagation, which we choose to protect. As our evaluation experiment shows, the results of thread SDC probability prediction are in close agreement with the results obtained from fault injection, with an average KL divergence of 0.53%. Moreover, our partial thread redundancy mechanism provides a good trade-off between reliability and overhead compared to full RMT.

References

1. Anwer, A.R., Li, G., Pattabiraman, K., Sullivan, M., Tsai, T., Hari, S.K.S.: GPU-trident: efficient modeling of error propagation in GPU programs. In: SC20: International Conference for High Performance Computing, Networking, Storage and Analysis. IEEE, November 2020. https://doi.org/10.1109/sc41405.2020.00092
2. Bakhoda, A., Yuan, G.L., Fung, W.W.L., Wong, H., Aamodt, T.M.: Analyzing CUDA workloads using a detailed GPU simulator. In: 2009 IEEE International Symposium on Performance Analysis of Systems and Software. IEEE, April 2009. https://doi.org/10.1109/ispass.2009.4919648
3. Chang, C.K., Lym, S., Kelly, N., Sullivan, M.B., Erez, M.: Evaluating and accelerating high-fidelity error injection for HPC. In: SC18: International Conference for High Performance Computing, Networking, Storage and Analysis. IEEE (nov 2018). https://doi.org/10.1109/sc.2018.00048
4. Che, S., Boyer, M., Meng, J., Tarjan, D., Sheaffer, J.W., Lee, S.H., et al.: Rodinia: A benchmark suite for heterogeneous computing. In: 2009 IEEE International Symposium on Workload Characterization (IISWC). IEEE, October 2009. https://doi.org/10.1109/iiswc.2009.5306797
5. Dimitrov, M., Mantor, M., Zhou, H.: Understanding software approaches for GPGPU reliability. In: Proceedings of 2nd Workshop on General Purpose Processing on Graphics Processing Units. ACM, March 2009. https://doi.org/10.1145/1513895.1513907
6. Gao, Y., Iqbal, S., Zhang, P., Qiu, M.: Performance and power analysis of high-density multi-GPGPU architectures: a preliminary case study. In: 2015 IEEE 17th International Conference on High Performance Computing and Communications, 2015 IEEE 7th International Symposium on Cyberspace Safety and Security, and 2015 IEEE 12th International Conference on Embedded Software and Systems. IEEE, August 2015. https://doi.org/10.1109/hpcc-css-icess.2015.68
7. Grauer-Gray, S., Killian, W., Searles, R., Cavazos, J.: Accelerating financial applications on the GPU. In: Proceedings of the 6th Workshop on General Purpose Processor Using Graphics Processing Units. ACM, March 2013. https://doi.org/10.1145/2458523.2458536
8. Gupta, M., Lowell, D., Kalamatianos, J., Raasch, S., Sridharan, V., Tullsen, D., et al.: Compiler techniques to reduce the synchronization overhead of GPU redundant multithreading. In: Proceedings of the 54th Annual Design Automation Conference 2017. ACM, June 2017. https://doi.org/10.1145/3061639.3062212

9. Kalra, C., Previlon, F., Li, X., Rubin, N., Kaeli, D.: PRISM: Predicting resilience of GPU applications using statistical methods. In: SC18: International Conference for High Performance Computing, Networking, Storage and Analysis. IEEE, November 2018. https://doi.org/10.1109/sc.2018.00072

10. Kalra, C., Previlon, F., Rubin, N., Kaeli, D.: ArmorAll: compiler-based resilience targeting GPU applications. ACM Trans. Archit. Code Optim. **17**(2), 1–24 (2020)

11. Khudia, D.S., Wright, G., Mahlke, S.: Efficient soft error protection for commodity embedded microprocessors using profile information. ACM SIGPLAN Not. **47**(5), 99–108 (2012)

12. Kim, J., Kim, H., Lakshmanan, K., Rajkumar, R.R.: Parallel scheduling for cyber-physical systems. In: Proceedings of the ACM/IEEE 4th International Conference on Cyber-Physical Systems. ACM, April 2013. https://doi.org/10.1145/2502524.2502530

13. Lu, Q., Pattabiraman, K., Gupta, M.S., Rivers, J.A.: SDCTune: a model for predicting the SDC proneness of an application for configurable protection. In: Proceedings of the 2014 International Conference on Compilers, Architecture and Synthesis for Embedded Systems. ACM, October 2014. https://doi.org/10.1145/2656106.2656127

14. Pouchet, L.N.: PolyBench: the polyhedral benchmark suite (2012). http://www.cs.ucla.edu/pouchet/software/polybench

15. Wadden, J., Lyashevsky, A., Gurumurthi, S., Sridharan, V., Skadron, K.: Real-world design and evaluation of compiler-managed GPU redundant multithreading. In: 2014 ACM/IEEE 41st International Symposium on Computer Architecture (ISCA). IEEE, June 2014. https://doi.org/10.1109/isca.2014.6853227

16. Wei, X., Jiang, N., Wang, X., Yue, H.: Detecting SDCs in GPGPUs through an efficient instruction duplication mechanism. In: Qiu, H., Zhang, C., Fei, Z., Qiu, M., Kung, S.-Y. (eds.) KSEM 2021. LNCS (LNAI), vol. 12817, pp. 571–584. Springer, Cham (2021). https://doi.org/10.1007/978-3-030-82153-1_47

17. Wei, X., Yue, H., Gao, S., Li, L., Zhang, R., Tan, J.: G-SEAP: analyzing and characterizing soft-error aware approximation in GPGPUs. Future Gener. Comput. Syst. **109**, 262–274 (2020). https://doi.org/10.1016/j.future.2020.03.040

18. Yang, L., Nie, B., Jog, A., Smirni, E.: Enabling software resilience in GPGPU applications via partial thread protection. In: 2021 IEEE/ACM 43rd International Conference on Software Engineering (ICSE). IEEE, May 2021. https://doi.org/10.1109/icse43902.2021.00114

19. Yang, L., Nie, B., Jog, A., Smirni, E.: Practical resilience analysis of GPGPU applications in the presence of single- and multi-bit faults. IEEE Trans. Comput. **70**(1), 30–44 (2021). https://doi.org/10.1109/tc.2020.2980541

20. Yue, H., Wei, X., Li, G., Zhao, J., Jiang, N., Tan, J.: G-SEPM: building an accurate and efficient soft error prediction model for GPGPUs. In: Proceedings of the International Conference for High Performance Computing, Networking, Storage and Analysis. ACM, November 2021. https://doi.org/10.1145/3458817.3476170

21. Yue, H., Wei, X., Tan, J., Jiang, N., Qiu, M.: Eff-ECC: protecting GPGPUs register file with a unified energy-efficient ECC mechanism. IEEE Trans. Comput. Aid. Des. Integr. Circ. Syst. **41**(7), 2080–2093 (2022)

22. Öz, I., Ömer, F.K.: Regional soft error vulnerability and error propagation analysis for GPGPU applications. J. Supercomput. **78**(3), 4095–4130 (2021). https://doi.org/10.1007/s11227-021-04026-6

FDRShare: A Fully Decentralized and Redactable EHRs Sharing Scheme with Constant-Size Ciphertexts

Zhichao Li[1], Zhexi Lu[1], Lingshuai Wang[2], Qiang Wang[1(✉)], and Che Bian[3]

[1] Software College, Northeastern University, Shenyang, China
`wangqiang1@mail.neu.edu.cn`
[2] School of Computer Science and Engineering, Northeastern University, Shenyang, China
[3] The Fourth Affiliated Hospital, China Medical University, Beijing, China

Abstract. Blockchain-based Electronic Health Records (EHR) sharing schemes can enable the owner to outsource the encrypted EHR to the powerful but untrusted cloud such that only authorized users can search or access EHRs and check whether the cloud returns valid results or not. However, these existing schemes suffer from three substantial shortcomings that limit their usefulness: (i) a centralized Trusted Authority (TA) is introduced to manage keys, which fully conflicts with the decentralized nature of blockchain; (ii) the size of the encrypted EHRs is linear to the attribute set size, which poses challenges for the blockchain network; (iii) the efficiency of search seriously impacts the user experience, which hinders the deployment in practice. To address these issues, we proposed FDRshare, a fully decentralized EHRs sharing scheme with constant-size ciphertexts.

Keywords: Searchable Encryption · Blockchain · Attribute-Based Encryption

1 Introduction

With the explosive increase of global information, cloud computing has been experiencing unprecedented development. To capture the market as much as possible, the cloud service providers have been rushing to launch their products, such as Amazon EC2 and S3, Microsoft Azure, and Google App Engine. Due to lower cost, higher reliability, better performance, and faster deployment, enterprises and individuals have been increasingly outsourcing their storage tasks to the cloud. As a typical and concrete application instance of cloud storage, the cloud-based electronic health record (EHR) has been playing a significant role in the healthcare industry. Unlike traditional paper-based health records, EHR can be shared among different institutions by outsourcing them to the cloud. It can vigorously facilitate personalized treatment plans, disease analysis, prediction,

Z. Tari et al. (Eds.): ICA3PP 2023, LNCS 14493, pp. 240–252, 2024.
https://doi.org/10.1007/978-981-97-0862-8_15

etc. However, past real-world incidents [1] and recent research [2] have shown that the cloud cannot be fully trusted and may expose sensitive data and forge the query result. To this end, several privacy-preserving EHR sharing schemes supporting integrity checks (PPShare) [3,4] have been proposed. In terms of privacy, the owners encode their EHR with some specified access control policies using attribute-based encryption (ABE) before uploading to the cloud. The users such as doctors or researchers with different attributes send the search tokens to make the cloud search the target EHR over all encrypted EHRs. The user can decode them if and only if the attribute set satisfies the policies embedded into ciphertexts. In terms of integrity, authenticated data structure (ADS) such as accumulator [5] or Merkel hash tree [6] is utilized to check the correctness of the query results with an overwhelming probability.

However, most of the existing PPShare schemes [7,8] rely on the centralized cloud to manage EHR and respond to the users' query requests. In such a centralized model, it inevitably suffers from DDoS attacks and single-point failure. To overcome this challenge, Hu et al. [9] proposed a novel blockchain-based scheme, which utilizes smart contracts to manage and search EHRs rather than the cloud. Given a search token, each consensus node has to faithfully execute search operations through the smart contract. If not, everyone can find his misbehavior. Due to its importance, several blockchain-based EHR sharing schemes (BBShare) [10–14] have been proposed over the past decade. However, these schemes still cannot be applied in practice. The main reasons are summarized as follows:

Conflict between Centralized TA and Blockchain: Most of the existing BBShare schemes [10,11,13,15] make use of ciphertext policy ABE (CP-ABE) [16] to facilitate fine-grained sharing of EHRs. However, CP-ABE depends on a centralized trusted authority (TA) for key management and attribute distribution. This seriously contradicts the decentralized nature of blockchain. In addition, the security of the schemes will be compromised if the fully trusted TA is corrupted. Hence, it is essential to achieve full decentralization.

Unbearable Storage Overhead: To protect EHR privacy, the existing BBShare schemes employ the CP-ABE to encode their EHR indexes and store the encrypted indexes on the blockchain. However, a common issue is that most of them rely on the traditional CP-ABE scheme, where the size of ciphertext increases at least linearly with the number of attributes involved in the access policy. This creates a conflict with the inherent storage limitations of blockchain, severely restricting its practicality. To address this issue, some schemes [12,14], such as MedShare and BBF, have been proposed to replace the traditional CP-ABE with CP-ABE with constant-size ciphertext. However, it is worth noting that these schemes still depend on a centralized TA as mentioned above. Hence, it is essential to design a fully decentralized CP-ABE with constant-size ciphertexts.

Search Efficiency Challenges: Most of the existing BBshare schemes [10–12] are interactive, where the user needs to interact with the cloud during the search phase. Multiple-round interactions impose a serious impact on the efficiency of search. To tackle this issue, Wang et al. [14] introduced the first non-interactive BBshare scheme called Medshare based on the work [17]. To the best of our knowledge, Medshare is the only non-interactive scheme achieving sublinear search performance, in line with the standard of plaintext information retrieval algorithms. However, it requires users to conduct $O(n \cdot q)$ exponentiation operations, where n is the number of matches of the least frequent keyword in the boolean search and q is the number of terms in the search. These exponentiation operations not only exhibit low efficiency but also cost high gas fees on the blockchain. Hence, it is essential to design a lightweight search method that eliminates computationally intensive operations such as exponentiation.

1.1   Our Results and Contributions

To resolve the problems mentioned above, we propose, FDRShare, a fully decentralized EHRs sharing scheme with constant-size ciphertexts. Our results and contributions can be summarized as follows:

- To achieve decentralized fine-grained access control with constant-size ciphertexts, we propose a decentralized constant-size CP-ABE (DCS-ABE) scheme, which employs multiple attribute authorities instead of one single TA for key management and utilizes an AND_m^* access control structure to generate the constant-size ciphertext.
- To improve search efficiency, we integrated a bloom filter into the non-interactive blockchain boolean search protocol. The Bloom filter is utilized to initially filter out the irrelevant search results, thereby preventing them from being processed by the search algorithm executed by the smart contract.

2   Related Work

Electronic health record (EHR) makes a transition from paper to digital medical records. This transition has accelerated health information sharing since it makes the use of them more flexible. To alleviate the maintenance burden of massive EHRs, the existing EHR sharing schemes commonly resort to the powerful cloud. Nevertheless, as the cloud is untrusted, it inevitably suffers from two major security challenges: privacy and integrity. To overcome these challenges, Nayak at el. initialed the first privacy-preserving EHR sharing scheme supporting integrity checks [18]. Following this pioneering work, plenty of schemes have been proposed. According to their architectures, it can be roughly categorized into two folds: centralized ones [19] and decentralized ones [10,14,20]. The former relies on the centralized cloud to manage EHR and handle query requests. In contrast, the latter is an enhanced version that incorporates security measures. It leverages blockchain techniques to resist DDoS attacks and single-point failure, which are potential vulnerabilities in the former. All search and upload operations over EHRs are performed through smart contracts rather than the cloud. Due to the limitation of space, we only focus on the decentralized privacy-preserving EHR sharing schemes supporting integrity checks (DPPShare) in the following.

Azaria et al. [20] utilize searchable symmetric encryption (SSE) to construct the first DPPShare scheme. Since SSE allows only the private key holder (i.e., the owner) to produce ciphertexts and to create trapdoors for search, the owners cannot share their EHRs with others. To overcome this challenge, [11,12] incorporated traditional CP-ABE with SSE schemes to achieve fine-grained access control. The users with different attributes send the search tokens to make the smart contract search the target EHR over all encrypted EHRs. However, traditional CP-ABE schemes, which introduce a TA for attribute authentication and key management, will make the system suffer from the single points of failure and compromise attacks. Additionally, the linear growth of CP-ABE ciphertext size with the attribute set's size imposes a significant storage burden on the blockchain.

To address these issues, we design a decentralized ABE scheme with a constant-size ciphertext, which supporting high-efficiency on-chain search. Our theoretical comparison with the most advanced model proposed by Wang [14] is presented in Table 1. Although our scheme shares the same time complexity with [14] in Search algorithm, our Search algorithm improves the search efficiency by avoiding time-consuming exponential or bilinear pairing operations.

Table 1. Comparison of Computation Cost

Complexities	MedShare [14]	FDRShare
ParameterSetup	$O(1)$	$O(1)$
IndexGen	$O(n)$	$O(n)$
KeyGen	$O(k)$	$O(k)$
STGen	$O(q^2)$	$O(q)$
Search	$O(n \cdot q)$	$O(n \cdot q)$

3 Preliminaries

3.1 Bilinear Pairing

Let $\mathbb{Z}_p$ be a finite field over prime p. Let $\mathbb{G}$, $\mathbb{G}_T$ be two cyclic multiplicative groups of the same prime order p. g is the generator of $\mathbb{G}$. Denote $\hat{e}$ a bilinear pairing map : $\mathbb{G} \times \mathbb{G} \rightarrow \mathbb{G}_T$ with the following properties:

1) Bilinear: $\forall a, b \in \mathbb{Z}_p$, $\hat{e}\left(g^a, g^b\right) = \hat{e}\left(g, g\right)^{ab}$.
2) Non-degenerate: $\hat{e}\left(g, g\right) \neq 1$.
3) Computable: There exists an efficient algorithm to compute $\hat{e}\left(g_1, g_2\right)$ for all $g_1, g_2 \in \mathbb{G}$.

3.2 DDH Assumption

Let $\mathbb{G}$ be a cyclic group of prime order p, the decisional Diffie-Hellman(DDH) problem is to distinguish the ensembles $\{(g, g^a, g^b, g^{ab})\}$ from $\{(g, g^a, g^b, g^z)\}$, where the elements $g \in \mathbb{G}$ and $a, b, z \in F_p$ are chosen uniformly at random. We say that the DDH assumption holds if for all efficient adversary $\mathcal{A}$, advantage

$$Adv_{\mathcal{A}}^{DDH}(1^\lambda) = |\Pr[\mathcal{A}(\mathbb{G}, p, g, g^a, g^b, g^c) = 1]$$
$$- \Pr[\mathcal{A}(\mathbb{G}, p, g, g^a, g^b, g^{ab}) = 1]|$$

is negligible.

3.3 Pseudo-Random Function

Let F be a function transforming the element $x \in X$ to an output $y \in Y$ with a secret seed $k \in K_{prf}$. We say F is a pseudo-random function (PRF), if for all efficient adversaries $\mathcal{A}$, advantage

$$Adv_{F,\mathcal{A}}^{prf}(1^\lambda) = \Pr[A^{F(k,\cdot)}(1^\lambda) = 1] - \Pr[A^{f(\cdot)}(1^\lambda) = 1]$$

is negligible, where f is a truly random function from X to Y.

4 System Definition

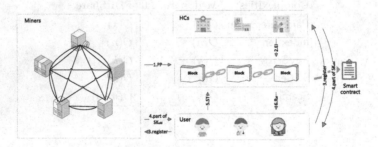

Fig. 1. The Architecture of FDRShare

The architecture of FDRShare is depicted in Fig. 1. To begin with, Miners initiate the public parameters denoted as PP and disseminate them across the blockchain network. Subsequently, HCs employ the same PP to generate encrypted indexes, denoted as EI , which are then deployed onto the smart contract.

After the system setup, users enter the system and initiate registration requests directed toward both Miners and HCs. Following the authentication of these requests, Miners and HCs provide the respective users with sets of secret keys. Armed with these secret key sets, users are able to generate encrypted search tokens and effectively engage with the smart contract to retrieve search results. Users can employ their secret key sets to decrypt the obtained search results. Successful decryption is contingent upon whether users' attributes fulfill the access policies integrated within the encrypted search results.

4.1   Threat Model

We make the following threat assumptions for three entities in our system:

- HCs(Healthcare Centers) are always honest. As the builders and sharers of EHR, they try their best to protect the confidentiality of EHR. They are responsible for generating and distributing authorized keyword search keys for users. Besides, they generate the encrypted indexes and deploy them to the smart contract according to the protocol.
- Miners are honest. We use the term "Miners" to refer to the consensus nodes in the blockchain. They are responsible for generating public parameters of the whole system. Meanwhile, they execute the tasks of authorizing attributes and issuing secret keys for users, as well as faithfully executing search contract based on a consensus algorithm.
- Users are potential adversaries. They are doctors, nurses, and so on, who are allowed to access the EHR. While they generate search tokens honestly according to the protocol, they intend to extract sensitive information from encrypted indexes, query transactions, and search results. Moreover, they attempt to collude with each other to gain access privileges beyond their identity.

4.2   System Model

Formally, the system model of FDRShare consists of the following four algorithms:

1) $(PP, MK) \leftarrow Setup\,(1^\lambda, DB)$: The system setup algorithm is run by HCs and Miners. Given the security parameter 1^λ and local EHR database DB as inputs, this algorithm outputs the public parameters PP and the system master key MK.
2) $SK_{uid} \leftarrow KeyGen\,(PP, MK, uid)$: The key generation algorithm is run by Miners and HCs jointly. Given the public parameters PP, the system master key MK, and the user identity uid as inputs, this algorithm outputs the users' secret key set SK_{uid}.
3) $ST \leftarrow STGen\,(SK_{uid}, Q, PP)$: The search token generation algorithm is run by Users and smart contract jointly. Given the users' secret key set SK_{uid}, the query set Q, and the public parameters PP as inputs, this algorithm outputs the search token ST.
4) $R_{id} \leftarrow Search\,(PP, ST)$: The smart contract runs the on-chain search algorithm. Given the public parameters PP and the search token ST as inputs, this algorithm outputs the result set R_{id}.

5   Building Block

Before delving into the detailed construction of FDRShare, we first introduce our detailed construction of DSC-ABE in this section. To resolve the tackles posed by TA and the unbearable storage overhead brought by traditional CP-ABE, we design a DCS-ABE scheme, which employs multiple authorities and has constant-size ciphertext, based on the MA-ABE scheme and constant-size CP-ABE scheme [21,22]. In EHR sharing application scenario, we consider that users have multiple attributes, and each of them has multiple values. We employ a set of pre-selected consensus nodes (miners) that serve as attribute authorities. Each Miner manages a set of mutually exclusive attributes. Let $A_k = \{i_1, i_2, ..., i_{n_{ki}}\}$ $(1 \leqslant k \leqslant n_{Miner})$ denote the set of attributes managed by the k-th Miner, and $i_k (1 \leqslant i_k \leqslant n_{ki})$ represent the i-th attribute managed by the k-th Miner. The set of attribute values managed by the k-th Miner is denoted as V_k , where $v_{k,n}$ represents the set of values for the n-th attribute managed by the k-th Miner. We specify j_{i_k} as the j-th value of the i-th attribute managed by the k-th Miner. The decentralizing constant-size CP-ABE (DCS − ABE) scheme consists of the following four algorithms:

1) $(mpk, msk) \leftarrow$ DSCABE.Setup (1^λ) : The DSC-ABE setup algorithm is run by Miners. Miners negotiate a secure parameter 1^λ as input to generate a collision-resistant hash function $H_0 : \{0,1\}^* \rightarrow \mathbb{Z}_p^*$, system attribute set A_k , attribute value set V_k , and an integer multiplication cyclic group $\mathbb{G}$. The cyclic group $\mathbb{G}$ has an order of p and generator g . Additionally, each miner independently chooses a pair of random numbers $\alpha_k, \beta_k \epsilon \mathbb{Z}_N$ as their private key and stores them locally, and computes

$$M_{i_k, j_{i_k}} = g^{-H(\alpha_k || i_k || j_{i_k})}, N_{i_k, j_{i_k}} = e(g, g)^{H_0(\beta_k || i_k || j_{i_k})}. \tag{1}$$

The ABE public key is $mpk = \left\langle g, \left\{ M_{i_k, j_{i_k}}, N_{i_k, j_{i_k}} \right\} \right\rangle$, and the ABE master key is $msk = \{\alpha_k, \beta_k\}, 1 \leqslant k \leqslant n_{Miner}$. Given access policy $\mathbb{A}$, we can aggregate mpk according to access policy $\mathbb{A}$:

$$\langle M_\mathbb{A}, N_\mathbb{A} \rangle = \left\langle \prod_{i_k \in \mathbb{A}} \bar{M}_{i_k}, \prod_{i_k \in \mathbb{A}} \bar{N}_{i_k} \right\rangle, \tag{2}$$

where $\bar{M}_{i_k} = M_{i_k, j_{i_k}}$ and $\bar{N}_{i_k} = N_{i_k, j_{i_k}}$. We can utilize this structure for ABE encryption then.

2) $ASK_{uid} \leftarrow$ DCSABE.KeyGen (msk, uid): The DSC-ABE key generation algorithm is run by Miners. Given the master key msk and the user identity uid as inputs, this algorithm generates the attribute secret key ASK_{uid} . When users join the system, they securely transmit uid to all Miners for identity verification. Once the verification is successful, each miner provides the corresponding attribute secret key. The attribute secret key from the k-th miner, with attribute i_k and attribute value j_{i_k}, is represented as:

$$ask_{i_k, j_{i_k}} = g^{H_0(\beta_k || i_k || j_{i_k})} H_2(uid)^{H_0(\alpha_k || i_k || j_{i_k})}. \tag{3}$$

The users' secret attribute secret key is $ASK_{uid} = \left\{ ask_{i_k, j_{i_k}} \right\}$.

3) $\mathbb{C} \leftarrow DCSABE.Enc\,(mpk, m, \mathbb{A})$: The DSC-ABE encryption algorithm is run by HCs. Given the master key mpk , message m, and access control policy $\mathbb{A}$ as inputs, this algorithm outputs the ABE ciphertext $\mathbb{C}$. In FDRShare, the file is encrypted by the AES algorithm and then uploaded to IPFS. The AES secret key is denoted as K_f . For each file upload, IPFS returns a unique file identifier id . HCs input mpk and $\mathbb{A}$ to generate an access control structure $\langle M_{\mathbb{A}}, N_{\mathbb{A}} \rangle$, which is used to encrypt $m = id \| K_f$. Finally, HCs select a random number $s \in \mathbb{Z}_p^*$ and encrypt the data:

$$\mathbb{C} = F_{id} = \langle C_0, C_1, C_2 \rangle, \tag{4}$$

where $C_0 = (id\,\|K_f) \cdot N_{\mathbb{A}}^s$, $C_1 = g^s$, $C_2 = M_{\mathbb{A}}^s$. Since attributes are aggregated into exponents, the ciphertext is constant-size.

4) $m \leftarrow DCSABE.Dec\,(uid, ASK_{uid}, \mathbb{C})$: The DCS-ABE decryption algorithm is run by Users. Given the user identity uid , attribute secret key ASK_{uid} , and ABE ciphertext $\mathbb{C}$ as inputs, this algorithm outputs the decrypted result $m = id\|K_f$ or an $\perp$. The user aggregates their attribute secret key $ASK_{uid} = \prod_{i_k \in L_{uid}} ask_{i_k, j_{i_k}}$, where L_{uid} denotes the user's attribute set, and decrypts:

$$m = id\,\|K_f = \frac{C_0}{\hat{e}\,(ASK_{uid}, C_1) \cdot \hat{e}\,(H_2\,(uid)\,, C_2)}. \tag{5}$$

If the user's attribute set L_{uid} satisfies the $\mathbb{A}$ embedded in F_{id} , or in other words, if $L \models \mathbb{A}$, the user can successfully decrypt and obtain the value $id\,\|K_f$. Otherwise, an $\perp$ will be returned. With result $id\,\|K_f$, the user can use the file identifier id to locate the corresponding file on IPFS and decrypt the file using the symmetric key K_f .

6   Detail Construction of FDRShare

The detailed construction of FDRShare is as follows:

1) $(PP, MK) \leftarrow Setup\,(1^\lambda, DB)$: The system setup algorithm consists of two sub-algorithms: ParameterSetup and IndexGen.
   - $(PK, MK) \leftarrow ParameterSetup\,(1^\lambda)$: The parameter setup algorithm is run by Miners and HCs. Given 1^λ as input, Miners invoke DCSABE.Setup to generate the ABE public key mpk and the ABE master key msk. Then, Miners choose two cyclic groups $\mathbb{G}$ and $\mathbb{G}_T$ with the same order p . g is the generator of $\mathbb{G}$. Based on the chosen groups, Miners selects the bilinear maps $e : \mathbb{G} \times \mathbb{G} \to \mathbb{G}_T$, $e_1 : \mathbb{G} \times \mathbb{G}_T \to \mathbb{G}$, and three hash functions $H_0 : \{0,1\}^* \to \mathbb{Z}_p^*$, $H_1 : \{0,1\}^* \to \{0,1\}^\lambda$, and $H_2 : \{0,1\}^* \to \mathbb{G}$. After that, Miners choose the pseudo-random function F and the invertible pseudo-random permutation function P .
   Receiving the public parameters, HCs randomly select a seed k and a generator $g_1 \in \mathbb{Z}_n^*$ for the pseudo-random function F . Then, HCs initiate

---

Algorithm 1:

Input: $IV(w)$, $FW(id)$, PK and the system *masker key MK*
Output: $EIindex\{AIindex, BFindex, PTindex\}$

1   Initialize set $AIindex$, $BFindex$, $PTindex$ to empty dict;
2   **for** *each $w \in DB$* **do**
3   Initialize counter int c = 0;
4     $w_{t_0} \leftarrow \{0,1\}^\lambda$;
5   **for** *all $id \in IV(w)$* **do**
6   Set the counter $c = c + 1$;
7   Generate a random nonce $t_c \leftarrow \{0,1\}^\lambda$;
8   Set $w_{t_c} \leftarrow P\left(t_c, w_{t_{c-1}}\right)$;
9   Calculate $stagw$ $(Eq.6)$;
10   Set $l \leftarrow H_1(stagw)$;
11   Calculate F_{id} $(Eq.4)$;
12   Calculate $\Gamma_{id} = F_{id}||t_c \oplus H_2(l||w_{t_c})$;
13   Set $\sigma \leftarrow H_1(stagw||w_{t_c})$;
14   Append $AIindex[\sigma] = \Gamma_{id}$;
15   **end**
16   Append $PTindex[l] = w_{t_c}||c$;
17   **end**
18   **for** *each id in DB* **do**
19   Initialize $\vec{V}_{BF}^{id}$ to empty vector;
20   Initialize I_{id} to empty list;
21   **for** *all $w \in FW(id)$* **do**
22       $\vec{V}_{BF}^{id} \leftarrow$ *storage w into* $\vec{V}_{BF}^{id}$;
23   calculate $I_w = g^{H_1(stagw)}$;
24   Add I_w to I_{id};
25   **end**
26   Append $BFilter[F_{id}] = \vec{V}_{BF}^{id}||I_{id} = \mathcal{B}_{id}$;
27   **end**
28   **Deploy** $AIindex$, $BFindex$, $PTindex$ **to smart contract**

---

the authorized keyword search keys $\mathbf{w} = \{w_1, w_2, ..., w_n\}$, where we map each keyword to a prime number w_k for efficient storage [23]. Part of the $\mathbf{w}$ will then be sent to authenticated Users for generating search tokens. The system public key is $\mathsf{PK} = (\mathsf{H_0, H_1, H_2, F, P, msk})$, and the system master key is denoted as $\mathsf{MK} = (\mathsf{p, g_1, k, msk, w})$. The order p and the ABE secret key msk are kept by Miners, while w, $\mathsf{g_1}$, and k are kept by HCs.

- $\mathsf{EI} \leftarrow \mathsf{IndexGen}(\mathsf{DB, PK, MK})$: The index generation algorithm is run by HCs. First, HCs process DB to obtain $\mathsf{FW(id)}$ and $\mathsf{IV}(w)$. Specifically, $\mathsf{IV}(w)$ establishes a mapping from the keyword w to all identifiers id of files containing w , while $\mathsf{FW(id)}$ establishes a mapping from the file identifier id to the keywords contained in the corresponding file. HCs generate the encrypted indexes according to Algorithm 1. The HCs then update all the new indexes on the blockchain.

The final output of the system setup algorithm is the system public key $PP = \{PK, EI\}$ and the system master MK.

Algorithm 2: Search token generation

Input: User private key SK_{uid}, Search query set Q, Partial Token index $PTindex$

Output: Search Token ST

1 Initialized Trapdoor T to an empty array.;
2 Initialized $\vec{V}_{BF}^{Q}$ to a vector.;
3 Calculate $stagw_l$; Set $l \leftarrow H_1(stagw_l)$;
4 Get $w_{t_c} \| c \leftarrow PTindex[l]$;
5 **for** *each* $w/\{w_l\}$ *in* Q **do**
6     $\vec{V}_{BF}^{Q} \leftarrow$ Store w to $\vec{V}_{BF}^{Q}$;
7   Calculate $I_w = g^{H_1(stagw)}$;
8   Add I_w to T;
9 **end**
10 **return** $ST\left\{\vec{V}_{BF}^{Q}, Trapdoor\ T, w_{t_c} \| c, stagw_l\right\}$

---

2) $SK_{uid} \leftarrow KeyGen(PP, MK, uid)$: The system pubic key PP can be parsed as $\{PK, EI\}$. Given the msk from MK and the user identity uid , which are sent to Miners and HCs by Users when registering, as inputs, Miners invoke DCSABE.KeyGen to generate the users secret attribute secret key ASK_{uid} . Subsequently, HCs assign authorized user keyword search keys $\mathbf{w}_{uid}$, where $\mathbf{w}_{uid} \in \mathbf{w}$, and a keyword private key set $WSK = g_1^{\frac{1}{\prod_{i=1}^{i=n} w_i}}$ to the user based on uid . The users' secret key set is $SK_{uid} = (k, \mathbf{w}_{uid}, WSK, ASK_{uid})$. Finally, Miners and HCs transmit the secret key set to the user through a secure channel.

3) $ST \leftarrow STGen(SK_{uid}, Q, PP)$: PP can be parsed as $\{PK, EI\}$. Given the users' secret key set SK_{uid}, boolean query Q , and the encrypted indexes EI as inputs, the user selects the least frequent word w_l from Q and calculates:

$$stagw = F\left(k, (WSK)^{\prod_{w \in \mathbf{w}_l/\{w_l\}} w}\right), \tag{6}$$

and submits it to the smart contract. Subsequently, the smart contract executes Algorithm 2 and returns the search token ST .

4) $R_{id} \leftarrow Search(PP, ST)$: The PP can be parsed as $\{PK, EI\}$. Given the encrypted indexes EI and the search token ST as inputs, the smart contract executes Algorithm 3.

Algorithm 3:

Input: Search Token $ST\left\{\vec{V}_{BF}^{Q}, Trapdoor\ T, w_{t_c}\|c, stagw_l\right\}$, $AIindex$, $BFindex$

Output: Result set R_{id}

1 **for** $i = c$ *to* 1 **do**
2   Set $\sigma = H_1(stagw_l\|w_{t_c})$;
3   Get $\Gamma_{id} \leftarrow AIindex.find(\sigma)$;
4     $F_{id}\|t_c \leftarrow \Gamma_{id} \oplus H_2(stagw\|w_{t_c})$;
5   Get $\left\{V_{BF}^{id}, I_{id}\right\} \leftarrow BFindex.find(F_{id})$;
6   **if** $V_{BF}^{Q} \cup V_{BF}^{id} \mathrel{!}= V_{BF}^{id}$ **then**
7   continue;
8   **else**
9   **if** $T \subseteq I_{id}$ **then**
10   Add F_{id} to result R_{id};
11   **end**
12   **end**
13   Set $w_{t_{i-1}} \leftarrow P^{-1}(t_i, w_{t_i})$;
14 **end**
15 **return** result R_{id};

---

7 Conclusion

In this paper, we present FDRShare, a new blockchain-based EHR sharing system with fully decentralized fine-grained access control. Compared to the existing solutions, our solution offers a more efficient privacy-protecting search service. Additionally, we have designed the DCS-ABE scheme based on blockchain characteristics, achieving further decentralization. To the best of our knowledge, our solution is currently the only one that addresses both of these aspects, making it highly practical for real-world applications. Future research directions include further refining the proposed DCS-ABE scheme and on-chain searchable encryption scheme to enhance their security. Additionally, introducing editability into the system represents a valuable avenue of investigation, as it can facilitate the broader adoption of the system. In the initial full version of the paper, we designed a scheme for an editable blockchain system, which is not presented in this article due to space constraints.

Acknowledgment. We thank the anonymous reviewers for their fruitful suggestions. This work was supported in part by the National Natural Science Foundation of China under Grant 62202090 and 62173101, by Liaoning Province Natural Science Foundation Medical-Engineering Cross Joint Fund under Grant 2022-YGJC-24, by Doctoral Scientific Research Foundation of Liaoning Province under Grant 2022-BS-077, and by the Fundamental Research Funds for the Central Universities under Grant N2217009.

References

1. Zhang, R., Liu, L.: Security models and requirements for healthcare application clouds. In: 2010 IEEE 3rd International Conference on Cloud Computing, pp. 268–275. IEEE (2010)
2. Hasan, M.Z., Hussain, M.Z., Mubarak, Z., Siddiqui, A.A., Qureshi, A.M., Ismail, I.: Data security and Integrity in Cloud Computing. In: 2023 International Conference for Advancement in Technology (ICONAT), pp. 1–5, January 2023
3. Reedy, B.E., Ramu, G.: A secure framework for ensuring EHr's integrity using fine-grained auditing and CP-ABE. In: 2016 IEEE 2nd International Conference on Big Data Security on Cloud (BigDataSecurity), IEEE International Conference on High Performance and Smart Computing (HPSC), and IEEE International Conference on Intelligent Data and Security (IDS), pp. 85–89 (2016)
4. Su, Y., Sun, J., Qin, J., Hu, J.: Publicly verifiable shared dynamic electronic health record databases with functional commitment supporting privacy-preserving integrity auditing. IEEE Trans. Cloud Comput. 10(3), 2050–2065 (2020)
5. Khedr, W.I., Khater, H.M., Mohamed, E.R.: Cryptographic accumulator-based scheme for critical data integrity verification in cloud storage. IEEE Access 7, 65635–65651 (2019)
6. Wang, Q., Wang, C., Ren, K., Lou, W., Li, J.: Enabling public auditability and data dynamics for storage security in cloud computing. IEEE Trans. Parallel Distrib. Syst. 22(5), 847–859 (2011)
7. Joshi, M., Joshi, K., Finin, T.: Attribute based encryption for secure access to cloud based EHR systems. In: 2018 IEEE 11th International Conference on Cloud Computing (CLOUD), pp. 932–935 (2018)
8. Li, M., Yu, S., Zheng, Y., Ren, K., Lou, W.: Scalable and secure sharing of personal health records in cloud computing using attribute-based encryption. IEEE Trans. Parallel Distrib. Syst. 24(1), 131–143 (2013)
9. Hu, S., Cai, C., Wang, Q., Wang, C., Wang, Z., Ye, D.: Augmenting encrypted search: a decentralized service realization with enforced execution. IEEE Trans. Dependable Secure Comput. 18(6), 2569–2581 (2021)
10. Xia, Q., Sifah, E.B., Asamoah, K.O., Gao, J., Du, X., Guizani, M.: MeDShare: trust-less medical data sharing among cloud service providers via blockchain. IEEE Access 5, 14757–14767 (2017)
11. Zhang, P., White, J., Schmidt, D.C., Lenz, G., Rosenbloom, S.T.: FHIRchain: applying blockchain to securely and scalably share clinical data. Comput. Struct. Biotechnol. J. 16, 267–278 (2018)
12. Wang, S., Zhang, Y., Zhang, Y.: A blockchain-based framework for data sharing with fine-grained access control in decentralized storage systems. IEEE Access 6, 38437–38450 (2018)
13. Wu, G., Zhu, B., Li, J.: BMKs: a blockchain based multi-keyword search scheme for medical data sharing. In: IEEE Symposium on Computers and Communications (ISCC), pp. 1–7. IEEE (2022)
14. Wang, M., Guo, Y., Zhang, C., Wang, C., Huang, H., Jia, X.: MedShare: a privacy-preserving medical data sharing system by using blockchain. IEEE Trans. Serv. Comput. 16(1), 438–451 (2023)
15. Yu, G., et al.: Enabling attribute revocation for fine-grained access control in blockchain-IoT systems. IEEE Trans. Eng. Manage. 67(4), 1213–1230 (2020)
16. Bethencourt, J., Sahai, A., Waters, B.: Ciphertext-policy attribute-based encryption. In: IEEE Symposium on Security and Privacy (SP 2007), pp. 321–334. IEEE (2007)

17. Cash, D., Jarecki, S., Jutla, C., Krawczyk, H., Roşu, M.-C., Steiner, M.: Highly-scalable searchable symmetric encryption with support for Boolean queries. In: Canetti, R., Garay, J.A. (eds.) CRYPTO 2013. LNCS, vol. 8042, pp. 353–373. Springer, Heidelberg (2013). https://doi.org/10.1007/978-3-642-40041-4_20

18. Nayak, S.K., Tripathy, S.: Privacy preserving provable data possession for cloud based electronic health record system. In: IEEE Trustcom/BigDataSE/ISPA 2016, pp. 860–867 (2016)

19. Chenthara, S., Ahmed, K., Wang, H., Whittaker, F.: Security and privacy-preserving challenges of e-health solutions in cloud computing. IEEE Access 7, 74361–74382 (2019)

20. Azaria, A., Ekblaw, A., Vieira, T., Lippman, A.: Medrec: using blockchain for medical data access and permission management. In: 2nd International Conference on Open and Big Data (OBD) 2016, pp. 25–30. IEEE (2016)

21. Lewko, A., Waters, B.: Decentralizing attribute-based encryption. In: Paterson, K.G. (ed.) EUROCRYPT 2011. LNCS, vol. 6632, pp. 568–588. Springer, Heidelberg (2011). https://doi.org/10.1007/978-3-642-20465-4_31

22. Zhang, Y., Zheng, D., Chen, X., Li, J., Li, H.: Computationally efficient ciphertext-policy attribute-based encryption with constant-size ciphertexts. In: Chow, S.S.M., Liu, J.K., Hui, L.C.K., Yiu, S.M. (eds.) ProvSec 2014. LNCS, vol. 8782, pp. 259–273. Springer, Cham (2014). https://doi.org/10.1007/978-3-319-12475-9_18

23. Sun, S.-F., et al.: Non-interactive multi-client searchable encryption: realization and implementation. IEEE Trans. Dependable Secure Comput. 19(1), 452–467 (2022)

An Efficient Fault Tolerance Strategy for Multi-task MapReduce Models Using Coded Distributed Computing

Zaipeng Xie[1,2(✉)], Jianan Zhang[2], Yida Zhang[2], Chenghong Xu[2], Peng Chen[2], Zhihao Qu[1,2], and WenZhan Song[3]

[1] Key Laboratory of Water Big Data Technology of Ministry of Water Resources, Hohai University, Nanjing, China
{zaipengxie,quzhihao}@hhu.edu.cn
[2] College of Computer and Information, Hohai University, Nanjing, China
{jianan_zhang,zhangyida,chxu,pengchen}@hhu.edu.cn
[3] Center for Cyber-Physical Systems, University of Georgia, Athens, GA 30602, USA
wsong@uga.edu

Abstract. MapReduce is a programming framework designed for processing and analyzing large volumes of data in a distributed computing environment. Despite its capabilities, it faces challenges due to silent data corruption during task execution, which can yield inaccurate results. Ensuring fault tolerance in the MapReduce framework while minimizing communication overhead presents considerable challenges. This study presents CDCFT (Coded Distributed Computing Fault Tolerance), a novel approach to fault tolerance within the MapReduce paradigm, combining the strengths of TMR (Triple Modular Redundancy) and CDC (Coded Distributed Computing). By leveraging task-level TMR and voting mechanisms, CDCFT robustly defends against silent data corruption. To further optimize, CDCFT employs intra-group broadcasts for relaying intermediate messages and has a finely-tuned node grouping combined with a strategic data and task allocation procedure. Through rigorous theoretical analysis, we establish that CDCFT's communication overhead during the Shuffle Stage is notably less than traditional CDC methods that rely on triple modular redundancy. Experimental results showcase the efficacy of CDCFT, signifying a substantial reduction in the overall communication overhead and execution time compared to the conventional fault-tolerant methods.

Keywords: MapReduce framework · Silent data corruption · Fault tolerance · Coding distributed computing · Communication load

---

Supported by The Belt and Road Special Foundation of the State Key Laboratory of Hydrology-Water Resources and Hydraulic Engineering under Grant 2021490811 the National Natural Science Foundation of China under Grant 62102131 Natural Science Foundation of Jiangsu Province under Grant BK20210361.

Z. Tari et al. (Eds.): ICA3PP 2023, LNCS 14493, pp. 253–271, 2024.
https://doi.org/10.1007/978-981-97-0862-8_16

1   Introduction

MapReduce [18] is a prominent distributed computing paradigm for processing large-scale computational tasks. Apache Spark [17] and Apache Hadoop YARN [32] are frameworks similar to MapReduce, with Spark offering high-performance processing engine capabilities and Hadoop YARN excelling in resource management and framework support. However, as the distributed clusters scale up, MapReduce faces challenges [23] due to system failures and data corruption caused by soft errors, leading to inaccurate or unreliable results.

As the number of nodes and the scale of data increase, the likelihood of failures rises. Dealing with a large number of potential failures and maintaining fault tolerance becomes more challenging [6,33]. The occurrence of Silent Data Corruption (SDC) [16] can substantially impact the accuracy and reliability of computations. SDC alters processed data, introducing erroneous information that can propagate through subsequent stages, ultimately compromising the overall validity and reliability [4]. With the growth of MapReduce systems, the communication overhead and Silent Data Corruption (SDC) escalates, intensifying the risks of inaccuracies and performance degradation. These challenges become particularly prominent in large and intricate distributed computing environments. Thus, it becomes crucial to implement effective measures and advanced techniques to mitigate these challenges successfully.

To achieve fault tolerance, MapReduce [6] makes use of speculative execution and task replication. Task replication [19] involves creating redundant copies of tasks on different nodes, where replication-based fault tolerance methods, such as Triple Modular Redundancy (TMR) [3], can be used. Speculative execution [6] involves launching multiple duplicate copies of slower-running tasks to expedite their completion. The result is taken from whichever copy finishes first, while the redundant copies are terminated. While speculative execution aims to improve performance, it also provides some fault tolerance if a copy fails. While these techniques provide fault tolerance, they also introduce substantial communication overhead. Task replication requires additional communication between nodes to coordinate the redundant copies. This overhead can impact the overall performance and efficiency of the MapReduce framework.

For example, the *Shuffle stage* in MapReduce can encounter a performance bottleneck due to increased communication load caused by assigning individual computing nodes to subtasks. Additionally, fault recovery mechanisms, such as redistributing tasks and data among functioning nodes, contribute to further communication overhead. The cumulative effect of these factors results in heightened network traffic and potential congestion during the *Shuffle Stage*, requiring effective management strategies to optimize communication and maintain efficient performance in MapReduce systems.

Coded Distributed Computing (CDC) [20] is a technique that introduces redundancy through data encoding and distributed task assignments to provide fault tolerance in distributed systems. CDC can enhance the efficiency and scalability in MapReduce, decreasing communication load during the *Shuffle Stage* by assigning repeated Map tasks to different servers and encoding message bits across keys and data blocks. However, integrating CDC into distributed fault tolerance in MapReduce can further increase the computation load at each node and present the challenge of effectively distributing and coordinating coding tasks across multiple nodes [26], potentially impacting system performance.

Silent data corruption (SDC) can significantly impact the accuracy of computation results in MapReduce by corrupting the encoded data during the coding process, potentially introducing errors into the final output and affecting subsequent stages of the MapReduce workflow. SDC can be triggered by various factors such as cosmic rays, hardware faults, software bugs, and environmental conditions. Detection and mitigation of SDC in a distributed setting are challenging [1], as conventional error detection mechanisms may not effectively identify silent corruptions. While task redundancy can provide some fault tolerance to SDC, it also incurs additional communication overheads that need to be managed, depending on the level of replication. Therefore, there's a growing interest in designing SDC-aware coding schemes that optimize reliability, resilience, and performance, especially when MapReduce frameworks operate in SDC environments. However, further research is essential to address SDC concerns effectively.

This study introduces CDCFT (Coded Distributed Computing Fault Tolerance), a fault-tolerance solution for the MapReduce framework that combines the benefits of TMR and CDC to achieve fault tolerance, including silent data corruption, at a low cost. CDCFT employs triple modular task redundancy and voting mechanisms for fault tolerance. It reduces communication overhead by using intra-group broadcasts for intermediate messages. CDCFT also includes an optimized node grouping scheme and data and task allocation scheme. The main contributions of this study are as follows:

- We propose a novel fault tolerance strategy, named CDCFT, that leverages the advantages of CDC and TMR for the MapReduce framework. This strategy distributes the workload between Solver and Helper nodes, enabling the realization of fault tolerance while minimizing the communication overhead.
- We theoretically prove that the communication overhead of CDCFT in the *Shuffle Stage* is always less than or equal to that of a vanilla CDC fault-tolerant method using triple modular redundancy.
- We evaluate the performance of distributed computing by implementing four fault tolerance algorithms for MapReduce running the Terasort tasks. Our experimental results demonstrate that CDCFT can achieve an 86% reduction in communication overhead and a 13.2% reduction in execution time when compared with the conventional TMR approach.

The rest of this paper is organized as follows: Sect. 2 outlines fault tolerance research in MapReduce and Coded Distributed Computing. Section 3 covers Preliminaries for understanding CDCFT. Section 4 presents the proposed framework, processes, and mathematical derivations related to CDCFT. Section 5 compares CDCFT with other fault tolerance methods through experimental analysis. Finally, Sect. 6 summarizes the key findings and contributions.

2   Related Work

MapReduce has emerged as a crucial technology for distributed data processing in large-scale computing environments [8,25,28]. However, when dealing with massive datasets, the risk of failures is inherent, necessitating the implementation of fault tolerance mechanisms to ensure seamless operation [23]. Silent data corruption (SDC) [2] denotes data corruption without explicit error detection or notification. It can originate from various factors, impacting data at rest, during transit, or during processing. In the context of MapReduce, SDC can introduce errors in intermediate or final computation results, potentially propagating them to subsequent stages. Detecting and mitigating SDC present challenges [4,29] as conventional error detection mechanisms may prove inadequate.

Various approaches [5,9,11,22,23,27,29–31] have been proposed to address fault tolerance in MapReduce. These include speculative execution, task replication, checkpoint-based recovery, and fault-tolerant coding. Speculative execution [18], designed primarily for performance enhancement, launches multiple task copies concurrently on different nodes and selects the result from the earliest completed instance, thereby alleviating straggling tasks' impact and offering some degree of fault tolerance. Task replication [23,28] involves creating redundant copies of tasks on different nodes. If a node fails, the tasks running on that node can be reassigned to other functioning nodes that hold copies of the required data. The redundancy ensures that the computation can continue without interruption, even if a node fails. Checkpoint-based recovery [23] in MapReduce involves periodically saving the computation state to checkpoints. In the event of a failure, the system can resume processing from the most recent checkpoint, reducing data loss and improving fault tolerance. Periodic system checkpoints allow for swift recovery by restarting from the latest checkpoint after failure, thereby improving fault tolerance [22]. Fault-tolerant coding [11] in MapReduce utilizes error-correcting codes to detect and correct errors during data transmission or storage, but it can introduce communication and computational overhead. With its stochastic nature, SDC poses a greater detection and control challenge compared to other errors [16]. Consequently, achieving effective and cost-efficient SDC fault tolerance with existing solutions remains challenging.

Among these solutions, Triple modular redundancy (TMR) [1,10] is a simple yet effective technique for managing silent data corruption (SDC). TMR replicates tasks three times and uses majority voting to determine the final result. To improve energy efficiency, Salehi et al. [24] propose LE-NMR which divides execution into indispensable and on-demand phases. Only a subset of task copies is executed during the indispensable phase, while the on-demand phase executes the remaining copies for majority voting. Mireshghallah et al. [19] present Reactive TMR, a system that can detect and deactivate malfunctioning components, reallocating their tasks to operational ones and distinguishing between permanent and transient faults. Through simulations, they show that Reactive TMR can provide substantial energy savings over conventional TMR while maintaining high reliability. However, TMR techniques may still incur significant communication overhead for large-scale systems.

Coded Distributed Computing (CDC) [13,15] aims to minimize communication overhead and enhance parallel processing in Multi-Stage Computation Tasks. A local implementation of a generalized CDC scheme is effective, whereas a global strategy for the whole task graph can significantly reduce bandwidth usage and response time. Li et al. [13] introduce Coded MapReduce to lessen the *Shuffle Stage* communication load in MapReduce settings. This is achieved by assigning repetitive Map tasks to various servers and using coding techniques for message bits, significantly reducing inter-server communication. However, fault tolerance is not addressed in Coded MapReduce. In a related work, Li et al. [14] unveil CodedTeraSort, a CDC-based distributed sorting algorithm, which leverages structured redundancy in data to overcome the data shuffling bottleneck of TeraSort. Ozfatura et al. [21] propose Coded Distributed Computation with Partial Recovery (CCPR) to address straggling workers in distributed computing. By having the number of subfiles reduced, a trade-off between computation and communication is enabled by CCPR, resulting in faster operations. Numerical simulations confirm the advantages of the proposed scheme in terms of the trade-off between computation accuracy and latency.

The CDC technique holds considerable potential for reducing communication overhead within the MapReduce framework. By leveraging coding methodologies, CDC effectively reduces data size, decreasing network bandwidth consumption and the overall communication burden. However, integrating CDC and TMR to enhance fault tolerance in the MapReduce framework remains relatively unexplored. Combining these techniques requires careful planning and synchronization, particularly when addressing challenges such as designing appropriate coding schemes and devising efficient decoding algorithms for redundant tasks. While this unified approach might introduce added complexity and overhead in computation, communication, and storage, it is evident that further research is crucial to formulate efficient algorithms and evaluate the trade-offs between fault tolerance, performance, and resource utilization in this merged strategy.

3   Preliminaries and Problem Formulation

3.1   The MapReduce Framework

The MapReduce framework has become a cornerstone for distributed data-intensive tasks [12,13] due to its capability to process vast datasets in a distributed manner. The intrinsic simplicity and scalability of MapReduce have led to its widespread adoption. The computational workflow in MapReduce is categorized into several stages: In the *Preparation Stage*, the *master node* breaks down the main task into numerous subtasks, delegating Map and Reduce functions to the worker nodes. During the *Map Stage*, these nodes tackle their specific subtasks and produce intermediate outcomes. In the *Shuffle Stage*, nodes exchange these outcomes, preparing data inputs for the Reduce functions. Lastly, in the *Reduce Stage*, nodes process the data, leading to the final outcomes which are then integrated by the system for the final output. Despite the built-in fault-tolerance features of MapReduce, Silent Data Corruption (SDC) introduces distinct challenges. Conventional fault-tolerance techniques may prove inadequate against SDC, highlighting the urgency for more tailored solutions.

Furthermore, the communication load within the MapReduce framework has been identified as one of the performance bottlenecks [7]. During the *Shuffle stage*, each worker node holds a list of key-value pairs, which then need to be redistributed among all worker nodes. The communication load L_{MR} of the MapReduce framework can be expressed as:

$$L_{MR} = 1 - \frac{r}{K}, \tag{1}$$

where r represents the number of nodes assigned to each Map function, and K is the total number of nodes.

In the MapReduce framework, task-level TMR can be realized by dispatching each task to three distinct nodes. Each of these nodes independently executes the task. Subsequently, their results are forwarded to a Voter node, which employs majority voting to determine the final result. This approach is predicated on assuming that the outputs from at least two nodes are correct.

3.2 Coded Distributed Computing

CDC [13] offers a solution to mitigate communication overhead in the MapReduce framework with multiple Reduce functions, particularly during the *Shuffle stage*. The efficiency of the system is enhanced by CDC through two primary strategies: First, distributing different subtasks across multiple nodes during the *Map stage*, ensuring redundancy in the intermediate results. Second, employing heterogeneous encoding techniques in the *Shuffle stage*. Instead of transmitting intermediate results individually, the CDC broadcasts encoded data throughout the cluster. Nodes subsequently decode the received data utilizing locally available redundant results, facilitating the retrieval of necessary intermediate outcomes.

These improvements have the potential to substantially reduce the communication overhead, especially during the *Shuffle stage*. The communication load for CDC, L_{CDC}, can be quantitatively represented as:

$$L_{CDC} = \frac{1}{r}\left(1 - \frac{r}{K}\right) = \frac{1}{r}L_{MR}. \tag{2}$$

Here, r denotes the number of nodes assigned to subtasks and each Map function. Despite nodes not naturally holding redundant information in their encoded data [20], CDC lacks inherent SDC fault tolerance, necessitating additional measures for addressing challenges related to SDC.

3.3 Problem Formulation

Consider the scenario where N input files are processed using the MapReduce framework to generate the results. In this scenario, the multi-Reduce function setup in MapReduce may result in multiple intermediate results corresponding to each input file. These intermediate results serve as input variables for different Reduce functions. The objective is to compute Q Reduce functions denoted as $R_q(\cdot)$, utilizing a distributed cluster consisting of K nodes and N input files w_n. It is important to note that N is greater than K. Here, q represents the index of the Reduce functions, n represents the file number, and $v_{q,n}$ represents the intermediate result of the q-th group corresponding to the n-th file.

The Map function is responsible for mapping an input file to an intermediate result denoted as $v_{q,n} = M_{q,n}(w_n)$. On the other hand, the Reduce function denoted as $R_q(\cdot)$ takes all the intermediate results of the q-th group and maps them to an output result $u_q = R_q(v_{q,1}, v_{q,2}, \ldots, v_{q,n})$. To quantify the degree of redundancy, we define the parameter μ as a function of the number of Reduce functions, Q, and the file redundancy r, and it can be given by

$$
\begin{cases}
\mu = \underset{r \in [0, Q-1]}{\arg\min} \, c_m \cdot \dfrac{r}{Q} + \dfrac{c_s \cdot (Q-r)}{Q \cdot (r+1)} \\
r = r^* + 1 \\
r^* = \max\{\mu\}
\end{cases}
\tag{3}
$$

where c_m denotes the computational node capacity, and $1/c_s$ signifies the communication capacity between nodes. The objective is to minimize μ, representing the combined communication and computation overhead. We define r^* as the maximum of μ, indicating that each file will be allocated to r^* *Solver nodes* and one *Helper node*. A relationship can be established between the variables K, N, and Q as follows:

$$
\begin{cases}
\gamma = \binom{r^*}{Q} \\
K = Q + \left\lceil \dfrac{Q}{r^*} \right\rceil \\
N = (K - Q) \cdot \gamma
\end{cases}
\tag{4}
$$

For the sake of description, the files deployed to *node z* are represented by the set F_z, while the Reduce functions computed by *node z* are denoted as W_z and the Map functions computed by *node z* are denoted as M_z.

In the typical MapReduce framework, we assume SDC events to be infrequent and specifically focus on their potential during the Map Stage, a critical stage of processing raw input into intermediate results. Our fault model concentrates solely on SDC within the Map operation's output, given that soft errors predominantly lead to silent data corruption with a relatively low occurrence. This model pertains to executing distributed tasks across Q Reduce operations, N files, and K nodes, maintaining a predefined redundancy level of r.

4   Fault Tolerance for Multiple Reduce Tasks Using Coded Distributed Computing

The overall process of CDCFT can be illustrated in Fig. 1. CDCFT involves four stages: Preparation, Map, Shuffle, and Reduce. This system is designed for enhanced error tolerance, accommodating SDC. This is achieved through node grouping, employing TMR strategies, and voting mechanisms. Furthermore, it efficiently minimizes communication load by using intermediate result coding and intra-group broadcasting.

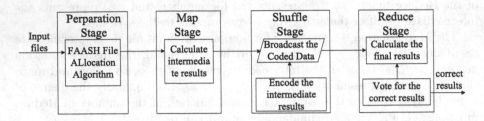

Fig. 1. CDCFT Flowchart that illustrates the four main stages of the CDCFT algorithm: Preparation, Map, Shuffle, and Reduce.

In our proposed CDCFT approach, nodes are categorized as either *Solver node* s_k, $k \in [1, Q]$, or *Helper nodes* p_j, $j \in [Q+1, K]$. The *Solver nodes* execute Map and Reduce functions, receive data from *Helper nodes* during the *Shuffle Stage*, and act as Voter nodes in the *Reduce Stage*, utilizing a majority voting mechanism to ensure fault tolerance and determine the final output. Conversely, *Helper nodes* carry out Map functions, encode intermediate outcomes into coded messages, and relay them to the respective *Solver nodes*.

4.1 Preparation Stage

The Preparation Stage is dedicated to establishing the distributed framework. This involves classifying nodes, distributing files, and designating Map and Reduce functions to specific nodes.

Initially, nodes are segregated into two primary categories: *Solver nodes* and *Helper nodes*. The number of *Solver nodes* matches the total count of Reduce functions, amounting to Q nodes. In contrast, the *Helper nodes* make up the difference, totaling $(K - Q)$ nodes. Files are then allocated to each node. For the *Solver nodes*, a selection mechanism pinpoints r^* nodes from the pool of γ options, determining which are most apt to receive a file. Meanwhile, every *Helper node* receives a file chosen from a set of $(K - Q)$ possibilities. This file allocation yields $(K - Q) \cdot \gamma$ unique configurations, optimizing the distribution of files between *Solver* and *Helper* nodes in the framework.

The rule set, defined in Eq. (5), integrates various allocation methods, symbolized as S. This is mathematically expressed as:

$$S = \{(i, A) \mid Q + 1 \leq i \leq K, |A| = r^*, A \subseteq x \mid 1 \leq x \leq Q, x \in \mathbb{Z}\}. \quad (5)$$

Here, each pair (i, A) signifies an allocation method, where i represents the *Helper node* number assigned to the file, and A defines the set of *Solver node* numbers designated to that file. Thus, the entire rule set S embodies $(K - Q) \cdot \gamma$ elements, which corresponds to the total file count. Consequently, every element (i, A) in the set S epitomizes the deployment strategy tailored for a particular file. In the construction of the rule set S, elements (i, A) are structured by consecutively populating node numbers in a cyclical pattern.

We can calculate the number of files assigned to each node using the following procedure. For each *Solver node* s_k, we iterate through each element (i, A) in S. If *Solver node* s_k belongs to set A, the file corresponding to that element is assigned to *Solver node* s_k. This process establishes the total number of files λ allocated to every *Solver node*, as depicted below:

$$\lambda = \frac{\binom{Q-1}{r-1} \cdot (K - Q)}{\gamma \cdot (K-Q)} = \frac{r^*}{Q}. \tag{6}$$

Similarly, for each *Helper node* p_j, we consider every element (i, A) in S, as defined in Eq.(5). When $i = j$, the file related to that element is allocated to *Helper node* j. Using this approach, the total files ψ for each *Helper node* are:

$$\psi = \frac{\gamma}{\gamma \cdot (K-Q)} = \frac{1}{K-Q}. \tag{7}$$

The file set at *node z* are given as F_z, with the n-th file represented by w_n. Based on this, we introduce the *File Allocation Algorithm for Solver and Helper nodes (FAASH)* in Algorithm 1.

Algorithm 1: FAASH Algorithm

Input : Q,K,N
Output: result F

1 **for** $n \leftarrow 1$ **to** N **do**
2 $\quad\mid$ allocate w_n to $S[n]$;
3 **end**
4 **for** *each Solver node* s_k **do**
5 $\quad\mid$ **if** $k \in A$ **then**
6 $\quad\quad\mid$ allocate corresponding w_n to F_k;
7 $\quad\mid$ **end**
8 **end**
9 **for** *each Helper node* p_j **do**
10 $\quad\mid$ **if** $j == i$ **then**
11 $\quad\quad\mid$ allocate corresponding w_n to F_j;
12 $\quad\mid$ **end**
13 **end**

Next, we proceed to assign Map functions to nodes based on file allocation. Specifically, when node z is assigned file w_i, we allocate to it the set of Map functions $\{M_{1,i}(\cdot), M_{2,i}(\cdot), \ldots, M_{Q,i}(\cdot)\}$, with each function utilizing w_i as its input. Subsequently, the algorithm assigns Reduce functions to the *Solver nodes*. We represent the Reduce function assignment scheme using a matrix $RM \in \mathbb{N}^{3 \times Q}$ as given by

$$RM = \begin{bmatrix} 1 & 2 & \cdots & Q-1 & Q \\ 2 & 3 & \cdots & Q & 1 \\ 3 & 4 & \cdots & 1 & 2 \end{bmatrix}, \tag{8}$$

where the elements in the matrix denote the Reduce function numbers. The k-th column of the matrix RM corresponds to the set of Reduce functions W_k that *Solver node* s_k is responsible for.

In summary, the Preparation Stage comprises four steps. First, nodes are categorized into *Solver* and *Helper* types. Next, the FAASH algorithm allocates files across nodes, ensuring an equitable distribution. Then, Map functions are designated to nodes anchored on their file ownership. Finally, Reduce functions are distributed using the RM matrix, streamlining task distribution. Collectively, these procedures set the stage, ensuring optimal file, Map function, and Reduce function allocation to *Solver* and *Helper* nodes, setting the groundwork for the subsequent stages of the CDCFT framework.

4.2   Map Stage

For each file w_n, *node z* applies the corresponding numbered Map function from each group, yielding an intermediate result set denoted by $\Omega_n(w_n) = \{v_{(q,n)} | q \in \mathbb{N} \text{ and } 1 \leq q \leq Q\}$. Here, $v_{(q,n)}$ symbolizes the intermediate result procured from the n-th Map functions of the q-th group, serving as input for the q-th numbered Reduce function related to file w_n. The collective intermediate results acquired by *node z* during the Map Stage is represented as $V_z = \{\Omega_n(w_n)\}$.

4.3   Shuffle Stage

Assuming each file is allocated to r redundancy nodes, the *Helper node* p_j encodes its local intermediate result set V_j into forwarded messages X_j, which are then broadcast to the entire node cluster. Then r *Solver nodes* decode these messages to acquire the intermediate results.

Consider G as the set of r^* *Solver nodes*. In the encoding phase, *Helper node* p_j generates encoded messages $x_{(j,G)}$ for nodes in G based on its intermediate results V_j. Given that each *Solver node* in G has three Reduce functions, the *Helper node* transmits three encoded messages to every node, corresponding to each Reduce function. Denote $F_j \subseteq w_n$ as the file set under helper p_j and W_G as the Reduce functions for solvers in G. The encoding functions are:

$$x_{(j,G)} = \left\{ \bigoplus_{j \in G} v_{(q,n)} \mid R_q(\cdot) \in W_G, n \in F_j \right\}, \tag{9}$$

where the symbol $\oplus$ denotes a heterogeneous encoding operation [20] enacted on the intermediate results. This operation activates when the group number in V_j of the *Helper node* aligns with the *Solver node* number within set G.

During the *Shuffle Stage*, *Helper node* p_j partitions its collection of intermediate results, V_j, into distinct subsets. Each subset is carefully assembled to meet the specific requirements of r^* distinct *Solver nodes*, ensuring no overlap or redundancy. After this categorization, the *Helper node* employs heterodyne encoding on each subset, producing the corresponding encoded message $x_{(j,G)}$. Consequently, the entire ensemble of these encoded messages is represented by X_j. The communication overhead associated with the *Shuffle Stage* can be encapsulated by the cumulative count of intermediate results set to be transferred, as delineated in Eq. (10):

$$L_f = 3 \cdot (K - Q) \cdot \binom{Q}{r^* + 1}. \tag{10}$$

It can be concluded that the communication overhead of CDCFT in the *Shuffle Stage* is always less than or equal to that of a CDC fault-tolerant method using the vanilla *Triple Modular Redundancy*, as stated in the theorem below.

Theorem 1. *Let CDCFT be a distributed algorithm with Q Reduce functions, N files, K nodes, and the redundancy set to r. We can conclude the communication overhead of CDCFT in the Shuffle Stage is always less than or equal to that of a CDC fault-tolerant method using the vanilla Triple Modular Redundancy.*

Proof. The CDC using Triple Modular Redundancy is achieved by running three instances of CDC. Each instance incorporates an additional $2r^*$ redundant files and $2K$ supplementary nodes, as elucidated in Eq. (3). The communication overhead in the *Shuffle Stage* is determined by the total number of intermediate results sent in the CDC using Triple Modular Redundancy. The calculation for the communication overhead in the *Shuffle Stage* is provided by Eq. (11):

$$L_c = 3 \cdot \left(1 - \frac{r}{K}\right) \cdot \frac{N}{r}, \tag{11}$$

where K represents the total number of nodes in the CDC system without TMR, N denotes the total number of files being processed, Q represents the total number of Reduce functions, and r indicates the degree of redundancy implemented. The communication overhead in the *Shuffle Stage* for both CDCFT and the CDC with TMR is determined by the total number of nodes used, denoted as K^*, under the same configurations of N, Q, and r. Drawing from Eq. (10) and Eq. (11), the expression for communication overhead can be delineated as follows:

$$\frac{L_f}{L_c} = \frac{r}{(r+1)} < 1. \tag{12}$$

As shown in Eq. (11) and Eq. (12), the communication overhead of CDCFT in the *Shuffle Stage* is always less than or equal to that of a CDC fault-tolerant method using the vanilla Triple Modular Redundancy, thereby proving the theorem. To demonstrate, let's examine the case where $r = 3$, the overhead of

CDCFT is at most 3/4 that of TMR. In summary, the theorem shows CDCFT is more communication efficient than vanilla TMR redundancy for fault tolerance in distributed computing for any redundancy configuration.

4.4 Reduce Stage

Solver node s_k, applies its specific decode function, $D_k^q(\cdot)$, to decode the encoded data and retrieve intermediate results. These intermediate results are then combined with the q-th set of local intermediate results to form the complete set of intermediate results $\Omega_n(w_n)$. *Solver nodes* repeat this process for their other Reduce functions to obtain their respective complete sets of intermediate results. By substituting the q-th set of complete intermediate results into the corresponding Reduce function, the output u_q is obtained.

Since *Solver node* s_k has multiple Reduce functions, it generates a set of outputs denoted as $U_k \subseteq \{u_q | q \in \mathbb{N} \cap 1 \le q \le Q\}$. Each *Solver node* hashes its output set U_k and sends the encoded result to a designated Voter node. Upon receiving the output sets from all *Solver nodes*, the Voter node selects the final output u_q by conducting a vote on the corresponding number of outputs and choosing the result with the highest number of occurrences.

5 Experiments

We evaluate the communication overhead of the *Shuffle Stage* in different approaches by using TeraSort [14] as a reference test program. TeraSort is a sorting task based on the MapReduce paradigm. In the *Map Stage* of TeraSort, each Map node partitions the data into R blocks. This ensures that the data in the $(i+1)$-th block is greater than the data in the i-th block. Additionally, each Map node applies a hash code to the data within each block. During the *Shuffle Stage*, the i-th Reduce node sorts the i-th data block received from all Map nodes. This ensures that the outcomes produced by the $(i+1)$-th Reduce node are greater than the outcomes of the i-th node. Finally, in the *Reduce Stage*, each Reduce node sorts all the received data. The sorting results from the first to the R-th Reduce nodes are then output sequentially, resulting in the final sorted output.

The principle of the TeraSort algorithm is depicted in Fig. 2. Each Map node divides its corresponding data into four blocks and sends them to the respective Reduce node. In the *Reduce Stage*, the Reduce node sorts the received data and outputs them in sequential order, producing the final sorted result.

5.1 Experiment Setups

We perform experiments to assess our proposed approach. The experiments utilize Docker containers to emulate a distributed computing environment with multiple nodes. Each node in the cluster is equipped with 4GB of memory and an Intel(R) Xeon(R) Gold 6126 CPU model. To emulate a real distributed network environment, we use Docker network plugins (TC-HTB and NetEM) to simulate various network conditions between containers. Each container is configured with a unique IP address on Docker's bridge network to facilitate communication. The cluster architecture consists of one Master node and multiple Worker nodes. The Master node handles data distribution and task allocation, while the Worker nodes execute the computational tasks.

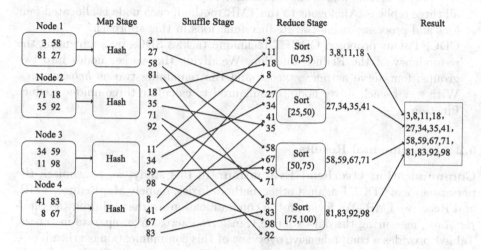

Fig. 2. Schematic of the TeraSort sorting algorithm demonstrating the distributed sorting of integers from 0 to 100 using four nodes.

For the experiment, we generate a 1.6GB file using *NumPy*, where each line in the file represents an integer ranging from 0 to 100. The experiment parameters are set as $N = 8$ and $Q = 4$. Here, $N = 8$ indicates that the input data was divided into eight smaller files, which are individually transmitted by the Master node to each Worker node.

Our proposed CDCFT algorithm is benchmarked against three other methods. We assess each algorithm based on two metrics: the duration of the MapReduce computation tasks and the communication overhead incurred during the Shuffle Stage. Below, we detail the specific configurations used for each method:

- MapReduce [7]: This approach leverages the MapReduce computing framework within Hadoop. It offers a straightforward and scalable programming model designed for processing vast amounts of data across large-scale clusters. The framework divides computational tasks into independent Map and Reduce stages, distributing data across multiple nodes for parallel processing. Each node is assigned four files in our experimental setup, and two Reduce functions are applied.
- TMR [1]: The MapReduce computation task is executed in triplicate, running concurrently in the TMR approach. Fault tolerance is established by

cross-referencing the final results from the three executions. In our specific experimental configuration for the TMR method, each node is assigned four files, and it uses two Reduce functions.

- Reactive TMR [4]: This method adopts a two-stage, three-module redundancy technique. Every computational task is executed in triplicate. Initially, the results of the first two replicas are assessed for consistency. If discrepancies emerge between these results, the third replica is then executed. The conclusive output is ascertained through a majority vote based on the outcomes of all three replicas. Analogous to the TMR method, each node is allocated four files and processes using two Reduce functions in this approach.
- CDCFT: Our proposed CDCFT technique tackles SDCs by augmenting the redundancy of the Reduce function. We divide the worker nodes into two groups: four serve as *Solver nodes* and the remaining two as *helper nodes*. With $r = 4$, each solver node is configured to calculate three unique Reduce functions.

5.2   Experimental Results

Communication Overhead Evaluation: In this study, we benchmark the performance of CDCFT against other methodologies, namely MapReduce, TMR, and Reactive TMR. We focus on the communication overhead during the *Shuffle Stage*, measuring the data volume transferred across five separate instances. Table 1 provides a comprehensive overview of this communication overhead, representing the average amount of data exchanged during the *Shuffle Stage*. Specifically, D_a represents the average volume of data transferred under fault conditions, whereas D_b depicts the average volume of data transferred in a faultless *Shuffle Stage*. All data volumes are represented in megabytes (MByte).

Table 1. Amount of data transmitted (MByte) during the Shuffle stage by each algorithm, with and without faults.

Algorithm	D_a	D_b
MapReduce	2369.38	2369.23
TMR	2494.94	2494.95
Reactive TMR	1769.05	1613.69
CDCFT	327.42	324.64

Table 1 reveals that CDCFT considerably reduces the volume of data transferred relative to its counterparts. The primary reason for this reduction is CDCFT's implementation of distributed encoding. In this method, helper nodes distribute encoded information, while solver nodes perform local decoding for their respective *Reduce* functions. Further, CDCFT mitigates unnecessary data exchanges amongst solver nodes, decreasing data transmissions during the *Shuffle Stage*. A noteworthy observation is the relatively uniform data transmission volumes in the CDCFT and TMR methods during the *Shuffle Stage*, irrespective of fault occurrences. Such consistency can be attributed to the proficiency of CDCFT and TMR in reliably detecting and rectifying discrepancies based on the acquired data. This holds even when unnoticeable errors emerge from individual *Map* function computations. On the other hand, the Reactive TMR approach exhibits an increase in data transfer during the *Shuffle Stage* when it identifies a silent error by contrasting results from two *Reduce* functions. This is because the erroneous detection prompts a third execution of the *Reduce* function, necessitating the *Map* node to retransmit intermediate results. This, in turn, leads to a greater data transfer volume during the *Shuffle Stage* compared to when no faults are present.

Execution Time Evaluation: Fig. 3(a) compares the performance of the execution times of CDCFT, MapReduce, TMR, and Reactive TMR across the map, shuffle, and reduce stages, examining their duration in these stages as well as their cumulative times under fault-free scenarios. Given the inherent requirement of CDCFT to maintain additional redundancy of intermediate results alongside task redundancy, it registers a longer span in the Map Stage. Nonetheless, during the *Shuffle Stage*, CDCFT notably surpasses TMR and Reactive TMR, slashing execution times by 27.3% and 19.6% compared to TMR and Reactive TMR, respectively. Moreover, CDCFT reduces the total task execution time by 17.5% and 10.6% compared to TMR and Reactive TMR, respectively. It is pivotal to observe that CDCFT's elongated Map Stage is due to its allocation of six files per solver node, while both TMR and Reactive TMR allocate four, causing CDCFT to process a larger volume of data than its peers.

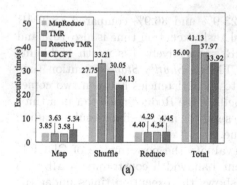 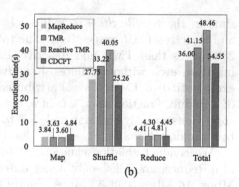

(a)   (b)

Fig. 3. Comparison of the execution time during the Map, Shuffle, Reduce stages and total execution time across different methods, (a) without faults, and (b) with faults.

Ablation Studies on Bandwidth Constraints: Fig. 3(b) illustrates the execution times for each stage of the algorithms in the presence of faults. Notably, CDCFT consistently outperforms TMR and Reactive TMR, regardless of whether faults exist. Specifically, when faults occur, CDCFT's execution

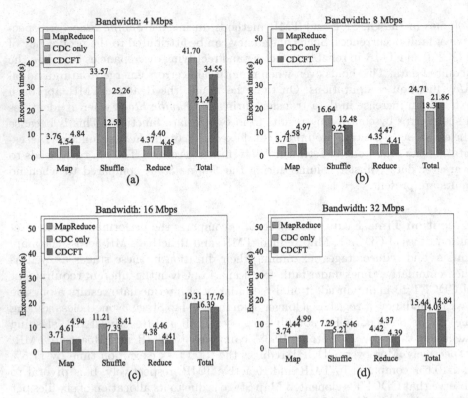

Fig. 4. Bandwidth ablation study: Execution time of each stage under varied bandwidth constraints: (a) 4 Mbps, (b) 8 Mbps, (c) 16 Mbps, and (d) 32 Mbps.

time in the *Shuffle Stage* is reduced by 23.9% and 36.9% compared to TMR and Reactive TMR, respectively. The total task execution time is also 16% and 28.6% less than TMR and Reactive TMR, respectively. It's worth highlighting that, even without failures, Reactive TMR's *Shuffle Stage* execution time exceeds that of TMR. This is attributed to inconsistencies between two copies of a Reduce function upon a failure, compelling the Reduce node to run a third copy. Consequently, the Map node has to resend the intermediate results, leading to a prolonged *Shuffle Stage* execution time.

We conduct ablation experiments to evaluate the performance of CDCFT, MapReduce, and CDC-only under different bandwidth constraints: 4 Mbps, 8 Mbps, 16 Mbps, and 32 Mbps. Figure 4 shows the execution times under the different bandwidth constraints. The findings indicate that MapReduce consistently exhibits the longest execution time. While CDCFT performs faster, it registers a slightly longer duration than CDC-only due to the additional task redundancy from the TMR mechanism. However, the superior fault tolerance of CDCFT relative to CDC-only justifies this minor increment in the execution time, particularly during the *Shuffle Stage*.

For instance, at a bandwidth of 4 Mbps, CDCFT lags by roughly 8 s in the *Shuffle Stage* compared to CDC-only. But as the bandwidth escalates, this gap narrows to about 2 s at 32 Mbps. This pattern suggests that with the advent of the 5G era and the increase in bandwidth availability, the benefits of CDCFT over CDC-only become even more noticeable.

6   Summary

MapReduce is widely recognized for its capacity to process distributed datasets efficiently, distributing tasks across multiple nodes to reduce costs. However, as cluster sizes expand, the emergence of silent data corruption emerges as a discernible concern in MapReduce. Traditional fault-tolerance strategies that rely on multiple replicas introduce significant communication overhead when applied to compute-intensive tasks.

In response to this challenge, this paper presents the CDCFT fault-tolerance method, which targets the frequent occurrence of silent data corruption in multi-Reduce function MapReduce applications. CDCFT employs CDC to diminish the communication load, offsetting the added overhead from task redundancy, and harnesses result voting to assure fault tolerance against silent data corruption. Experimental results demonstrate that CDCFT significantly reduces the execution time of the *Shuffle Stage* relative to traditional fault-tolerant techniques, concurrently slashing the communication overhead substantially.

Future research can explore applying CDCFT in heterogeneous clusters where nodes have different capabilities. Node allocation and file deployment strategies can be developed to balance processing times across heterogeneous nodes.

References

1. Benoit, A., Cavelan, A., Cappello, F., et al.: Coping with silent and fail-stop errors at scale by combining replication and checkpointing. J. Parallel Distrib. Comput. **122**, 209–225 (2018)
2. Charyyev, B., Alhussen, A., Sapkota, H., et al.: Towards securing data transfers against silent data corruption. In: 2019 19th IEEE/ACM International Symposium on Cluster, Cloud and Grid Computing (CCGRID), pp. 262–271. IEEE (2019)
3. Deveautour, B., Traiola, M., Virazel, A., et al.: Reducing overprovision of triple modular redundancy owing to approximate computing. In: 2021 IEEE 27th International Symposium on On-Line Testing and Robust System Design (IOLTS), pp. 1–7. IEEE (2021)
4. Dixit, H.D., Pendharkar, S., Beadon, M., et al.: Silent data corruptions at scale. arXiv preprint arXiv:2102.11245 (2021)
5. Dong, Y., Tang, B., Ye, B., Qu, Z., Lu, S.: Intermediate value size aware coded mapreduce. In: 26th IEEE International Conference on Parallel and Distributed Systems, (ICPADS), Hong Kong, December 2–4, 2020. pp. 348–355. IEEE (2020)
6. Gandomi, A., Movaghar, A., Reshadi, M., et al.: Designing a MapReduce performance model in distributed heterogeneous platforms based on benchmarking approach. J. Supercomput. **76**, 7177–7203 (2020)

7. Glushkova, D., Jovanovic, P., Abelló, A.: MapReduce performance model for Hadoop 2.x. Inf. Syst. **79**, 32–43 (2019)
8. Khader, M., Al-Naymat, G.: Density-based algorithms for big data clustering using MapReduce framework: a comprehensive study. ACM Comput. Surv. (CSUR) **53**(5), 1–38 (2020)
9. Krishnan, R.M., Zhou, D., Kim, W.H., et al.: TENET: memory safe and fault tolerant persistent transactional memory. In: 21st USENIX Conference on File and Storage Technologies (FAST 23), pp. 247–264 (2023)
10. Li, C., Wang, Y.P., Tang, H., et al.: Dynamic multi-objective optimized replica placement and migration strategies for SaaS applications in edge cloud. Future Gener. Comput. Syst. **100**, 921–927 (2019)
11. Li, C., Zhang, Y., Tan, C.: Fault-tolerant computation meets network coding: optimal scheduling in parallel computing. IEEE Trans. Commun. **71**(7), 3847–3860 (2023)
12. Li, P., Guo, S., Yu, S., et al.: Cross-cloud MapReduce for big data. IEEE Trans. Cloud Comput. **8**(2), 375–386 (2015)
13. Li, S., Maddah-Ali, M.A., Avestimehr, A.S.: Coded MapReduce. In: 53rd Annual Allerton Conference on Communication, Control, and Computing (Allerton), pp. 964–971. IEEE (2015)
14. Li, S., Supittayapornpong, S., Maddah-Ali, M.A., et al.: Coded TeraSort. In: 2017 IEEE International Parallel and Distributed Processing Symposium Workshops (IPDPSW), pp. 389–398 (2017)
15. Li, S., Yu, Q., Maddah-Ali, M.A., et al.: Coded distributed computing: fundamental limits and practical challenges. In: 50th Asilomar Conference on Signals, Systems and Computers, pp. 509–513. IEEE (2016)
16. Li, Z., Menon, H., Maljovec, D., Livnat, Y., Liu, S., et al.: SpotSDC: revealing the silent data corruption propagation in high-performance computing systems. IEEE Trans. Visual Comput. Graphics **27**(10), 3938–3952 (2021)
17. Luo, C., Cao, Q., Li, T., et al.: Mapreduce accelerated attribute reduction based on neighborhood entropy with apache spark. Expert Syst. Appl. **211**, 118554 (2023)
18. Maleki, N., Rahmani, A.M., Conti, M.: MapReduce: an infrastructure review and research insights. J. Supercomput. **75**, 6934–7002 (2019)
19. Mireshghallah, F., Bakhshalipour, M., Sadrosadati, M., et al.: Energy-efficient permanent fault tolerance in hard real-time systems. IEEE Trans. Comput. **68**(10), 1539–1545 (2019)
20. Ng, J.S., Lim, W.Y.B., Luong, N.C., et al.: A comprehensive survey on coded distributed computing: fundamentals, challenges, and networking applications. IEEE Commun. Surv. Tutor. **23**(3), 1800–1837 (2021)
21. Ozfatura, E., Ulukus, S., Gündüz, D.: Coded distributed computing with partial recovery. IEEE Trans. Inf. Theory **68**(3), 1945–1959 (2022)
22. Saadoon, M., Hamid, S.H.A., Sofian, H., et al.: Experimental analysis in Hadoop MapReduce: a closer look at fault detection and recovery techniques. Sensors **21**(11), 3799 (2021)
23. Saadoon, M., Hamid, S.H.A., Sofian, H., et al.: Fault tolerance in big data storage and processing systems: a review on challenges and solutions. Ain Shams Eng. J. **13**(2), 101538 (2022)
24. Salehi, M., Ejlali, A., Al-Hashimi, B.M.: Two-phase low-energy n-modular redundancy for hard real-time multi-core systems. IEEE Trans. Parallel Distrib. Syst. **27**(5), 1497–1510 (2016)
25. Saleti, S., Subramanyam, R.B.V.: A MapReduce solution for incremental mining of sequential patterns from big data. Expert Syst. Appl. **133**, 109–125 (2019)

26. Woolsey, N., Chen, R.R., Ji, M.: Cascaded coded distributed computing on heterogeneous networks. In: IEEE International Symposium on Information Theory (ISIT), pp. 2644–2648. IEEE (2019)

27. Xu, D., Chu, C., Wang, Q., et al.: A hybrid computing architecture for fault-tolerant deep learning accelerators. In: 2020 IEEE 38th International Conference on Computer Design (ICCD), pp. 478–485. IEEE (2020)

28. Xu, H., Liu, Y., Lau, W.C.: Multi resource scheduling with task cloning in heterogeneous clusters. In: Proceedings of the 51st International Conference on Parallel Processing, (ICPP), Bordeaux, France, 29 August 2022–1 September 2022, pp. 41:1–41:11 (2022)

29. Yakhchi, M., Fazeli, M., Asghari, S.A.: Silent data corruption estimation and mitigation without fault injection. IEEE Can. J. Elect. Comput. Eng. **45**(3), 318–327 (2022)

30. Yang, N., Wang, Y.: Predicting the silent data corruption vulnerability of instructions in programs. In: 25th IEEE International Conference on Parallel and Distributed Systems, (ICPADS), Tianjin, China, December 4–6, 2019, pp. 862–869 (2019)

31. Zhang, G., Liu, Y., Yang, H., et al.: Efficient detection of silent data corruption in HPC applications with synchronization-free message verification. J. Supercomput. **78**(1), 1381–1408 (2022)

32. Zhang, J., Lin, M.: A comprehensive bibliometric analysis of apache Hadoop from 2008 to 2020. Int. J. Intell. Comput. Cybern. **16**(1), 99–120 (2023)

33. Zhu, Y., et al.: Fast recovery MapReduce (FAR-MR) to accelerate failure recovery in big data applications. J. Supercomput. **76**(5), 3572–3588 (2020)

Key-Based Transaction Reordering: An Optimized Approach for Concurrency Control in Hyperledger Fabric

Haoliang Ma[1,2(✉)], Peichang Shi[1,2], Xiang Fu[1,2], and Guodong Yi[3]

[1] National Key Laboratory of Parallel and Distributed Computing,
College of Computer Science, National University of Defense Technology,
Changsha 410073, China
[2] Key Laboratory of Software Engineering for Complex Systems,
College of Computer Science, National University of Defense Technology,
Changsha 410073, China
{mhliang0640,pcshi}@nudt.edu.cn
[3] Xiangjiang Lab, Changsha 410073, China

Abstract. As blockchain technology garners increased adoption, permissioned blockchains like Hyperledger Fabric emerge as a popular blockchain system for developing scalable decentralized applications. Nonetheless, parallel execution in Fabric leads to concurrent conflicting transactions attempting to read and write the same key in the ledger simultaneously. Such conflicts necessitate the abortion of transactions, thereby impacting performance. The mainstream solution involves constructing a conflict graph to reorder the transactions, thereby reducing the abort rate. However, it experiences considerable overhead during scenarios with a large volume of transactions or high data contention due to capture dependencies between each transaction. Therefore, one critical problem is how to efficiently order conflicting transactions during the ordering phase. In this paper, we introduce an optimized reordering algorithm designed for efficient concurrency control. Initially, we leverage key dependency instead of transaction dependency to build a conflict graph that considers read/write units as vertices and intra-transaction dependency as edges. Subsequently, a key sorting algorithm generates a serializable transaction order for validation. Our empirical results indicate that the proposed key-based reordering method diminishes transaction latency by 36.3% and considerably reduces system memory costs while maintaining a low abort rate compared to benchmark methods.

Keywords: Hyperledger Fabric · Reordering Algorithm · Concurrency Control · Transaction Conflicts

1 Introduction

Originating from Nakamoto's Bitcoin whitepaper [12], Bitcoin only supports cryptocurrency. Ethereum [1] was then developed to facilitate Turing-complete smart contracts, thus enabling arbitrary data processing logic. Consequently, the blockchain evolved from merely a cryptocurrency platform to a distributed transaction system. Traditional blockchain systems, such as Bitcoin and Ethereum, employ an Order-Execute(OE) model, whose sequential transaction execution characteristic restricts performance, as evidenced in an analysis of seven blockchain systems [14]. In contrast, Hyperledger Fabric leverages an Execute-Order-Validate (EOV) model to enhance performance: transactions submitted are first executed by the endorsing peers, then ordered and batched by the ordering services, and finally validated by the validating peers. Fabric exploits today's multi-core architecture to facilitate transaction processing by supporting parallel processing of transactions [2]. It overcomes the limitations of the OE model by providing parallelism of transaction execution on different endorsing peers.

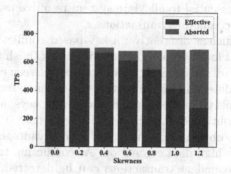

Fig. 1. Effective and aborted throughput under vaying skewness

However, the delay between execution and commitment of a transaction increases the probability of conflicting transactions, which are subsequently rejected by peers during the verification stage, thus creating a scalability bottleneck. The Fabric uses an optimistic concurrency control (OCC) mechanism, terminating conflicting transactions to ensure the consistency of the ledger under concurrent updates. However, this measure comes with a substantial transaction abort rate exceeding 40% [3] due to many inter & intra-block conflicts, amplified particularly under high contention workload characterized by a large number of conflicting transactions. Various degrees of data race conditions can be simulated by adjusting the skew parameters implying Zipfian distribution. It reveals that higher skewness corresponds to an increased percentage of conflicting transactions, e.g., $skew = 0$ represents uniform access, and $skew = 2.0$ represents extremely skewed access. Figure 1 reports Fabric's throughput under varying skewness [15], with its blue and red components, respectively, demonstrating effective and aborted throughput. The raw throughput remains consistent despite the workload type and requests skewness. But with higher skewness, a larger proportion of transactions are aborted for serializability.

Current studies [15,16] employ conflict graph construction, with Tarjan's and Johnson's algorithms [18] used for cycle detection and removal to decrease discarded transactions during the sorting phase of transaction reordering. However, the overhead associated with conflict graph construction is significant due to the need to map dependencies between every transaction pair, especially when large transaction volumes or considerable conflicting transactions are present. As transactions increase, so do blocks that need to be processed, implying more conflict graphs need construction and processing. Heightened data contention amplifies this issue as each transaction potentially conflicts with a larger number of other transactions, leading to an increase in edges that may trigger out-of-memory issues. Additionally, Tarjan's and Johnson's algorithms require complex operations on the graph to identify strongly connected components or cycles, demanding considerable computational and memory resources. This approach can result in substantial delays and potential system failures. Therefore, an efficient algorithm or strategy for conflict graph construction and transaction processing and order is essential to alleviate system resource usage and manage an increased number of conflicting transactions.

We propose an efficient alternative: a key-based conflict graph (KCG) construction method that leverages key dependency to establish a global transaction order instead of capturing conflicting relationships between each pair of transactions in the conflict detection of transaction dependency-based strategies. This key dependency reveals transaction order on different keys, and more dependent transactions can be obtained on each key. Subsequently, a transaction sorting method is adopted to obtain a commit order. The advantage of our solution lies in its efficiency under high data contention. As conflicting transactions increase on each key, more dependent transactions can be detected, thus reducing system resource overhead. Utilizing the key-based reordering method, we can get a submission of non-conflicting transactions and achieve higher performance under concurrency conflicts on Fabric. The contributions of our paper are as follows:

- We present a theoretical classification of various types of concurrency-related transaction conflicts in Fabric and formulate the problem that our study aims to address.
- We introduce a key-based conflict graph construction approach, leveraging key dependency instead of the conventional transaction dependency, to efficiently resolve concurrency update conflicts in Hyperledger Fabric. Our solution proves particularly suitable for large transaction volumes under data contention. Notably, our method remains functional even when the CG method crashes due to Out-of-Memory (OOM) errors.
- We evaluate the performance of our solution and compare it with methods employed by vanilla Fabric and Fabric++/FabricSharp. Additionally, we conduct a sensitivity analysis to study the impact of different workload parameters on performances. Compared to these existing methods, our model diminishes transaction latency by 36.3% and considerably reduces system memory costs while maintaining a low abort rate.

The remainder of this paper is organized as follows. Section 2 reviews the related work. Section 3 categorizes transaction conflicts and formulates the problem. In Sect. 4, we present the system model and propose our approach. Performance evaluations are provided in Sect. 5, followed by a conclusion and future work in Sect. 6.

2   Related Work

Efficient handling of concurrency conflicts is a hot research topic in distributed databases, and conflicting transactions are also existing in Hyperledger Fabric which is a distributed system. Many studies have proposed the optimization of the performance for processing conflicting transactions. In this section, we will introduce these works along three categories according to the Fabric lifecycle: optimization for endorsement, ordering, and validation.

2.1   Endorsement Phase Optimization

Xu et al. [21] propose a lock mechanism to create a temporary database index for conflicting transactions, with the subsequent merging of the newly created index with the original index after the transaction is verified. However, in asynchronous blockchain systems, lock services are required to create and merge database indexes synchronously, resulting in substantial communication costs. Minsu et al. [9] introduce a read and write transactions separating method to accelerate transaction processing. Consequently, the transaction endorsement latency is reduced by 60% compared to the traditional Fabric network. Trabelsi et al. [19] offer a methodology to maintain a cache for conflict transaction detection at this stage and, based on this, compares three different cache storage strategies.

2.2   Ordering Phase Optimization

For the ordering phase, FastFabric [7] redesigns the ordering service to operate only with transaction IDs. By separating the transaction header from the payload, the process for determining transaction order is expedited, thus boosting throughput. Sharma et al. [16] introduce a reordering step immediately before block formation but after consensus, analyzes transaction conflicts by constructing a conflict graph, reorders and selectively discard transactions that cannot be serialized to determine a conflict-free transaction sequence, and eliminates Multi-version Concurrency Control (MVCC) Read Conflicts. Although Fabric++ reduces the number of conflicting transactions in a block, it does not apply a straightforward discarding strategy for cross-block transactions, limiting its reordering effect. Subsequently, Ruan et al. [15] consider transaction cross-block conflicts and varying conflict types based on the work of Fabric++. They proposed FabricSharp, a method capable of handling conflicts in a more fine-grained manner. However, the reordering algorithm has problems in usability and security [17]. In high-concurrency scenarios, the conflict graph becomes complex, and solving it can become a performance bottleneck and potentially even cause system crashes. To mitigate the overhead incurred by cycle detection and removal, Dickerson et al. [4] and FastBlock [11] introduce a happen-before graph for transaction execution and employ assumptions about software and hardware configurations to detect conflicts. Nevertheless, this reliance is not supported by all blockchain nodes. The transaction reordering method is also adopted in other distributed transaction processing systems. Furthermore, this method is utilized for improving OCC in online transaction processing systems [5]. Xiao et al. [20] employ the key-based concurrency control method to resolve conflicting transactions in directed acyclic graph (DAG)-based blockchains.

2.3   Validation Phase Optimization

Multiple articles propose the parallel execution of the validation process (syntax verification, endorsement policy verification, MVCC validation) to accelerate block validation [6–8]. Gorenflo et al. [6] advocate for the XOX transaction process. He believes that if a transaction is only marked as invalid due to conflicts in the verification phase, there is no trust problem in the entire execution process of the transaction, so conflicts can be found during the verification phase. Then the node executes the transaction locally to get the latest result. However, this method ignores the trust problem that still exists in the alliance chain built by Fabric, and different nodes may maliciously write wrong data, resulting in ledger data errors. FabricCRDT [13] focuses on automatically merging conflicting transactions using CRDT techniques without rejecting them. However, this approach is only suitable for use cases that can be modeled with CRDT. Skipping MVCC verification makes FabricCRDT lose the ability to detect "double spend attacks."

3   Problem Definition

3.1   Types of Transaction Conflicts

There are three categories of transaction conflicts in Fabric [3]:

3.1.1   Endorsement Failure Conflicts: All transactions need to be endorsed in the execution phase. The reasons for endorsement failure include invalid endorsement signatures or other reasons such as configuration or network errors. In this article, we only focus on endorsement policy failures caused by read-write set mismatches. Every peer independently maintains a ledger using a key-value store, which will update independently by each peer in the validation phase. Therefore, transient world state inconsistencies between peers are possible. Moreover, in the execution phase, the tail delay of block propagation makes it impossible for each endorsement node in the organization to obtain the latest ledger status for the first time. Due to the inconsistency of the world state of peers, the error of read/write set mismatch is called endorsement failure conflict.

As illustrated in the Table 1, when two different endorsement nodes Peer1, Peer2 $\in$ P endorse the same transaction $T_i \in T$, the version numbers of the same value Key in the read-write set RWSet generated by Peer1 and Peer2 are inconsistent, and an error occurs in the endorsement phase.

Table 1. Example of Endorsement Failure Conflicts

Peers	Execution Phase		World State	
	Transaction from Client	Generated Read/Write Set	Key	Version
Peer 1	$T_1[R(A), W(A)]$	R(A, Version 1), W(A)	A	1
Peer 2	$T_2[R(A), W(A)]$	R(A, Version 2), W(A)	A	2

3.1.2 MVCC Read Conflicts: MVCC Read Conflicts arise in the transaction verification phase. MVCC is a low-cost optimistic concurrent access processing method widely utilized in database systems. Its core principle involves creating historical snapshots for read transactions, and for write transactions, a new version snapshot is created instead of overwriting original data.

During the verification process, each peer node examines the transactions within the current block sequentially, and compares the version number of each transaction's read set with the current world state. The peer ensures that the current ledger state is consistent with the state achieved by transaction simulation. If any key's version number in the read set doesn't match the present world state, the transaction is considered as invalid. MVCC Read Conflicts can happen under two circumstances:

Condition 1: A read-after-write conflict, where a transaction's read operation takes place after another transaction's write operation.

Condition 2: A stale read conflict happens because a node can be either a committing peer or an endorsing peer. During a transaction's transition from the execution to the validation phase, other transactions could get validated and committed to the chain, thereby updating the world state. Therefore, the ledger update turns the data read by the transaction into stale data.

Table 2. Example of an MCVV Read Conflict

Transactions	Validation Phase			World State	
	Transaction from Ordering Service	Read Set Version Matches World State	Status	Key	Version
1	$T_1[R(A, Version 1)]$	Yes	Success	A	1
2	$T_2[W(A, Version 1)]$	/	Success	A	1
3	$T_3[R(A, Version 1)]$	No	Fail	A	2
4	$T_4[R(B, Version 1)]$	No	Fail	B	2

A typical instance of MVCC read conflict is depicted in Table 2. Transaction 1 (T1) reads key A, whose world state version is the same as the one in the transaction's read set. Therefore, the read set contains the latest value of Key A. Transaction 2 (T2) modifies Key A's value, giving it a new version 2. For Transactions 3 and 4 (T3 and T4), that read Key A and Key B respectively, the world state and read set host different versions. This implies that T3 and T4 are accessing an older key version, hence, they fail. Specifically, T3 fails due to condition 1 and T4 fails owing to condition 2.

3.1.3 Write-Write Conflicts:
In traditional databases, "write-write" concurrency conflicts primarily arise when multiple requests attempt to modify the same database index concurrently. Similarly, on blockchain platform like Fabric, when multiple transactions seek to modify the same ledger data simultaneously, it creates a similar concurrency problem. Although the data written later will overwrite the previous ones, these transactions in Fabric will eventually be submitted successfully and not marked as invalid. However, this process consumes system resources.

3.2 Problem Formulation

The problem we are focusing on is solving MVCC Read conflict with lower system overhead and acceptable latency. This is in contrast to the current problem resolved by Fabric++, which uses a CG of consuming a lot of resources. On EOV blockchain systems like Hyperledger Fabric, in the simulation execution phase, multiple peer nodes execute transactions in parallel to obtain read and write sets, which are then sent by the client to the ordering service for sorting and packaging and then verification. For the Hyperledger Fabric blockchain platform, its essence as a distributed database also has concurrency problems. We analyzed the concurrency problems in Sect. 3.1 and defined three types of transactions that cause transaction conflicts. In order to avoid MVCC Read Conflicts caused by the order of transactions in the validate phase, a transaction sequence that satisfies serialized execution can be obtained through a CG reordering method based on transaction dependencies instead of original first-in-first-out (FIFO) ordering, thus can reduce the aborted rate of transactions. CG reordering method is used to guide the ordering of conflicting transactions adopting transactions as vertices and transaction dependencies as edges. However, a transaction dependency only indicates the order between two transactions.

When constructing a conflict graph, its memory usage is closely related to the conflict graph relationship of transactions. As the capacity of transactions within a block or the skewness of transactions escalate, so does the count of conflicting transactions. As the skewness increases, the access pattern tends to concentrate on a small number of hotkeys. Smallbank workload corresponds to frequent asset update operations on a small number of accounts. In this case, the potential conflict between transactions will increase because they may more frequently access the same keys. This will increase the number of nodes and edges of the conflict graph, thereby increasing the complexity of the conflict graph, which in turn increases memory usage. Johnson's algorithm, which is used for identifying cycles in a strongly connected subgraph, can be done in linear time in $O((N + E)(C + 1))$, where N is the number of nodes and E is the number of edges, C is the number of cycles in the graph. Meanwhile, it uses a recursive algorithm called depth-first-search (DFS) for cycle detection, substantially increasing the system resources overhead required for cycle detection and transaction processing latency. In extreme cases, an Out-Of-Memory (OOM) might occur, which can potentially lead to a system crash. That is why CG is not suitable for data contention situations. Hence, it motivates us to find a more efficient way to generate a commit order.

4 System Design

4.1 System Model

Supposing the client sends two transaction requests (T_u) and (T_v), and the endorsement node gets the transaction read-write set after execution: $RS(T_u)$, $WS(T_u)$, $RS(T_v)$, $WS(T_v)$. Assume that T_u and T_v are executed in parallel, and when any of the following conditions (1)(3) is true, T_u and T_v conflict with each other. The difference is that conflicting transactions that satisfy condition (1) will be marked as invalid transactions during the verification phase, while conflicting transactions that satisfy condition (3) are valid transactions, that is, transactions whose write sets are finally successfully applied to the state database.

(1) $WS(T_u) \cap RS(T_v) \neq \phi$

(2) $RS(T_u) \cap WS(T_v) \neq \phi$

(3) $WS(T_u) \cap WS(T_v) \neq \phi$

Definition 1. *Transaction Dependency*. *Given two transactions T_u and $T_v(u < v)$, when T_u is verified before T_v, a transaction dependency $T_u{\to}T_v$ exists if condition (1) is met, $T_u \xrightarrow{rw} T_v$, that is, read-write dependency, or when (3) is satisfied, $T_u \xrightarrow{ww} T_v$, that is, write-write dependency.*

Definition 2. *Conflict Graph*. *A conflict graph, denoted as CG, is a directed graph that consists of a set of vertices $V = \{T_1, T_2, ..., T_N\}$ and a set of edges $E = \{(T_u, T_v)|1 \leq u \neq v \leq N, T_u \to T_v\}$. In this graph, $|V| = N$.*

Based on the captured transaction dependencies, a conflict graph that takes transactions as vertices and dependencies as edges is build. CG can guide the ordering of transactions to reduce over-aborting transactions that are still serializable.

Definition 3. *Key Dependency*. *Let's consider two distinct keys K_i and $K_j(i \neq j)$. We can say that K_i is dependent on K_j (notated as $K_i{\dashrightarrow}K_j$) if there exists a transaction T_v such that T_v^W belongs to RW_i and T_v^R is part of RW_j. Here, T_v^R and T_v^W denote the read and write units of transaction T_v, respectively.*

Definition 4. *Key-based Conflict Graph*. *A key-based conflict graph, denoted as $KCG = (V, E)$, is a directed graph where $V = \{RW_j|j = 1, 2, ..., n\}$, and $E = \{(RW_i, RW_j)|1 \leq i \neq j \leq n, \exists v \in [1, N_e], T_v^W \in RW_i \wedge T_v^R \in RW_j\}$. Here, n represents the number of keys being accessed.*

The key dependency is identified between the read and write units of a transaction. Contrasting with transaction dependency, we methodically map these read and write units to the associated key queues and position them accurately within this sequence. Using the captured key dependency, we build a directed edge between the write-read units of each transaction across different keys. We employ these edges to organize the read/write sets of all keys into a novel conflict graph called KCG.

Table 3. Four concurrent transactions

Transaction	T_1	T_2	T_3		T_4
R/W Operation	R	W	R	W	W
Accessed Key	K_1	K_1	K_2	K_1	K_2
Total order			$T_1 \Rightarrow T_2 \Rightarrow T_3 \Rightarrow T_4$		

Due to the increased conflicts per key, as shown in Table 3, more dependent transactions can be obtained on each key. Such dependency speeds up the processing of all writes and reads by incorporating a relatively small yet fast solution compared with a transaction dependency.

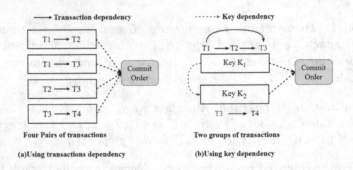

Fig. 2. Example of obtaining a commit order

Figure 2 presents a concrete example. It shows that employing the transaction dependency requires four pairs of dependent transactions to obtain the total order. Instead, it only requires two groups of dependent transactions detected by key K1 and K2 by relying on key dependency. Specifically, there are four transactions from T1 to T4 waiting ordering. In transaction dependency, if there is a dependency between two transactions, an edge is added, and at the same time, the transaction acts as a node. In key dependency, the read and write units of transactions that operate on the same key are stored in the same queue. We can see that employing the transaction dependency requires four pairs of dependent transactions to obtain the total order. Instead, it only requires two groups of dependent transactions detected by K1 and K2 by relying on key dependency. Hence, comparing to transaction dependency, key dependency is more suitable for large number of transaction conflicts.

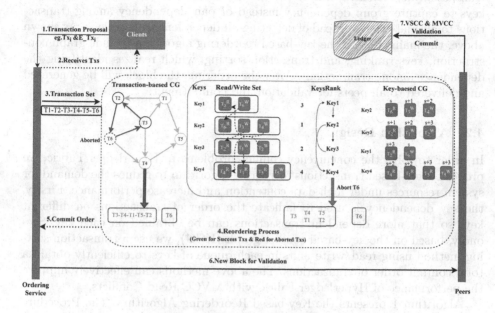

Fig. 3. Overview of System Model

To solve the concurrency conflict problem in Hyperledger Fabric, we propose a new transaction processing optimization method using a key-based transaction reordering algorithm. The main workflow of the system is shown in Fig. 3. The main goal is to get a serialized sorting after ordering service, which produces the least transaction discard and latency under the lowest system resources overhead.

The system workflow is as follows: the proposal is first signed and executed by the endorsing peers before it reaches the client. Then, the client will assemble endorsements into a transaction that contains the read/write sets, the endorsing peers' signatures, and the channel ID. Then send them to the Ordering Service to package and propagate. In the original version, the Ordering Service does not read the transaction details; it simply receives transactions from all channels in the network, orders them chronologically by channel, and creates blocks of transactions per channel. Therefore, you can take advantage of different ordering strategies to order the transactions inside a block to reduce the production of invalid transactions, such as transaction-based CG and key-based CG.

The main goal of the reordering algorithm is to generate a serialized sort in the middle of serialized transactions, ensuring that the least amount of transaction discards occur while consistent state transitions occur. The CG reordering algorithm using transaction dependency mainly includes five steps. They have directed conflict graph construction, subgraph division, cycle detection and removal, cycle-free conflict graph construction, and topological sorting. With the increase of conflicting transactions, we discover that each key may exhibit more transaction dependencies due to the increased conflicts per key. So, we utilize keys to capture group dependency instead of pair dependency among transactions to alleviate the overhead of detecting all dependent transactions. As shown above, the main steps of the key-based reordering algorithm include graph construction, keys ranking, and transaction sorting, which realizes identifying the dependence among transactions. After the ordering, the block will be generated and delivered to all peers for validation and commitment.

4.2   Algorithm Design

In order to solve the concurrency conflict problems in Hyperledger Fabric, we propose a key-based transactions reordering algorithm to reduce the demand for system resources under high data contention and increase performance. Firstly, the key dependency is used to indicate the order of transactions on different keys so that more dependent transactions can be obtained on each key. Secondly, based on the key-based conflict graph (KCG), we use a transaction sorting method using read/write units in each queue of keys to efficiently obtain a total commit order of transactions. The above methods can effectively improve the performance of Hyperledger Fabric with MVCC Read Conflicts.

Algorithm 1 presents the Key-based Reordering Algorithm. The Procedure CreateGraph creates a graph whose nodes contain queues of read and write units, and the data structure of rwNodes in each node is create to record read/write sets. During the graph construction, the edges of the graph is first created, and then in each read-write node (rwNodes), according to its read-write set, (RWSet) generates queues for the keys and stores these queues in a list. Since each queue of keys maintains all dependent transactions that read and write to it, we can obtain a partial order between transactions on each key. As to transactions that read and write to multiple keys, we need to get their order using key dependency.

After building the KCG, the next step is to determine the specific order of each transaction. Based on key dependency, we can obtain the sorting priority of keys. Procedure KeysRank ranks vertices in a graph based on in-degree and out-degree. There may exist cycles among keys. This phenomenon is caused by unserializable transactions, which will drop in the final sorting process. After that, Procedure TransactionSorting is used to generate a commit order based on KCG. Inspired by Lamport's logical clock [10], we assign a unified sequence number for each read/write unit in the queue to represent their sequence in the total order. After removing unserializable transactions, we got a conflict-free transaction order by switching transactions with the same sequence number to a serial order for deterministic state transfer.

Algorithm 1. Key-based Reordering Algorithm

```
 1: procedure CREATEGRAPH
 2:     Initialize edges, queueArray, and queues as empty dictionaries
 3:     Construct a list of read/write nodes named rwNodes from set of transactions S
 4:     for each rw in rwNodes do
 5:         Create an edge with rw and an id
 6:         Add the new edge to edges
 7:         for each node n in rw do
 8:             append n to the list of nodes associated with the string key in queueArray
 9:         end for
10:     end for
11:     for each key in queueArray do
12:         Sort the nodes associated with the key into rSlice and wSlice
13:         Create a new queue with rSlice and wSlice, and add it to queues
14:     end for
15:     return a new QueueGraph with queues and edges
16: end procedure
17: procedure SORTINGRANKDIVISION
18:     Initialize a sequence seq to represent sorting ranks
19:     if G.vertices == ∅ then
20:         return
21:     end if
22:     Find the minimum in-degree in G, assign it to min
23:     for each K_j in G do
24:         if A_j.inDegree == min then
25:             Select A_j and append it to seq, then break
26:         end if
27:     end for
28:     if min > 0 then
29:         Find the first keys with maximum out-degree in minAddrs, append it to seq
30:     end if
31:     Remove the vertex and edges of the selected vertex from G
32:     Recursively call SORTINGRANKDIVISION with the updated G
33: end procedure
34: procedure TRANSACTIONSORTING
35:     Initialize initialSeq with seq from SORTINGRANKDIVISION
36:     Find read units with sequence numbers in RW_j, assign it to sortedRSet
37:     if sortedRSet is empty then
38:         Assign initialSeq to sequence in RW_j and update maxRead to initialSeq
39:     else
40:         Find the minimum and maximum sequence numbers in sortedRSet, assign maxRead to maxSeq
    and update sequence of remaining units in RW_j to minSeq
41:     end if
42:     Find write units with sequence numbers in RW_j, assign it to sortedWSet
43:     if sortedWSet contains a unit whose read unit exists in RW_j then
44:         Increment sequence number
45:     end if
46:     for each unit in sortedWSet do
47:         if its sequence is less than maxRead then
48:             Abort the unit
49:         end if
50:     end for
51:     for remaining units in RW_j do
52:         while writeSeq is assigned do
53:             Increment writeSeq and then assign writeSeq to their sequence
54:         end while
55:     end for
56: end procedure
```

---

5   Experimental Evaluation

In this section, we present a comprehensive evaluation of our key-based reordering method. We first describe the experimental setup for our prototype and the workload in our experiments. Then, we evaluate the performance of KCG against FIFO adopted by vanilla Fabric and CG used by Fabric++.

5.1   Experimental Setup

Table 4. Experiment Configuration

Parameters	Values
Number of users	10,000
World State Database	LevelDB
Number of transactions per block(block size)	100
Probability for picking a read transactions (Pr)	50%
s-value of Zipfian distribution	0.8

Environment: This paper designs a blockchain prototype system implemented in GO 1.19 including simulated execution, sorting and verification of the Fabric. During the simulation phase, we adopt LEVM (Little Ethereum Virtual Machine) to provide an execution environment for our smart contracts written in Solidity. The open source panjf2000/ants library is used to provide the ability to manage and recycle a massive number of goroutines to simulate multiple peer nodes executing transactions in parallel. In the sorting phase, three ordering algorithms, FIFO, CG and KCG, are implemented. Table 4 respectively presents the various parameters that we have configure for system and workload.

Workloads: We use SmallBank as the workload, which simulates typical asset transfer scenarios. It provides 6 types of transactions for operating the data, including 5 update transactions and 1 read transaction. The read transaction is selected with probability P_r, while one of the five update transactions is chosen with a probability $1 - P_r$. The degree of skewness influences the distribution of read/write operations among the 10,000 available accounts. A higher skewness indicates a greater concentration of read/write operations on a smaller subset of accounts, leading to an increased potential for conflicts (Table 5).

Table 5. Transaction Types in SmallBank Workload

Transaction	Implication
CreateAccount	Initialize random funds for each customer's checking and savings accounts
TransactSavings	Add a certain amount of money to a savings account
DepositChecking	Add a certain amount of money to a checking account
WriteCheck	Indicates the removal of an amount from a customer's checking account
SendPayment	Transfer funds between two checking accounts
Amalgamate	Transfer all funds from a savings account to a checking account
Query	Query about the amount of a customer's savings or checking account

5.2 Results

This section presents a comprehensive analysis of the performance metrics of blockchain systems, specifically focusing on the influence of various parameters and different ordering strategies. We parameterize three key factors: 1) the skew parameter of the Zipfian distribution, 2) the block size, and 3) the percentage of read transactions. Our benchmarking analysis primarily considers latency and abort rate as the key metrics. It is worth mentioning that the ordering method employed in Fabric is abbreviated as FIFO, while the conflict graph utilized in Fabric++ and Fabricsharp is referred to as CG. Our proposed solution is denoted as KCG. For all experiments, we conduct ten runs for each parameter, and the reported results are the average values of these sum runs.

5.2.1 Impact of Block Size:
To evaluate the impact of block size on performance, we increase the number of transactions in each block from 50 to 200. We set the percentage of read transactions as 50% and the skew parameter as 0.8. For other parameters, refer to Table 4.

Figure 4(a) shows the transaction abort rate under different strategies. The results indicate that all three systems experience a drop in transaction rates ranging from 5% to 15% when the block size is set to 50 and 100. However, the CG strategy encounters memory-related failures when the block size increases from 100 to 150. This is attributed to the increasing number of transactions within a block, leading to a higher likelihood of conflicts and, consequently, a more complex conflict graph. This is attributed to the increasing number of transactions within a block, leading to a higher likelihood of conflicts and, conse-

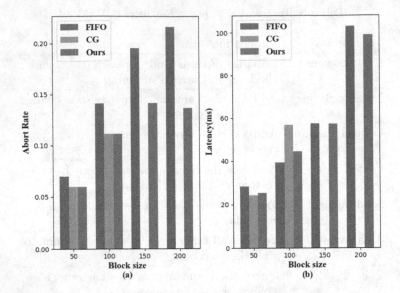

Fig. 4. Impact of the block size on transaction (a) aborting rate, (b) latency of FIFO, CG and ours.

quently, a more complex conflict graph. In the CG strategy, the DFS algorithm is utilized for searching, and when numerous cycles are present, memory consumption becomes significant as a new object is created for each cycle detected in a strongly connected subgraph. Our proposed KCG strategy maintains a transaction abort rate that is lower compared to FIFO and equal to or lower than CG. Moreover, unlike CG, our method can handle larger block sizes such as 150, 200, or even larger. This is possible because KCG resolves conflicting transactions by leveraging key dependency rather than transaction dependency. Key dependency denotes the order of transactions on different keys, allowing for faster processing of all writes and reads through the incorporation of a relatively small yet efficient solution.

Figure 4(b) shows the average latency under different block sizes. It is observable that the average latency escalates with the increase in block size, while FIFO always maintains a low latency. These results are expected since both CG and our KCG require additional reordering processing to resolve conflicting transactions in the ordering phase. Additionally, more transactions included in a block result in longer processing time, thus leading to higher latency. But the latency in KCG is always lower than CG and is comparable with FIFO. This is because the KCG method does not require the time-consuming cycle detection and removal stages under each strongly connected component within the graph. Furthermore, non-serializable transactions are directly discarded. Consequently, compared to FIFO, the number of transactions necessitating verification within KCG is reduced, thereby mitigating the delay. Hence, KCG proves to be a highly efficient method in high contention scenarios with concurrency conflicts.

5.2.2   Impact of Transaction Contention: Next, we conduct a comprehensive examination of transaction contention and its effects on overall performance. We set varying degrees of transaction contention by adjusting the skew parameter of the Zipf distribution from 0.2 to 1.2 in steps of 0.2.

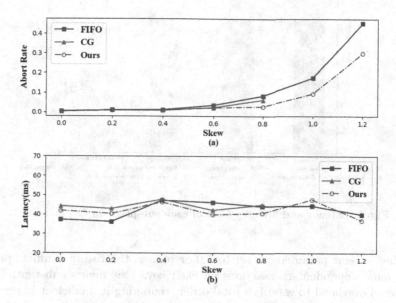

Fig. 5. Impact of skew parameter on transaction (a) aborting rate, (b) latency of FIFO, CG, and ours.

Figure 5(a) summarizes the results of the transaction abort rate under various skew parameters. We can see that for a small skew parameter (< 0.6), the transaction abort rate of all three systems is relatively low because the number of potentially conflicting transactions is small. However, an increase in the transaction abort rate of FIFO is witnessed with higher skew parameters (> 0.6), attributed to the fact that high data skewness in the Smallbank workload leads to a large number of potentially conflicting transactions. What's worse, the CG process fails due to exhausted memory when the skew exceeds 0.8. On the contrary, KCG's abort rate is always lower than FIFO. When the number of conflicting transactions is small (with skew below 0.8), this can be resolved by CG adopted by Fabric++. However, when it is set to 1.0 or higher, CG is prone to failure due to memory exhaustion, and OOM occurs. Contrarily, our KCG allows the system to operate normally for all skew degrees. KCG is not sensitive to the increase of contention degree and demands fewer system resources due to avoid building an edge between each pair of dependent transactions. Specifically, KCG assigns each transaction to the corresponding keys. Unique keys housing dependent transactions construct a novel scheduling graph, known as the key-based conflict graph (KCG), where edges capture key dependencies.

288   H. Ma et al.

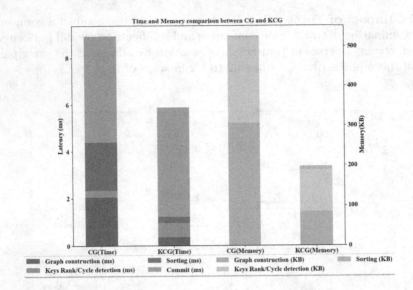

Fig. 6. Latency and Memory Cost of each sub-phase in CG and KCG.

When a skew parameter is set to 1.0 or higher, the rise in conflicts per key yields more dependent transactions on each key. This incurs substantial computational overhead to establish total order, rendering it inefficient in scenarios with high contention and a large volume of conflicting transactions. The result demonstrates KCG improves the transactions abort rate compared to FIFO. We compare the average latency in Fig. 5(b). We observe that our KCG latency is always lower than CG. Moreover, when the skew surpasses 0.8, the CG method becomes inapplicable. This is due to the increase in conflicting transactions leading to a larger transaction processing time in CG. In contrast, it has a relatively minor influence on our KCG system, where the rise in conflicts results in a limitation of accessed keys.

As presented in Fig. 6, we evaluate the latency and memory cost of each subphase in CG and KCG. The first and second columns represent the utilization of time. Overall, the latency of KCG is 36.3% lower than that of CG, where the time excluding the transaction commit stage is nearly equivalent across all systems. However, latency for other time periods in KCG is markedly lower than the corresponding stage latency in CG. Particularly under high data contention scenarios with a skew rising to 1.0, the latency for cycle detection and removal in CG substantially increases. This change is due to the recursive Johnson's algorithm, which has a time complexity of $O(((|V|+|E|) \cdot (C+1))$, and consumes a significant amount of memory when the number of cycles is large. Turning to the third and fourth columns, they indicate the utilization of memory resources. The sorting phase in both systems only constitutes a small fraction, with the KCG method outperforming CG in the other two stages. We can see that the graph construction and keys rank/cycle detection occupy a large portion of memory due to the overhead involved in establishing an edge between each pair of dependent transactions and the object overhead created for each cycle during the process of cycle detection in strongly connected components.

5.2.3 Impact of Read Transaction Percentage: This section presents an evaluation of how the percentage of read transactions affects the performance of simulated blockchain systems. We vary the percentage of read transactions from 0.1 to 0.9 in steps of 0.2. Figure 7(a) illustrates the impact of the percentage of read transactions on the transaction abort rate. It can be observed that as the percentage of read transactions increases, the transaction aborting rate decreases. This can be attributed to fewer conflicting transactions occurring when a higher proportion of read transactions is present, resulting in fewer updates to transaction records within a short time slot. Similar to the previous case, the CG method is not applicable when there are too many cycles in the conflict graph, while the proposed KCG method remains effective under any read transaction rate. Hence, the transaction abort rate of KCG outperforms that of FIFO and CG.

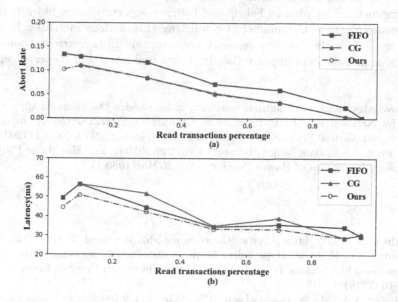

Fig. 7. Impact of the read transaction percentage on (a) aborting rate, (b) latency of FIFO, CG, and ours.

Figure 7(b) depicts the average latency of the three systems. As the proportion of read transactions increases and the proportion of write transactions decreases, the average latency shows a decreasing trend. Specifically, KCG demonstrates lower latency compared to FIFO and CG. However, when the read transaction proportion is as low as 10%, there is a possibility of CG experiencing errors or taking significantly longer than expected, primarily due to the time-consuming cycle detection and removal. To summarize, the aforementioned experiments demonstrate that the key-based CG construction method enables efficient concurrency control and yields few transaction aborts in scenarios with considerable transactions and conflicts. This method proves to be more suitable for high contention scenarios than the compared schemes.

6   Conclusion and Future Work

This paper focuses on enhancing the efficiency of handling concurrency conflicts in Hyperledger Fabric. We propose a key-based transaction reordering algorithm called KCG, which effectively resolves conflicts with minimal overhead. We first adopted a key-based transaction conflict graph construction method to replace transaction dependency to support parallel transaction processing. Then, the keys rank and sorting method is used to generate a transaction commit order. Through evaluations conducted on a real workload using Smallbank, we demonstrate that the key-based ordering method outperforms both the original and CG ordering method adopted by Fabric and Fabric++ permissioned blockchain systems, particularly as the number of conflicting transactions increases. In future work, we will further analyze resource consumption in concurrent transaction processing and strive to improve the abort rate and latency of Hyperledger Fabric.

Acknowledgement. The authors gratefully acknowledge the financial support provided by National Key R&D Program of China (No. 2022ZD0115302), in part by the National Natural Science Foundation of China (No. 62202479, No. 61772030), the Major Program of Xiangjiang Laboratory (No. 22XJ01004) and the Major Project of Technology Innovation of Hunan Province (No. 2021SK1060-1).

References

1. Ethereum (2023). https://www.ethereum.org/zh/. Accessed 25 May 2023
2. Androulaki, E., et al.: Hyperledger fabric: a distributed operating system for permissioned blockchains. In: Proceedings of the Thirteenth EuroSys Conference, pp. 1–15 (2018)
3. Chacko, J.A., Mayer, R., Jacobsen, H.A.: Why do my blockchain transactions fail? A study of hyperledger fabric. In: Proceedings of the 2021 International Conference on Management of Data, pp. 221–234 (2021)
4. Dickerson, T., Gazzillo, P., Herlihy, M., Koskinen, E.: Adding concurrency to smart contracts. In: Proceedings of the ACM Symposium on Principles of Distributed Computing, pp. 303–312 (2017)
5. Ding, B., Kot, L., Gehrke, J.: Improving optimistic concurrency control through transaction batching and operation reordering. Proc. VLDB Endow. **12**(2), 169–182 (2018)

6. Gorenflo, C., Golab, L., Keshav, S.: XOX fabric: a hybrid approach to blockchain transaction execution. In: 2020 IEEE International Conference on Blockchain and Cryptocurrency (ICBC), pp. 1–9. IEEE (2020)
7. Gorenflo, C., Lee, S., Golab, L., Keshav, S.: Fastfabric: scaling hyperledger fabric to 20 000 transactions per second. Int. J. Network Manage **30**(5), e2099 (2020)
8. István, Z., Sorniotti, A., Vukolić, M.: StreamChain: do blockchains need blocks? In: Proceedings of the 2nd Workshop on Scalable and Resilient Infrastructures for Distributed Ledgers, pp. 1–6 (2018)
9. Kwon, M., Yu, H.: Performance improvement of ordering and endorsement phase in hyperledger fabric. In: 2019 Sixth International Conference on Internet of Things: Systems, Management and Security (IOTSMS), pp. 428–432. IEEE (2019)
10. Lamport, L.: Time, clocks, and the ordering of events in a distributed system. In: Concurrency: the Works of Leslie Lamport, pp. 179–196 (2019)
11. Li, Y., et al.: FastBlock: accelerating blockchains via hardware transactional memory. In: 2021 IEEE 41st International Conference on Distributed Computing Systems (ICDCS), pp. 250–260. IEEE (2021)
12. Nakamoto, S.: Bitcoin: a peer-to-peer electronic cash system. Decentralized business review, p. 21260 (2008)
13. Nasirifard, P., Mayer, R., Jacobsen, H.A.: FabricCRDT: a conflict-free replicated datatypes approach to permissioned blockchains. In: Proceedings of the 20th International Middleware Conference, pp. 110–122 (2019)
14. Reijsbergen, D., Dinh, T.T.A.: On exploiting transaction concurrency to speed up blockchains. In: 2020 IEEE 40th International Conference on Distributed Computing Systems (ICDCS), pp. 1044–1054. IEEE (2020)
15. Ruan, P., Loghin, D., Ta, Q.T., Zhang, M., Chen, G., Ooi, B.C.: A transactional perspective on execute-order-validate blockchains. In: Proceedings of the 2020 ACM SIGMOD International Conference on Management of Data, pp. 543–557 (2020)
16. Sharma, A., Schuhknecht, F.M., Agrawal, D., Dittrich, J.: Blurring the lines between blockchains and database systems: the case of hyperledger fabric. In: Proceedings of the 2019 International Conference on Management of Data, pp. 105–122 (2019)
17. Sun, Q., Yuan, Y.: GBCL: reduce concurrency conflicts in hyperledger fabric. In: 2022 IEEE 13th International Conference on Software Engineering and Service Science (ICSESS), pp. 15–19. IEEE (2022)
18. Tarjan, R.: Depth-first search and linear graph algorithms. SIAM J. Comput. **1**(2), 146–160 (1972)
19. Trabelsi, H., Zhang, K.: Early detection for multiversion concurrency control conflicts in hyperledger fabric. arXiv e-prints arXiv:2301.06181 (2023). https://doi.org/10.48550/arXiv.2301.06181
20. Xiao, J., Zhang, S., Zhang, Z., Li, B., Dai, X., Jin, H.: NEZHA: exploiting concurrency for transaction processing in DAG-based blockchains. In: 2022 IEEE 42nd International Conference on Distributed Computing Systems (ICDCS), pp. 269–279. IEEE (2022)
21. Xu, L., Chen, W., Li, Z., Xu, J., Liu, A., Zhao, L.: Solutions for concurrency conflict problem on hyperledger fabric. World Wide Web **24**, 463–482 (2021)

Decentralized Self-sovereign Identity Management System: Empowering Datacenters Through Compact Cancelable Template Generation

Junwei Yu(✉) , Shaowen Li, Yepeng Ding , and Hiroyuki Sato

The University of Tokyo, Tokyo, Japan
{yujw,li-shaowen879,youhoutei,schuko}@satolab.itc.u-tokyo.ac.jp

Abstract. Digital identity management functions as a critical infrastructure for various information and communications technologies. However, traditional centralized systems are raising security concerns due to their reliance on trusted intermediaries, which prompts the development of self-sovereign identity (SSI). However, SSI still face challenges regarding network pressures, blockchain costs, and security vulnerabilities. In this paper, we propose Coconut, a novel system leveraging a decentralized SSI management architecture to facilitate the establishment of secure, localized digital identity and credential verification mechanisms, while obviating the necessity for reliance on trusted intermediaries and blockchain technologies. Coconut reduces the storage overhead by minimizing the responsibility of data centers and enabling them to solely store public keys. On the end-user side, individuals retain the prerogative to store their verifiable credentials within local environments. Besides, we introduce a compact cancelable template generation algorithm to enhance security and efficiency. Additionally, our experiments demonstrate the effectiveness and performance of Coconut.

Keywords: Self-sovereign identity · Biometric-based cryptography · Privacy preservation

1 Introduction

In today's landscape of information and communications technologies, the management of digital identities has gained paramount importance. Traditional centralized identity management systems [1] have long been plagued by vulnerabilities. The realm of digital identity information management within data centers raises a triad of crucial concerns encompassing storage capacity, computational power expenditure for identity verification, and the paramount aspect of user privacy.

This research was partially supported by KAKENHI (Grant-in-Aid for JSPS Fellows) 21J21087.

Conventional paradigms often necessitate considerable storage real estate to accommodate the voluminous identity information. Simultaneously, the computational outlay for verification processes adds a layer of complexity to the operational dynamics. Low flexibility results in high costs for simple adjustments in digital identity management. And, the revelation of sensitive identity data during verification procedures can intrude upon user privacy and potentially lead to adversarial consequences.

These multifaceted challenges underscore the imperative to employ a solution that addresses these intricacies comprehensively. In this context, the adoption of Self-Sovereign Identity (SSI) within the digital identity management system of data centers [2] emerges as a salient and promising strategy.

However, while SSI brings the advantage of decentralization, it introduces its own set of challenges, notably heightened network communication demands, substantial blockchain costs, and the intricate task of securely binding digital identities to physical entities to thwart identity theft [3] and fraud [4].

To address this complex array of challenges, a fresh approach emerges: the integration of biometric measures into the system of SSI. By establishing a tangible and verifiable link between digital identities and the individuals they represent, biometric integration presents a compelling solution to the limitations and vulnerabilities in current SSI models.

The heart of our solution lies in a meticulously crafted compact cancelable template generation algorithm. This algorithm serves as a robust shield for sensitive biometric information, preserving privacy while allowing for secure verification. Most notably, this algorithm's implementation effectively mitigates the memory-intensive demands on data centers.

Moreover, our approach significantly alleviates the computational strain during authentication procedures. By embedding biometric information within verifiable credentials, we enable seamless and secure decentralized verification processes via trusted edge computing. This strategic distribution of computation minimizes the computational power needed at any single point, ensuring efficient and flexible authentication while conserving energy.

2 Related Work

2.1 Self-sovereign Identity

Digital identity management is critical in the pervasive adoption of digital services. The majority of existing digital identity systems are centralized and rely on a central authority to manage and verify identity information. This centralized approach poses several security and privacy concerns, such as data breaches and identity theft [5]. As an emerging paradigm, SSI [1,6] allows individuals to manage, share, and verify their identity information securely and transparently, which enhances system resilience and security. In addition, the ability to record consent for data sharing provides transparency and enables individuals to make informed decisions about sharing their identity information [7].

Decentralized Identifiers (DIDs) are a crucial component of the self-sovereign identity model. The first version of the DID specification was published in 2019 by the World Wide Web Consortium (W3C) [8]. Instead of a centralized authority, DIDs rely on decentralized infrastructure such as blockchain to uniquely associate an identifier with its owner. DIDs are further leveraged by verifiable credentials (VCs) to bind entities with their credentials. Numerous DID systems have been proposed and developed, such as uPort [9], Sovrin [10], Blockstack [11], and Microsoft ION [12]. These systems are built on blockchains. However, the adoption of blockchains can potentially increase costs and network communication demands, as well as limit scalability, interoperability, and sustainability [6].

2.2 Biometric-Based Key Generation

Biometric-based key generation methods offer promising solutions for secure authentication, access control, and data transmission. These methods are characterized by their respective modality and revocation approaches. "Modality" refers to the type or mode of biometric data used for key generation, while "Revocation" pertains to the ability to revoke or invalidate generated keys in case of compromise or unauthorized access. We summarize these works in Table 1.

Table 1. Comparison of Biological-based Key Generation Algorithms

Algorithm	Key Generation	Modality	Revocation
BIBE [13]	Identity-based	Multiple	Yes
FIBE [14]	Identity-based	Single	Yes
CBT-KG [15]	Template-based	Multiple	Yes
PBKG [16]	Physiological-based	Single	No
CBAKG [17]	Behavioral-based	Single	No
MBKG [18]	Multi-modal	Multiple	No
RBKG [19]	Fuzzy commitment-based	Single	Yes
HBKG [20]	Homomorphic-based	Multiple	No

However, challenges such as the requirement for precise and reliable biometric sensors, protection of biometric data from theft or misuse, and ensuring user privacy need to be addressed.

2.3 Fingerprint Matching Algorithm

There are several types of biometric features that can be utilized for authentication purposes, such as face, fingerprint, and palm print. Compared with others, fingerprint biometrics are more accurate, easily acceptable to users, and more convenient to collect.

Fingerprint recognition algorithms can be classified into three categories: Minutiae-based, Correlation-based, and Pattern recognition-based algorithms. Minutiae-based algorithms extract minutiae points, such as ridge endings, bifurcations, and dots, to identify unique features of a fingerprint [21]. Correlation-based algorithms compare an input fingerprint with a pre-stored database of fingerprints using a correlation-based matching method [22], while Pattern recognition-based algorithms use advanced techniques to identify unique features, such as texture and orientation, of the fingerprint [23].

Fingerprint recognition, owing to its convenience and low cost, has been widely used and promoted. However, decentralized fingerprint verification is an area that has received relatively little attention in research. To address this gap, further exploration is needed to identify suitable algorithms that can ensure data security and privacy in a decentralized context. Additionally, the size of fingerprint templates and the potential trade-offs between template size and recognition accuracy to optimize the performance of fingerprint recognition systems should be considered. By addressing these challenges, the use of fingerprints can be further expanded and enhanced in various fields.

3 Coconut

3.1 Overview

Our goal is to create a system that seamlessly integrates biometric information into verifiable credentials, establishing a strong and reliable link between digital identities and the physical individuals they represent. By doing so, we anticipate overcoming the limitations and vulnerabilities of current SSI technologies, paving the way for a more secure and privacy-preserving identity management solution.

In pursuit of the aforementioned objectives, we put forth an innovative ecosystem centered around the utilization of Coconut ID, a secure and privacy-enhancing identifier, to realize the desired outcomes. Our proposed ecological system aims to effectively address the challenges associated with existing SSI technologies, leveraging the integration of biometric measures.

Figure 1 illustrates the key components and interactions within the Coconut ecosystem. The user, as the holder, plays a central role by generating a local Coconut ID using the Compact Cancelable Template Generation algorithm. This algorithm guarantees the creation of a unique and unlinkable identifier, protecting the user's privacy while maintaining the necessary security.

1. The user, as the holder, initiates the system by locally generating a unique Coconut ID using the Compact Cancelable Template Generation algorithm. This algorithm ensures the creation of a secure and privacy-preserving identifier.
2. The holder has the option to request a verifiable credential from the issuer. To obtain this credential, the holder's identity is verified by the issuer. Once the verification is completed, the issuer digitally signs the verifiable certificate, incorporating the Coconut ID, and transmits it to the user.

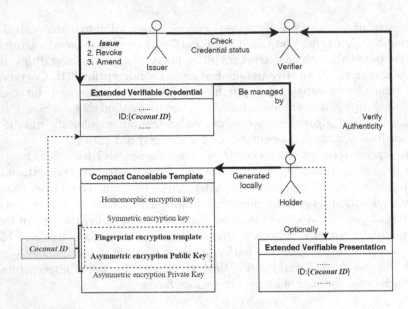

Fig. 1. Overview of Coconut System.

3. The holder maintains a local repository for managing the received verifiable credentials. Optionally, the holder can generate a verifiable presentation that encapsulates one or more verifiable credentials. This presentation serves as a concise and structured representation of the holder's credentials.
4. The holder presents the verifiable credentials, either individually or within a verifiable presentation, to a verifier. The presentation enables the holder to provide relevant and trustworthy information to the verifier in a secure manner.
5. The verifier undertakes the task of verifying the authenticity and integrity of the received verifiable credentials or verifiable presentation. Additionally, the verifier verifies the user's digital identity by using suitable verification mechanisms. This process includes verifying the credential status to ensure the non-revocation of the verifiable credentials as provided by the issuer.

Next, we will delve into an in-depth introduction that delves into the fundamental concepts and technologies that form the core of the Coconut ecosystem.

3.2 Compact Cancelable Template Generation

Fingerprint Enhancement and Minutiae Extraction. In the phase of fingerprint enhancement, we mainly employ Grayscale Conversion, Contrast Limited Adaptive Histogram Equalization, and Gaussian Blur Filter, which can enhance the contrast of the fingerprint and reduces unwanted artifacts in the image.

The original fingerprint may contain artifacts and noise that can affect the accuracy of fingerprint recognition systems. After undergoing the enhanced algorithm, which has effectively removed noise and enhanced ridge details to improve the clarity and fidelity of the fingerprint image.

Then, we have selected to use the Scale-Invariant Feature Transform (SIFT) [24] algorithm for minutiae extraction in enhanced fingerprint images, utilizing a Laplacian filter response instead of the Hessian determinant [25]. The SIFT algorithm is used for detecting and describing local features in images.

$$D(x, y, \theta) = \sum_{i=1}^{nBins} w(i) \cdot h_i(x, y, \theta)$$

In this formula, D(x, y, theta) represents the SIFT descriptor for a keypoint at location (x, y) with orientation theta. The descriptor is computed by concatenating nBins orientation histograms, h_i(x, y, theta), weighted by the values w(i) (Fig. 2).

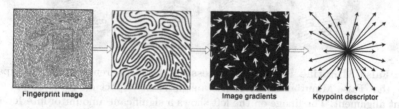

Fig. 2. Creation of Keypoint and Descriptor. Compute the gradient magnitude and orientation at each image sample point within a region around the keypoint location. The resulting samples are weighted by a Gaussian window and used to generate orientation histograms that summarize the contents over subregions.

The SIFT algorithm is widely used in fingerprint recognition due to its ability to detect and describe distinctive local features that are invariant to changes in scale, rotation, and illumination. In this implementation, the SIFT algorithm is utilized to detect and compute keypoints and descriptors from the input image. The resulting keypoints and descriptors are expected to provide a robust representation of the fingerprint image.

False Minutiae Removal. The filtering process involves the identification and removal of various types of keypoints that are not indicative of unique and discriminative features of the fingerprint. These types of keypoints include islands, short ridges, bifurcations with sharp angles, endings near the edge of the fingerprint, and spur or delta features.

The algorithm employs a set of criteria to ascertain the selection of relevant keypoints to eliminate from the given list, which is based on OpenCV (Open Source Computer Vision Library). It checks for "islands" in the fingerprint, which are small areas where the ridges converge and then diverges again. Keypoints with a size of less than 4 are identified as islands and are removed. The function then checks for short ridges by analyzing the keypoint's response. If the response is less than 0.02, the keypoint is removed. Additionally, the algorithm checks for ending points that are too close to the edge of the fingerprint, and removes them if they are present. Then, the function checks for spur or delta features, which are other types of interference in the fingerprint. If a keypoint's class ID is equal to 3 or 4, the keypoint is removed (Fig. 3).

The removal of these undesirable keypoints helps to increase the accuracy and efficiency of subsequent fingerprint matching and recognition tasks by reducing the noise and redundancy in the extracted feature set.

Fig. 3. False Minutiae Removal processing. It is apparent from the comparison images that False Minutiae Removal has a significant impact on the accuracy of fingerprint alignment. The image on the left shows a significant amount of interference in the form of non-discriminative keypoints, while the image on the right demonstrates the effects of removing these undesirable keypoints.

Following a thorough empirical analysis of the dataset, it was discerned that the dimensionality of the descriptor linked to 80 standard fingerprint images underwent a False-Minutiae-Removal process, culminating in an average reduction to 21.03% of their initial magnitude, which is a significant achievement. The enhanced robustness and precision of fingerprint matching obtained through this method increase the reliability of fingerprint-based authentication systems, which assume a paramount role in safeguarding secure digital identities.

Cancelable Biometrics. Cancelable biometrics is a technique that provides a means of revocability and update ability in biometric authentication systems [26]. It allows the user to update or cancel their biometric template when it is compromised or no longer trusted.

To attain Cancelable Biometrics, we have presented a Homomorphic Obfuscation Encryption (HOE) algorithm. HOE is a class of encryption schemes that are based on the idea of obfuscating the data in such a way that it can still be processed without being decrypted [27].

Furthermore, the use of HOE in cancelable biometrics enables users to use a single password to generate multiple template variations, thereby providing additional security against attacks such as dictionary attacks [28] and brute-force attacks [29]. The resulting scrambled descriptors have the same statistical distribution as the original descriptors, which ensures that the matching results are consistent.

In our algorithm, referred to as Homomorphic Obfuscation Encryption Scrambling Descriptor (HOE-SD), we introduce the process of generating the scrambled descriptor vector, which involves the following steps:

Hashing: We apply the hash function $hash$ to the password p to obtain the salted hash value h. Seed Generation: From the first eight hexadecimal characters of h, we derive an integer value that serves as the seed for a pseudorandom number generator. Pseudorandom Permutation: Utilizing the pseudorandom number generator initialized with the seed value, we generate a random permutation of the elements in the descriptor vector v. This step results in the creation of the scrambled descriptor vector sv. Output: Finally, the function returns the scrambled descriptor vector sv.

By applying the HOE-SD algorithm, we achieve obfuscation and homomorphic encryption of descriptor vectors, allowing for secure processing while maintaining the confidentiality of the underlying data.

To further enhance the security and usability of the generated cancelable biometric templates, we also employed a combination of compression and symmetric encryption techniques. Specifically, the widely-used and trusted zlib-deflated compression algorithm [30] and AES mode CBC encryption mode [31] were utilized to efficiently compress and secure the fingerprint feature descriptor data. These techniques provide high levels of protection for the fingerprint feature descriptor data.

To recapitulate, the aforementioned algorithm enables the generation of a compact fingerprint template that integrates multi-layer encryption and Cancelable mechanisms.

3.3   Experiment

Fingerprint Database. FVC 2004 DB3 B is a widely utilized fingerprint database in the research community. This database is a part of the Fingerprint Verification Competition 2004 and consists of 400 grayscale fingerprint images from 100 subjects, with four samples per subject. The fingerprints were captured using a digital camera with a resolution of 500 dpi and a bit depth of 8 bits per pixel [32]. This makes it an ideal tool for assessing the performance of feature extraction and matching algorithms and comparing the effectiveness of different fingerprint recognition techniques. In this study, we use the FVC 2004 DB3 B database to evaluate the performance of our proposed algorithm under various acquisition conditions, including variations in image quality, sensor resolution, and finger placement.

Experimental Data. The present study evaluated the performance of a proposed fingerprint-matching algorithm, and the experimental results demonstrate

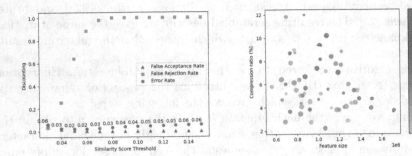

Fig. 4. (Left): Effects of Similarity Score Boundary on Error Rates. The Figure shows the effects of the similarity score boundary on the false acceptance rate (FAR), false rejection rate (FRR), and overall error rate (ER) of a fingerprint recognition system. **(Right): Relationship between Compression Rate and Size of Features.** The Figure presents the byte size of fingerprint features for multiple samples alongside the corresponding compression ratios.

its high success rate. As Fig. 4 (left) shows, at a similarity score boundary of 0.08, out of a total of 2000 attempts, 1935 successful matches were achieved, resulting in an impressively low error rate of 0.0325. Notably, the false acceptance rate (FAR) was found to be 0, indicating that no impostor fingerprints were erroneously identified as legitimate. These findings suggest that the proposed algorithm is highly accurate and effective in distinguishing between genuine and impostor fingerprints.

Figure 4 (right) reveals a lack of discernible correlation between the size of fingerprint features and their corresponding compression rates, primarily due to inherent variations in the fingerprint features themselves. The average size of fingerprint features is determined to be 1020433.55 bytes, while the average compression rate achieves 5.86%. Notably, the compression algorithm employed attains a notable performance, with the worst compression rate reaching 11.98%. These findings underscore the successful implementation of the encryption compression algorithm, which effectively balances the compression effect while maintaining a high level of matching accuracy for fingerprint recognition.

Experimental Analysis. The results of the present study indicate that the proposed algorithm exhibits a high success rate and an impressively low error rate, which exhibits the advantageous characteristic of a compact cancelable template. These findings suggest that the proposed algorithm is a promising solution for fingerprint recognition tasks.

In summary, our developed Compact Cancelable Template Generation algorithm for fingerprint matching successfully achieves our primary goal. The algorithm effectively combines compactness and high-security measures, ensuring its suitability for deployment in industrial-grade applications. By utilizing our algorithm, enhanced accuracy in fingerprint matching can be achieved while simultaneously providing robust privacy protection in real-world scenarios. Notably, our algorithm excels in its ability to compress fingerprint features and generate cancelable templates, further enhancing its practicality and versatility in various applications.

3.4   Evaluation

The proposed algorithm outlines a template-based fingerprint matching algorithm that culminates in the creation of a compact and cancelable template. This template is bound to the public key through a digital signature as the Coconut ID, which is the core of the entire system.

The datacenter stores IDs as unique identifiers for user entities for effective access control and authorization. Through standard interactions, the datacenter can authenticate the VC or VP managed by the user. This approach empowers the user with a degree of self-governance concerning identity management, contributing to a streamlined responsibility system for the data center. Storage considerations primarily revolve around the Coconut ID, minimizing storage-related concerns. Verification, in contrast, capitalizes on client-side computational resources, alleviating the computational burden on the data center effectively.

Evident in this architectural design is the principle of Minimalist Responsibility, whereby the data center's obligations are judiciously reduced. Storage obligations narrow down to the singular Coconut ID, while the onus of verification is shifted to the client-side computational infrastructure. Such an arrangement significantly mitigates the data center's computational load. Moreover, this configuration augments the potential for scaling and flexibility within the access control system. The data center's role predominantly entails the formulation of pertinent protocols and specifications, thus ensuring the secure and seamless expansion of functionalities.

In sum, this system heralds a sophisticated approach, leveraging a template-based fingerprint matching algorithm to yield a malleable and secure entity, intrinsically tied to a Coconut ID via digital signatures. The intricate web of identity relationships is adeptly managed through refined access control mechanisms, empowering both users and data centers with optimized identity governance strategies.

4   Conclusion

In this paper, we present Coconut, a self-sovereign identity management system, which offers a novel approach to address the challenges associated with identity management. We introduce the Compact Cancelable Template Generation algorithm, enabling users to securely and affordably manage their identities and claims independently, eliminating the reliance on third parties and blockchains.

The data center identity management system, integrated by Coconut, adheres to the principle of Minimalist Responsibility. In comparison to conventional identity management systems, Coconut requires less storage and computing power while offering enhanced privacy protection mechanisms and greater flexibility.

Through extensive experimentation, we have successfully validated the performance of Coconut. Our results indicate an impressively low error rate of 0.0325 for one-to-one matching of the encrypted fingerprints, with a false acceptance rate of 0. Additionally, our Compact Cancelable Template Generation algorithm achieves a noteworthy average compression ratio of 5.86%. These findings underscore the efficiency and reliability of our proposed solution.

To summarize, our study introduces a robust and cost-effective solution for self-sovereign identities and claims. Coconut, empowered by the Compact Cancelable Template Generation algorithm, exhibits exceptional performance and addresses critical concerns in current identity management practices.

References

1. Mühle, A., Grüner, A., Gayvoronskaya, T., Meinel, C.: A survey on essential components of a self-sovereign identity. Comput. Sci. Rev. **30**, 80–86 (2018). ISBN 1574-0137
2. Ding, Y., Sato, H., Machizawa, M.G.: Leveraging self-sovereign identity in decentralized data aggregation. In: 2022 Ninth International Conference on Software Defined Systems (SDS), Paris, France, pp. 1–8. IEEE (2022)
3. Newman, G.R., McNally, M.M., et al.: Identity theft literature review (2005)
4. Willox, N.A., Regan, T.: Identity fraud: providing a solution. J. Econ. Crime Manage. **1**(1), 1–15 (2002)
5. Anderson, K.B., Durbin, E., Salinger, M.A.: Identity theft. J. Econ. Perspect. **22**(2), 171–192 (2008)
6. Ding, Y., Sato, H.: Self-sovereign identity as a service: architecture in practice. In: 2022 IEEE 46th Annual Computers, Software, and Applications Conference (COMPSAC), Los Alamitos, CA, USA, pp. 1536–1543. IEEE (2022)
7. Tobin, A., Reed, D.: The inevitable rise of self-sovereign identity. Sovrin Found. **29**(2016), 18 (2016)
8. W3C: Decentralized identifiers (DIDs) v1.0. W3C Recommendation, May 2019. Accessed 12 Mar 2023
9. Lundkvist, C., Heck, R., Torstensson, J., Mitton, Z., Sena, M.: UPORT: a platform for self-sovereign identity (2017). https://whitepaper.uport.me/uPort_whitepaper_DRAFT20170221.pdf
10. Khovratovich, D., Law, J.: Sovrin: digital identities in the blockchain era. Github Commit by Jasonalaw October **17**, 38–99 (2017)
11. Ali, M., Nelson, J., Shea, R., Freedman, M.J.: Blockstack: a global naming and storage system secured by blockchains. In: 2016 USENIX Annual Technical Conference (USENIX ATC 2016), pp. 181–194 (2016)
12. Microsoft ION. https://github.com/decentralized-identity/ion. Accessed Mar 2023
13. Sarier, N.D.: A new biometric identity based encryption scheme. In: 2008 the 9th International Conference for Young Computer Scientists, pp. 2061–2066. IEEE (2008)

14. Sahai, A., Waters, B.: Fuzzy identity-based encryption. In: Cramer, R. (ed.) EURO-CRYPT 2005. LNCS, vol. 3494, pp. 457–473. Springer, Heidelberg (2005). https://doi.org/10.1007/11426639_27

15. Lee, C., Choi, J.-Y., Toh, K.-A., Lee, S., Kim, J.: Alignment-free cancelable fingerprint templates based on local minutiae information. IEEE Trans. Syst. Man Cybernet. B (Cybernet.) **37**(4), 980–992 (2007)

16. Miao, F., Bao, S.-D., Li, Y.: Biometric key distribution solution with energy distribution information of physiological signals for body sensor network security. IET Inf. Secur. **7**(2), 87–96 (2013)

17. Sitová, Z., et al.: HMOG: new behavioral biometric features for continuous authentication of smartphone users. IEEE Trans. Inf. Forensics Secur. **11**(5), 877–892 (2015)

18. Evelyn Brindha, V., Natarajan, A.M.: Multi-modal biometric template security: fingerprint and palmprint based fuzzy vault. J. Biom. Biostat. **3**(3), 100–150 (2012)

19. Zhang, L., Sun, Z., Tan, T., Hu, S.: Robust biometric key extraction based on iris cryptosystem. In: Tistarelli, M., Nixon, M.S. (eds.) ICB 2009. LNCS, vol. 5558, pp. 1060–1069. Springer, Heidelberg (2009). https://doi.org/10.1007/978-3-642-01793-3_107

20. Gomez-Barrero, M., Maiorana, E., Galbally, J., Campisi, P., Fierrez, J.: Multi-biometric template protection based on homomorphic encryption. Pattern Recogn. **67**, 149–163 (2017)

21. Jain, A., Ross, A., Prabhakar, S.: Fingerprint matching using minutiae and texture features. In: Proceedings 2001 International Conference on Image Processing (Cat. No. 01CH37205), vol. 3, pp. 282–285. IEEE (2001)

22. Nandakumar, K., Jain, A.K.: Local correlation-based fingerprint matching. In: ICVGIP, pp. 503–508 (2004)

23. Cole, S.A.: History of fingerprint pattern recognition. Automatic fingerprint recognition systems, pp. 1–25 (2004)

24. Lindeberg, T.: Scale invariant feature transform (2012)

25. Lowe, D.G.: Distinctive image features from scale-invariant keypoints. Int. J. Comput. Vision **60**, 91–110 (2004)

26. Patel, V.M., Ratha, N.K., Chellappa, R.: Cancelable biometrics: a review. IEEE Signal Process. Mag. **32**(5), 54–65 (2015)

27. Sahai, A., Waters, B.: How to use indistinguishability obfuscation: deniable encryption, and more. In: Proceedings of the Forty-Sixth Annual ACM Symposium on Theory of Computing, pp. 475–484 (2014)

28. Pinkas, B., Sander, T.: Securing passwords against dictionary attacks. In: Proceedings of the 9th ACM Conference on Computer and Communications Security, pp. 161–170 (2002)

29. Florêncio, D., Herley, C., Coskun, B.: Do strong web passwords accomplish anything? HotSec **7**(6), 159 (2007)

30. Gailly, J., Adler, M.: Zlib compression library (2004)

31. Vaidehi, M., Justus Rabi, B.: Design and analysis of AES-CBC mode for high security applications. In: Second International Conference on Current Trends In Engineering and Technology-ICCTET 2014, pp. 499–502. IEEE (2014)

32. Maio, D., Maltoni, D., Cappelli, R., Wayman, J.L., Jain, A.K.: FVC2004: third fingerprint verification competition. In: Zhang, D., Jain, A.K. (eds.) ICBA 2004. LNCS, vol. 3072, pp. 1–7. Springer, Heidelberg (2004). https://doi.org/10.1007/978-3-540-25948-0_1

Low-Latency Consensus with Weak-Leader Using Timestamp by Synchronized Clocks

Yue Ni[1,2], Guangping Xu[1,2(✉)], and Yi Tian[1,2]

[1] School of Computer Science and Engineering, Tianjin University of Technology, Tianjin, China
[2] Tianjin Key Laboratory of Intelligence Computing and Novel Software Technology, Tianjin, China
xugp@email.tjut.edu.cn

Abstract. Weak leader algorithms can improve the efficiency of reaching consensus by reducing the number of communications for distributed network services. However, they generate a large number of conflicts during operation, which can lead to expensive cross-region communications and make them difficult to adapt to WAN environments. So how to efficiently resolve conflicts becomes a key challenge. In this paper, we propose an approach applied to weak-leader algorithms, which effectively reduces the happens of conflicts and provides low-latency and high-throughput consensus in WAN systems, called the Low-Conflict Consensus method (LCC). Our proposed LCC uses timestamps generated by synchronized clocks to reduce conflicts. We present how LCC determines the delayed time in message processing and adopts some rules to sort the received messages. We validate and evaluate LCC through extensive experiments, which show that LCC can effectively reduce conflicts and the latency to achieve consensus.

Keywords: Weak Leader · Timestamp · Conflict

1 Introduction

The characteristics of distributed service computing systems [15,16] determine that they will have many nodes with wide distribution; at the same time, node crashes and network failures [17,18] may make system failures normal. The availability and fault tolerance of distributed systems can be implemented by deploying the replicated data in multiple data centers across geographical locations [11–13]. Data consistency is the most important question in data replication, and it means that multiple replicas agree on the same data or on the same order of multiple data. Various distributed consensus algorithms aim to solve data consistency problem [10]. However, the problem of high consensus latency seriously affects the performance of the system.

When such algorithms are deployed in a wide area network (WAN), message transmission between replicas will take much long time. That is, consensus latency is mainly determined by the communication time and the number of communications for message transmission. In WAN networks, some works, such as Fast Paxos [4] and EPaxos [3], try to reduce the number of messages communicated across regions if we want to reduce the consensus latency. These algorithms are the variants of Paxos [1] which have no leader or a weak leader, but different from Multi-Paxos [2] and Raft [25] which have a strong leader. Both the no-leader and weak-leader algorithms reduce the communication time by reducing the number of communications between replicas to the leader by weakening the performance of the leader. However, because of this, the Acceptor will process messages from different Proposers, thus creating conflicts that affect the performance of the algorithm. The no-leader algorithm completely offloads the leader and requires the Acceptor to sort the messages on its own when resolving conflicts. Taking EPaxos as an example, it easily receives the influence of complex dependency models and generates deadlocks when resolving conflicts. The leader in the weak-leader algorithm can still assume the role of message ordering and only needs to recollect the information and resend the Prepare message to resolve the conflict. Taking Fast Paxos as an example, it only needs the leader to re-execute the Classic Round to resolve the conflict, which is relatively simple. Therefore, we conduct research on low consensus delay algorithms for the weak-leader algorithm like Fast Paxos. Because the leader requires many additional communication steps in resolving conflicts, and greatly increases the consensus latency of the algorithm. We hope to propose methods for simple conflict resolution to reduce communication rounds to accommodate WAN.

We observe that conflicts occur mainly because replicas process operations at different times and in different orders, and that setting rules for the order in which operations are processed is a key point in resolving conflicts. We propose LCC, which uses timestamps and synchronized clocks on the replicas to impose intentional delays on the replicas, and the replicas try to process operations in order at the same time to reduce conflicts. In Clock-RSM [5], TOQEPaxos [19,24], and Nezha [23], all propose methods to order the commands based on clock synchronization while ensuring the correctness of the algorithm, which shows that it is feasible to use synchronized clocks in consensus algorithms and to sort messages in consensus algorithms based on their recorded times. We analyze and evaluate LCC and compare it with Fast Paxos and Multi-Paxos. Experiments show that LCC can reduce conflict rates by 15% and latency by 10% in a high concurrency state.

Briefly, this paper makes the following contributions:

- We propose an effective consensus algorithm, called LCC, which is a way to reduce conflicts in Fast Paxos using synchronized clocks and timestamps.
- We analyze the conflict rate and execution latency of Fast Paxos and LCC and show the latency advantage of LCC.
- We compare and evaluate the throughput of LCC, Fast Paxos, and Multi-Paxos, and show the throughput advantage of LCC.

The remainder of this paper is organized as follows: Sect. 2 provides some backgrounds, Sect. 3 describes the implementation of LCC, Sect. 4 evaluates LCC, Sect. 5 discusses related work, and Sect. 6 provides a conclusion.

2 Background and Motivation

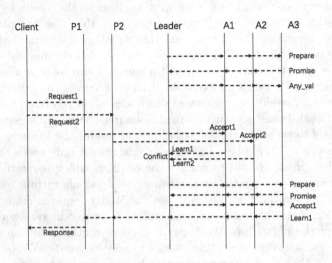

Fig. 1. Fast Paxos conflict resolution example.

We address the conflict problem using Fast Paxos as a representative of weak-leader algorithms, which is characterized by optimal performance by simplifying the communication process. However, the conflict problem of Fast Paxos can cause its advantages to be lost and even have higher consensus latency. In the rest of this section, we will describe the conflict generation of Fast Paxos and the conflict resolution idea.

2.1 Conflicts

The execution process of Fast Paxos is divided into Classic Round and Fast Round. Classic Round works the same as in classic Paxos. The Fast Round allows consensus to be reached in one round-trip communication, and an Acceptor can communicate directly with a Proposer if the leader sends Any-val message in the Fast Round. Each Acceptor will decide on its own which Accept message to take as the accept value. Therefore, Fast Paxos will create many conflicts. As shown in Fig. 1, the Proposer and Acceptor can communicate directly after the leader sends an Any-val message. P1 and P2 send Accept messages with different values. Due to the different distances between P1 and P2 arriving at A1, A2, and A3, the Acceptor accepts the first message that arrives. Then, A1 and A2 accept the Accept messages with different values and create a conflict. To ensure the correctness of the algorithm, the leader will rejoin the communication and execute Classic Round to resolve the conflict. In high concurrency scenarios, the probability of conflict occurrence and the cost of conflict resolution can be very high. Note that reducing the probability of conflict is the key for the algorithm to achieve as low as possible for consensus latency.

2.2 Motivation

Based on the above analysis of conflict formation, we can impose an intentional delay on the replicas to reduce the conflict rate. The replica closer to the Proposer delays the processing of the Accept message until the farthest replica accepts this message. Our basic idea is that the replica processes the Accept messages from the Proposer in an inert manner. After a replica receives Accept messages from Proposers, it needs to wait for some time and then process them in a certain order. We want to minimize the generation of conflicts through the way that replicas process Accept messages in the same order.

2.3 Synchronized Clock

LCC is proposed on the base of synchronizing the clocks of the replicas in the cluster. About the clock synchronization, we run NTP to keep the physical clock at each replica synchronized with a nearby public NTP server [9]. NTP is a protocol used to synchronize the time of computers. It allows computers to synchronize to their servers or clock sources, providing highly accurate time correction.

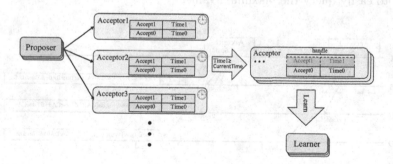

Fig. 2. System Overview

3 Proposed Approach

3.1 Overview

LCC uses timestamps and synchronized clocks to reduce conflicts, and it enables weak leader algorithms to reduce consensus latency by minimizing communication steps while ensuring correctness. We implement LCC based on Fast Paxos.

Figure 2 shows the system overview of the LCC. First, synchronize the clocks of all replicas in the cluster. We use the Network Time Protocol (NTP) clock synchronization [9] to synchronize the clocks of multiple replicas in the system; Second, replicas calculate the time recorded by the timestamp and sort the Accept messages that replicas receive based on the calculated time; Finally, multiple replicas can process messages in the same order, and reduce the impact of conflicts about the consensus algorithm.

We divide the design of the method into two parts: calculating the one-way delay time between replicas and constructing the order to process the messages after the replicas receive them.

3.2   Calculation of One-Way Delay Time

LCC requires each replica to estimate the one-way delay time of message transmission to other replicas. It is difficult to directly calculate the time difference of message transmission among multiple replicas across geographic regions. In this paper, timestamps are attached to messages to calculate the one-way delay between replicas. As shown in Fig. 3, the timestamp records the SendAt time when A sends an Accept message and the ReceiveAt time when B receives this Accept message. B records the ReceiveAt time when it receives an Accept message from A. B can calculate the one-way delay time between B and A. The one-way delay time between other replicas is calculated in the same way. Meanwhile, each replica maintains a time set that writes the one-way delay time between itself and all other replicas, and the existence of the time set allows the replica to easily query the maximum value.

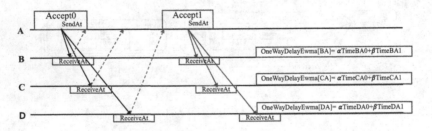

Fig. 3. Calculation of one-way delay time.

In order to accurately reflect the trend of recent time changes, we use a weighted moving average method to update the time data in the set. Different weights are assigned to the original data and the updated data, and the moving average is calculated based on these different weights. As shown in Fig. 3, after the Accept0 message is transmitted between replicas, the one-way delay time from other replicas to replica A is recorded. When the Accept1 message begins to transmit, the recorded time in each replica needs to be updated. Let the time weight of the Accept0 message be α and the time weight of the Accept1 message be β. The updated one-way delay time is αTimeBA0+βTimeBA1. See Algorithm 1 for implementation details.

3.3   Processing of Replica Messages

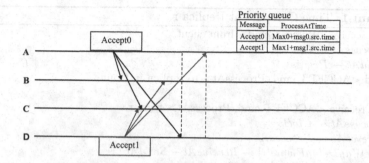

Fig. 4. Sorting the messages received by the replica by priority.

After the replicas have recorded the one-way delay time, they need to sort the received Accept messages in a certain order. The replica first needs to maintain a priority queue when sorting messages. After receiving an Accept message, the replica puts this message and its corresponding ProcessAtTime into the priority queue. The smaller ProcessAtTime is given a higher priority. As shown in Fig. 4, there will be two columns of messages in the priority queue of B: Accept0 and its corresponding ProcessAtTime0; and Accept1 and its corresponding ProcessAtTime1. In the priority sequence, the ProcessAtTime1 recorded in B is larger than ProcessAtTime0. Then in the priority sequence of B, the Accept0 message is first as the queue head message out of the queue for execution, and on the other replicas as well. See Algorithm 2 for implementation details.

Table 1. Definition of symbols used in the algorithm.

Symbols	Definitions
Config	All surviving replicas in the current configuration
Clock	Current time
Log	Command logging to stable storage
ClockSyncPQ	Priority Queue
PeakPriority	The head element of the priority queue
OneWayDelayEwma [ ]	Array holding one-way delay times

3.4 Implementation

Algorithms 1 and 2 provide the pseudo-codes for LCC. Table 1 defines the symbols used in the algorithm.

One-Way Time Determination: The first part of the method is given in Algorithm 1, which explains how to calculate one-way delay time.

Algorithm 1. Time Compute at Replica r

1: **upon** receive <Request cmd> from client
2:    $ProcessAt \leftarrow Clock$
3:    $SendAt \leftarrow ProcessAt$
4:    send <ACCEPT cmd, ProcessAt> to replica in Config
5:
6: **upon** receive <ACCEPT cmd, ProcessAt> from r
7:    $ProcessAt \leftarrow Clock$
8:    $ReceiveAt \leftarrow ProcessAt$
9:    $OneWayDelayEwma[r] \leftarrow ReceiveAt - SendAt$
10:   $ProcessAtTime \leftarrow ComputTime()$
11:   append <ACCEPT cmd, ProcessAtTime > to ClockSyncPQ
12:
13: **function** $ComputTime()$
14:   **return** $SendAt + max(OneWayDelayEwma[r])$

Sorting Steps: The second part of the method is given in Algorithm 2, which describes the steps of message execution by priority.

Algorithm 2. Message Execution

1: **if** $TimestampSorting = true$ & $ClockSyncPQ! = null$ **then**
2:    $PeakPriority = ClockSyncPQ.Top()$
3:    $CurrentTime \leftarrow Clock$
4: **end if**
5:
6: **if** $CurrentTime >= PeakPriority.Time$ **then**
7:    send Accept< ACCEPT cmd, ProcessAtTime > to Log
8:    remove Accept< ACCEPT cmd, ProcessAtTime > from ClockSyncPQ
9: **end if**

---

4 Evaluation

In the performance evaluation, the variations of the number of conflicts generated in the execution of the algorithm is quantified and analyzed, as well as the variations of the execution latency and the throughput of the algorithm are analyzed under different levels of message interference.

4.1   Experimental Setup

We use Centos 7 operating system with replicas. The commands submitted on the replicas are processed by Fast Paxos and LCC separately, comparing their latency time during execution. Replicas use the NTP protocol to keep each clock synchronized of replicas with a nearby public NTP server. For the message interference case, we simulate the level of concurrency of multiple clients sending messages by modifying the percentage of Shuffle in the experiment.

4.2   Conflict Rate

The experiment mainly depends on the percentage of Shuffle to affect the conflict rate. As shown in Fig. 5, For the Fast Paxos, the high the percentage of Shuffle, the more conflicts are generated. For LCC, the advantage becomes more and more obvious as the conflict rate increases, and it can effectively reduce the conflict rate.

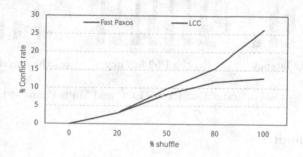

Fig. 5. Comparing the conflict rate of Fast Paxos and LCC with different Shuffle percentages.

4.3   Latency

In the above experiments, it is shown that LCC can effectively reduce the conflict rate. However, our ultimate goal of reducing the conflict rate is to reduce the latency, so the latency time also needs to be evaluated.

Average Latency: In order to observe the effect of LCC in comparison with the Fast Paxos in terms of latency, we evaluated the average latency of the replicas during the execution of the algorithm. Figure 6(a) shows the average latency of each replica, and it can be seen that LCC provides a better average latency than the Fast Paxos.

P99 Latency: To more completely evaluate the latency aspect of LCC, we measure the P99 latency of the replica during the execution of the algorithm. Figure 6(b) shows the P99 latency of the replicas, and LCC has a lower P99 latency than the Fast Paxos. Because the Fast Paxos will generate many conflicts in high concurrency scenarios, LCC can effectively reduce conflicts and reduce message communication between replicas in high concurrency scenarios. And just waiting for a period of time before executing an Accept message, the waiting time is relatively short than the time generated by message communication. Therefore, LCC has smaller P99 latency than the Fast Paxos.

Execution Latency: We also compare and evaluate LCC, Fast Paxos, and the now widely used Multi-Paxos. As shown in Fig. 6(c), for Multi-Paxos, as Shuffle increases, the leader creates a bottleneck problem and the execution delay of Multi-Paxos increases significantly. For Fast Paxos and LCC, the execution latency of LCC is similar to that of Fast Paxos during execution with less interference. However, the conflict rate increases significantly when the message interference becomes greater, and the advantage of LCC increases significantly.

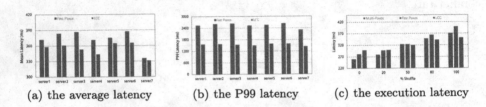

(a) the average latency	(b) the P99 latency	(c) the execution latency

Fig. 6. Comparing the latency of Fast Paxos, LCC and Multi-Paxos at the seven sites.

4.4   Throughput

The request message volume in the experiments is the number of Requests initiated by the client. Figure 7 shows the throughput comparison of Multi-Paxos, Fast Paxos, and LCC. The change in throughput for low conflict rates is shown in Fig. 7(a). At low conflict rates, the advantage of LCC is not obvious. At high conflict rates, as shown in Fig. 7(b). Fast Paxos will reach the throughput bottleneck quickly as the message volume increases significantly during the execution.

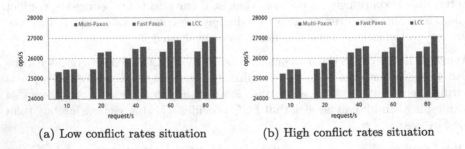

(a) Low conflict rates situation	(b) High conflict rates situation

Fig. 7. Comparing the throughput of Multi-Paxos, Fast Paxos, and Timestamp Sorting Method for low and high conflict cases.

LCC can reduce many conflicts and has a more obvious advantage in terms of throughput.

5   Related Work

As distributed systems are more and more widely deployed in WAN, the problem of high consensus latency is frequently mentioned. Optimization of the fundamental consensus algorithm has been frequently proposed in order to achieve low latency [6,14,20,21]. FPaxos [7,22] provides a more flexible quorum selection. It reduces the consensus latency caused during communication by reducing the number of nodes in the system that need to communicate across regions. WPaxos [8] uses dynamic partitioning and local commits to regions to reduce the consensus latency during algorithm execution. Clock-RSM [5] is a low latency timestamp-based replication algorithm for inter-data center state machines. Sequencing the instances depending on the timestamp of the synchronized clocks allows state machine replication to be done in less time. Which advantage lies mainly in the fact that the replica broadcasts an answer to all other replicas, which is also possible for Fast Paxos.

6   Conclusion

In this paper, we take the Fast Paxos algorithm as an example to analyze the impact of conflicts on performance in weak leader algorithms, and then propose the LCC method, which uses synchronized clocks and timestamps to reduce the conflicts and consensus delay in Fast Paxos. Most weak leader algorithms reduce the consensus delay by reducing the number of communications between each replica and the leader, but result in many conflicts that make them unsuitable for the WAN. LCC allows the replicas to impose intentional delays on incoming messages and to order the messages to process, which allows most replicas to execute messages in the same order, which effectively reduces the number of conflicts and lowers the delay to reach consensus.

Acknowledgement. This work was supported in part by grants from National Natural Science Foundation of China (Project number: 61971309).

References

1. Lamport, L.: Paxos made simple. In: ACM SIGACT News (Distributed Computing Column), pp. 18–25 (2001)
2. Lamport, L.: The part-time parliament. ACM Trans. Comput. Syst., 133–169 (1998)
3. Moraru, I., Andersen, D.G., Kaminsky, M.: There is more consensus in egalitarian parliaments. In: Proceedings of the Twenty-Fourth ACM Symposium on Operating Systems Principles, pp. 358–372 (2013)
4. Lamport, L.: Fast paxos. In: Distributed Computing, pp. 79–103 (2006)

5. Du, J., Sciascia, D., Elnikety, S., Zwaenepoel, W., Pedone, F.: Clock-RSM: low-latency inter-datacenter state machine replication using loosely synchronized physical clocks. In: 2014 44th Annual IEEE/IFIP International Conference on Dependable Systems and Networks, pp. 343–354 (2014)

6. Arun, B., Peluso, S., Palmieri, R., Losa, G., Ravindran, B.: Speeding up consensus by chasing fast decisions. In: 2017 47th Annual IEEE/IFIP International Conference on Dependable Systems and Networks (DSN), pp. 49–60 (2017)

7. Howard, H., Malkhi, D., Spiegelman, A.: Flexible Paxos: quorum intersection revisited. In: International Conference on Principles of Distributed Systems, pp. 1–14 (2016)

8. Ailijiang, A., Charapko, A., Demirbas, M., Kosar, T.: WPaxos: wide area network flexible consensus. IEEE Trans. Parallel Distrib. Syst. **31**, 211–223 (2020)

9. The network time protocol (2014). http://www.ntp.org/

10. Burrows, M.: The Chubby lock service for loosely-coupled distributed systems. In: Symposium on Operating Systems Design Implementation, pp. 335–350 (2006)

11. Schneider, F.B.: Implementing fault-tolerant services using the state machine approach: a tutorial. ACM Comput. Surv. **22**, 299–319 (1990)

12. Baker, J., Bond, C., Corbett, JC., et al.: Megastore: providing scalable, highly available storage for interactive services. In: Conference on Innovative Data Systems Research, pp. 223–234 (2011)

13. Calder, B., Wang, J., Ogus, A., et al.: Windows azure storage: a highly available cloud storage service with strong consistency. In: Symposium on Operating Systems Principles, pp. 143–157 (2011)

14. Zieliński, P.: Low-latency Atomic Broadcast in the presence of contention. In: Dolev, S. (ed.) DISC 2006. LNCS, vol. 4167, pp. 505–519. Springer, Heidelberg (2006). https://doi.org/10.1007/11864219_35

15. Corbett, C., Dean, J., Epstein, M., et al.: Spanner: Google's globally-distributed database. ACM Trans. Comput. Syst. **31**, 1–22 (2013)

16. Baker, J., Bond, C., Corbett, C., et al.: Megastore: providing scalable, highly available storage for interactive services. In: Conference on Innovative Data Systems Research, pp. 223–234 (2011)

17. Waldo, J., Wyant, G., Wollrath, A., Kendall, S.: A note on distributed computing. In: Vitek, J., Tschudin, C. (eds.) MOS 1996. LNCS, vol. 1222, pp. 49–64. Springer, Heidelberg (1997). https://doi.org/10.1007/3-540-62852-5_6

18. Rotem-Gal-Oz, A.: Fallacies of distributed computing explained[EB/OL] (2006). http://www.rgoarchitects.com/Files/fall/

19. Geng, Y., et al.: Exploiting a natural network effect for scalable, fine-grained clock synchronization. In: Proceedings of the 15th USENIX Conference on Networked Systems Design and Implementation, pp. 81–94 (2018)

20. Seo, H., Park, J., Bennis, M., Choi, W.: Communication and consensus co-design for distributed, low-latency, and reliable wireless systems. IEEE Internet Things J. **8**, 129–143 (2021)

21. Wang, Y., Hu, H., Qian, W., Zhou, A.: Migratable Paxos. In: Nah, Y., Cui, B., Lee, S.-W., Yu, J.X., Moon, Y.-S., Whang, S.E. (eds.) DASFAA 2020. LNCS, vol. 12112, pp. 296–304. Springer, Cham (2020). https://doi.org/10.1007/978-3-030-59410-7_20

22. Howard, E., Mortier, R.: Relaxed Paxos: quorum intersection revisited (again). In: Proceedings of the 9th Workshop on Principles and Practice of Consistency for Distributed Data, pp. 16–23 (2022)

23. Jinkun, G., Sivaraman, A., Prabhakar, B., Rosenblum, M.: NEZHA: deployable and high-performance consensus using synchronized clocks. In: Proceedings of VLDB Endowment, pp. 629–642 (2022)

24. Tollman, S., Jin Park, S., John, O.: EPaxos revisited. In: Symposium on Networked Systems Design and Implementation, pp. 613–632 (2021)

25. Ongaro, D., Ousterhout, J.: In search of an understandable consensus algorithm. In: 2014 USENIX Annual Technical Conference (USENIX ATC 2014), pp. 305–319 (2014)

AOPT-FL: A Communication-Efficient Federated Learning Method with Clusterd and Sparsification

Danlei Zhang, Geming Xia$^{(\boxtimes)}$, and Yuxuan Liu

College of Computer Science and Technology, National University of Defense Technology, Changsha 410000, China
546143539@qq.com, xiageming@163.com

Abstract. Federated Learning is a distributed machine learning technique that allows multiple devices to learn a shared model collaboratively without exchanging their data. It can be used to improve model accuracy while preserving user privacy. But traditional Federated Learning incurs significant communication overhead and does not perform well when the training data are not independent and identically distribute (Non-IID). Therefore, a Federated Learning algorithm based on adaptive Top-k sparsification and OPTICS method is proposed, which solves the problem that Federated Learning has low accuracy and high communication overhead on Non-IID data. Compared to existing Federated Learning algorithm, our algorithm has improved the accuracy of the model and reduced communication overhead.

Keywords: Cluster · Federated Learning · Sparsification · Residual

1 Introduction

With the rapid development of the Internet of Things and edge computing technologies, there is an increasing need for devices and organizations to share data and train machine learning models collaboratively [1]. However, traditional centralized machine learning methods face important challenges, such as the need to centralize data in one place for training, which can lead to issues related to data leakage and privacy concerns [2]. In addition, cross-domain data exchange between devices and organizations also presents a number of technical challenges.

To address these issues, Google first proposed the Federated Learning theory in 2016 [1]. Federated learning allows multiple devices or organizations to conduct joint model training through data sharing without centralizing raw data in one place. This decentralized model training approach can effectively protect data privacy and security while achieving more efficient model training and resource utilization [3].

However, one of the main problems in Federated Learning is the excessive cost of communication. The model size of a simple four-layer convolutional neural network is about 800KB, and training on large datasets and multiple clients at the same time can easily increase communication overhead to more than 1PB [4]. At the same time, the edge environment itself has the characteristics of small communication bandwidth, and if methods are not taken to reduce communication overhead, the application scenarios of Federated Learning will be greatly limited.

Another problem with Federated Learning is that it does not perform well on data that is Non-IID [5]. Training data in Federated Learning is directly generated by the data owner, so the distribution characteristics of the data are no longer controllable [6]. When client data is not independent and identically distribute, model training will be directly affected.

To solve the above problems, this paper proposes an efficient Federated Learning method that combines gradient compression and clustering methods. The main contributions of this paper are as follows:

- An Adaptive Sparse Ternary Compression (ASTC) algorithm is proposed. Based on the original STC algorithm [2], according to the convergence characteristics of machine learning, Newton's law of cooling is used to fit the convergence curve. Use the depreciation rate λ to optimize residuals. Experimental results on different datasets and models have shown that our method effectively improves the accuracy of Federated Learning.
- Based on the idea of multitasking learning, we use OPTICS method in the Federated Learning process to improve the accuracy of Federated Learning on Non-IID data. The current DBSCAN algorithm [7] is sensitive to original parameters. OPTICS solves this problem. Compared to DBSCAN, our algorithm has better performance in most cases.

2   Related Work

Although Mcmahan et al. [1] claim that FedAvg can handle Non-IID data to some extent, numerous studies have shown that Degradation of Federated Learning accuracy on Non-IID data is almost inevitable [8]. The performance degradation is mainly due to the divergence of the weights of the local model due to Non-IID data.

Sparsification is a commonly used method for reducing communication overhead. It limits parameter updates to a small portion. Storm [9] proposed a method that only sends elements that exceed the threshold while keeping the rest in the residue. They proved that 1-bit quantization is sufficient and carries no significant degradation in neither accuracy nor convergence speed. Golomb-Rice [10] coding the gaps reduces the average size per weight update to 10–11 bits, representing an additional 3x compression. However, the threshold is difficult to determine, so Aji et al. [11] proposed using sparsity p to select a portion of each gradient. Their method has little impact on accuracy. SignSGD alleviates communication overhead by transmitting only the symbols of each batch's random gradient. Bernstein et al. [12] proved that it can get the best of both worlds: compressed gradients and SGD-level convergence rate. Sattler et al.

[2] applied the sparsization to both upstream and downstream communication in Federated Learning. Their method requires both fewer gradient evaluations and communicated bits to converge to a given target accuracy and performs well on Non-IID data.

Cluster belongs to the category of unsupervised learning in machine learning, which refers to the process of dividing a data set into multiple subsets of data with high similarity according to a certain strategy, and the divided data subsets are called clusters. Avishek Ghosh *et al.* [13] used one-time clustering and cluster-to-cluster dissharing to reduce the impact of Non-IID data. However, it is necessary to specify the number K of clusters in advance. And the distance-based clustering algorithm does not perform well in practical applications. Clustering using the K-means method cannot exclude the interference of outliers, and may be attacked by malicious clients [13]. Y.Kim *et al.* [14] proposed a three-stage data clustering algorithm namely: generative adversarial network-based clustering, cluster calibration, and cluster division. Cluster calibration handles dynamic environments by modifying clusters. Cheng Xi*et al.* [15] aiming at the problem of data heterogeneity, an iterative joint clustering algorithm is proposed. IFCA divides clients into many clusters and lets clients in the same cluster optimize the shared model. However, in IFCA, clusters are not overlapping, which leads to inefficient use of local information because only one cluster uses the client's knowledge in each round.

The OPTICS clustering method is a density-based clustering algorithm, which can find arbitrarily shaped clusters in noisy spaces and find noise points compared to distance-based algorithms such as K-means. Compared with DBSCAN [16], it is not sensitive to input parameters, and it is difficult to determine the characteristics of data in advance in Federated Learning, so the method in this paper is more adaptable and widely used.

3   Preliminaries

In this section, we introduce building blocks of our approach and explain how to implement our algorithm.

3.1   Adaptive Top-K Sparsification

A common method to improve communication efficiency in Federated Learning is sparsification. Sparsification reduces the communication overhead of Federated Learning by limiting the range of parameter updates. The client only sends gradients that are above the threshold. All other gradients are accumulated in a residual.

According to Sattler Felix *et al.*'s research [2], Top-k sparsification shows the most promising performance in distributed learning environments with Non-IID data. Based on this conclusion, we design an adaptive top-k sparsification method to improve the communication efficiency of Federated Learning.

Top-k Sparsification: Top-k sparsification can be written as:

$$G^{t+1} = \frac{1}{n} \sum_{i=1}^{n} topk(G_i^{t+1} + A_i^t) \tag{1}$$

$$A^{t+1} = A^t + G_i^{t+1} - topk(G_i^{t+1} + A_i^t) \tag{2}$$

G_i^t means the gradient parameter of client i when the communication rounds is t. And A_i^t means the residual.

In the early stages of model training, the model changes rapidly. The gradient g is large enough that the low sparsity rate p is acceptable. In the later stages of model training, the gradient g tends to flatten, and the initially harsh sparsity rate p will affect the speed of model convergence. To solve this problem, we propose applying Newton's cooling law to fit the gradient descent curve and dynamically adjust the value of a sparsity rate p. The processing of sparsity p can be written as:

$$p^t = p \times e^{-\alpha \times (E - E^t)} + p \tag{3}$$

where E^t is current communication round, E is maximum communication round, α is the adjustment coefficient that defaults to 0.1.

Based on the method proposed in [4] and [2], combining sparsification and Binarization, we propose an adaptive sparse compression algorithm. In contrast to the previous algorithm, our approach is able to adjust the sparsity dynamically throughout the training process, which enhances the overall training efficiency.

The algorithm is formalized in Algorithm 1.

In the adaptive sparse ternary compression(ASTC) algorithm, first, we use Newton's cooling law to obtain sparsity p^t based on the current epoch. The parameter k is obtained by sparsity p^t, which is an important parameter of Top-k algorithm and is used to limit the number of saved parameters. Then, we find the smallest of the largest k parameters which is v and the mean of these k parameters which is u. Finally, iterate through each dimension of tensor T, setting values greater than v to u, values less than $-v$ to $-u$, and the rest to 0. This gives us a ternary tensor T with only u, $-u$, and 0.

Figure 1 shows the accuracy of RNN, VGG9 and CNN model when trained with and without adaptive algorithm. It can be seen that the accuracy of ASTC exceeds that of STC in all models, especially on CIFAR10, which is 11.61% higher than STC. On MNIST, the accuracy of ASTC increased by 5.81%. Performance is not significant on the EMNIST dataset because accuracy is already high and difficult to improve.

Algorithm 1. Adaptive Sparse Ternary Compression

Input :Tensor T, sparsity rate p, parameter α, current epoch E^t
Output :Tensor T
$p^t \leftarrow p \times e^{-\alpha \times (E - E^t)} + p$
$k \leftarrow np^t$
$v \leftarrow min(T_{topk})$
$u \leftarrow \frac{1}{k} \sum_{i=0}^{k} topk\{|T[i]|\}$
for i in range$(0, n)$ **do**
  if $T[i] > v$ **then**
    $T[i] = u$
  else if $T[i] < -v$ **then**
    $T[i] = -u$
  else
    $T[i] = 0$
  end if
end for
Return T

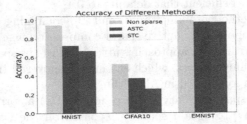

Fig. 1. Accuracy of Different Methods on Three Non-IID Datasets

Adjusting the sparsity p will increase the impact of outdated gradient on model convergence. The combination of ASTC and residual processing method can significantly improve the accuracy of model.

Residual Processing: During the sparsization process, using residual A to save the unused gradients can avoid the loss of gradient information. However, outdated gradient can deviate from the current gradient and be detrimental to model training. Therefore, it is necessary to reduce the weight of outdated gradient based on gradient's unuploaded rounds.

We introduce depreciation rate λ to achieve it. So the update rule for top-k sparsified communication can be written as:

$$G^{t+1} = \frac{1}{n} \sum_{i=1}^{n} topk(G_i^{t+1} + \lambda A_i^t) \tag{4}$$

$$A^{t+1} = A^t + G_i^{t+1} - topk(G_i^{t+1} + \lambda A_i^t) \tag{5}$$

When $\lambda = 1$, the algorithm is the same as Eq. (1) (2). It should be noted that the number of communication rounds in Federated Learning is very large. The residuals decrease exponentially, and a depreciation rate that is too small can lead to a loss of information after a few rounds with the residuals too small. After many experiments, we chose the depreciation rate $\lambda = 1$ to start at 0.99.

Figure 2 shows the accuracy of RNN, VGG9 and CNN model when trained at different depreciation rates with and without adaptive algorithm. In the case of Non-IID, each client holds only two random labels. Setting the depreciation rate to 0.99 and 0.999, it can be seen that on MNIST, the accuracy increased 39.9% when the dataset is Non-IID; on CIFAR10, the accuracy was improved by 22.93% when the dataset is Non-IID. It can be seen that the accuracy decreases at the later stage of training, mainly due to the deviation of the convergence direction caused by outdated gradient accumulation. Using depreciation rate to reduce the weight of outdated gradients can solve this problem. Especially in the case of Non-IID dataset, the depreciation rate is particularly significant in improving accuracy. Because the gradient accumulated during training on Non-IID dataset is more complex than in the case of IID dataset.

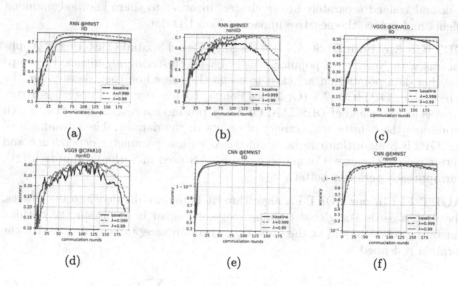

Fig. 2. The accuracy of the model under different depreciation rates.

Deflate Encode: Deflate Encode is a popular lossless compression algorithm commonly used to compress data such as files, images, and videos. This algorithm was developed by Phil Katz [17], the author of Zlib software library, in 1993 and is widely used in a variety of applications. In the compression section, Deflate Encode uses two techniques: LZ77 algorithm and Huffman encoding to reduce file size. LZ77 [18] algorithm achieves data compression by finding duplicate data blocks, while Huffman encoding encodes different data frequencies to achieve more efficient data storage.

In our algorithm, there are only three symbols $\{-v, 0, v\}$ for gradients, and most of them are 0. Huffman encoding is based on the probability of symbol occurrence. Symbols with a high probability of occurrence use shorter encodings, while symbols with a low probability of occurrence use longer encodings. This feature is very suitable for our sparse data. Subsequent experiments will demonstrate that Deflate Encode can effectively reduce communication overhead in our method.

3.2 Clustered Federated Learning

The advantage of Federated Learning is that it can effectively use distributed data. However, due to the characteristics of edge environments, datasets from different clients are likely to be heterogeneous. This leads to difficulty or even non convergence of the model. Therefore, clustered Federated Learning based on the idea of multi-task learning is proposed to solve this problem. Partitioning clients into different clusters according to certain rules can effectively reduce the impact of Non-IID data. Clients in the same cluster are considered a learning task and trained separately in the cluster, in order to share knowledge among clients and reduce the negative impact of Non-IID data.

OPTICS Method: DBSCAN [19] (Density-Based Spatial Clustering of Applications with Noise) is a popular density-based clustering algorithm. It groups together data points that are close to each other based on their density. On the other hand, OPTICS [20] (Ordering Points To Identify the Clustering Structure) is an extension of DBSCAN. OPTICS produces a reachability plot, which represents the density connectivity of points in the dataset. The advantages of the OPTICS algorithm can be summarized as less parameter dependence and better noise resistance. Compared to DBSCAN used in Sarhad's approach [21], our approach performs better.

AOPT-FL: In our AOPT-FL algorithm, it is mainly divided into two stages, the first stage is the global training stage, the goal is to obtain the optimal clustering solution, and at the same time train an acceptable global model, the formula is defined as:

$$\min_{w \in \mathbb{R}^d} f(w) \quad where \quad f(w) = \frac{1}{N} \sum_{i=1}^{N} f_i(w) \tag{6}$$

In the second stage, the local training stage, the goal is to train a global model in the C_i cluster. The server no longer performs global aggregation, and the client only trains an intra-cluster model with other clients in the cluster.

$$\min_{w \in C_i} f_{C_i}(w) \quad where \quad f_{C_i}(w) = \frac{1}{N_{C_i}} \sum_{i \in C_i}^{N_{C_i}} f_i(w) \tag{7}$$

N_{C_i} is the number of cluster C_i.

Algorithm 2 gives the entire process of AOPT-FL. First, the server performs OPTICS clustering on the parameter $model(w_1, w_2, ...w_k)$ uploaded by the client for the first time, divides the client into C clusters. Then in global aggregation stage, before each round of aggregation begins, the server calculates the Silhouette Coefficient of the current cluster, and if it is less than the input threshold, it recluster all clients. Otherwise, server decodes the received sparse gradient and aggregates them. When it comes to in-cluster aggregation stage, the server divides clients into several clusters, and each cluster trains a model separately. The client's operation is relatively simple, accepting the gradient from the server, and sparsing, encoding, and uploading the generated gradient after local training.

Algorithm 2. AOPT-FL

Input :initial paramenters w, set of clients C, Silhouette coefficient threshold α
while not converged **do**
  Server does:
  $\alpha_t \leftarrow \text{OPTICS}(w_1^0, w_2^0, ..., w_i^0)$
  if $epoch \leq globalepoch$ **then**
    if $\alpha_t \leq \alpha$ **then**
      $\alpha_t \leftarrow \text{OPTICS}(g_1^t, g_2^t, ..., g_i^t)$
    end if
    for each client i **do**
      $Msg \leftarrow \text{ClientUpdate}(k, Encode(\hat{g}_i^t))$
      $\hat{g}_i^t \leftarrow Decode(Msg)$
    end for
    $g^{t+1} \leftarrow \sum_{i=1}^n \frac{1}{n} \hat{g}_i^t$
    return g^{t+1} to each Client i
  else
    for each client i **do**
      $Msg \leftarrow \text{ClientUpdate}(k, Encode(\hat{g}_i^t))$
      $\hat{g}_i^t \leftarrow Decode(Msg)$
    end for
    for each Class C_k **do**
      $g_{C_k}^{t+1} \leftarrow \sum_{i=1}^{n_k} \frac{1}{n_k} \hat{g}_i^t$
    end for
    return $g_{C_k}^{t+1}$ to each Class C_k
  end if
  Client i does:
  receive g^t from server
  $g_i^t \leftarrow \text{local train}(g^t + w_i^{t-1})$
  $\hat{g}_i^t \leftarrow \text{Adaptive Sparse Compression}(g_i^t)$
  $Msg \leftarrow encode(\hat{g}_i^t)$
  return Msg to server
end while

---

4 Experimental

4.1 Experimental Environment

The experimental environment of this paper is AMD Ryzen 7 5800H with Radeon Graphics @3.20 GHz processor and NVIDIA GeForce RTX 3060 Laptop GPU. The experiment simulated a server and 50 local client nodes. Non-IID data was constructed based on the MNIST, EMNIST and CIFAR10 datasets. In the case of NON-IID data, each client was randomly assigned two kinds of labels. Models parameters are shown in Table 1.

Table 1. Details of datasets and models

Dataset	#Records	#Features	#Classes	Model	#Parameters
MNIST	70000	784	10	RNN	24714
EMNIST	814255	784	62	CNN	206922
CIFAR10	60000	1024	10	VGG9	3491530

4.2 Accuracy

Figure 3 shows the accuracy of our method on MNIST, CIFAR10, and EMNIST datasets. The accuracy of our method on MNIST is 14.09% higher than that of STC when $p = 0.001$, 4.08% higher than that of STC when $p = 0.01$ and 2.99% higher than that of STC when $p = 0.1$. For CIFAR10, the accuracy of AOPT is 7.46% higher than that of STC when $p = 0.001$, 5.17% higher than that of STC when $p = 0.01$ and 10.24% higher than that of STC when $p = 0.1$. For EMNIST, the accuracy of AOPT is 0.39% higher than that of STC when $p = 0.001$, 0.17% higher than that of STC when $p = 0.01$ and 0.45% higher than that of STC when $p = 0.1$. The accuracy of SignSGD on RNN models is 60.12%, but it does not converge on VGG9 and CNN models.

As shown in Fig. 3, our algorithm generally has higher accuracy than STC. Especially when the baseline performs poorly or the model is complex, the improvement is more significant. Compared to no-comp line, it is worth noting that there has been a significant improvement in accuracy when $p = 0.1$. The reason is that the greater the sparsity, the more gradients are uploaded. Therefore, the clustering effect is better, so the accuracy is significantly improved compared to no-comp line. Correspondingly, communication overhead will increase, as will be shown in the next section.

4.3 Communication Overhead

Table 2 shows the amount of communication overhead required to train model for the different methods in megabytes. Nan means that the model has not converged. On the MNIST learning task AOPT at a sparsity rate of $p = 0.001$ only

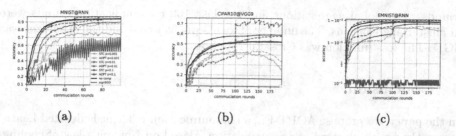

Fig. 3. The accuracy of our proposed method on different datasets and models. (a) is based on MNIST and CNN model. (b) is based on CIFAR10 and VGG9 model. (c) is based on EMNIST and CNN model.

communicates 2.84 MB worth of data, which is a reduction in communication by a factor of ×81.5 as compared to the baseline with requires 231.6 MB. And when is comes to CIFAR10, communication overhead has been reduced by a factor of ×562.1 as compared to the baselinet a sparsity rate of $p = 0.001$. On EMNIST learning task AOPT's communication overhead has been reduced by a factor of ×307.1 as compared to the baseline at a sparsity rate of $p = 0.001$. On CIFAR10 and EMNIST learning tasks, SignSGD did not converge within the communication rounds. At the same time, we can note that compared to the STC, the communication overhead of AOPT is lower at a sparsity rate of $p = 0.1$ and the model accuracy has been greatly improved, especially in complex model. Due to the large sparsity, AOPT can converge faster and have lower communication overhead.

Table 2. Bits required to train models on different learning tasks in a Non-IID learning environment. A value of "Nan" in the table signifies that the method has not converged within the iteration budget.

	RNN@MNIST	VGG9@CIFAR10	CNN@EMNIST
Baseline	231.6 MB	133253.81 MB	3961.30 MB
SignSGD	23.04 MB	Nan	Nan
STC $p = 0.001$	2.66 MB	235.55 MB	12.80 MB
STC $p = 0.01$	1.86 MB	762.77 MB	29.86 MB
STC $p = 0.1$	2.95 MB	3883.01 MB	132.20 MB
AOPT $p = 0.001$	2.84 MB	237.09 MB	12.96 MB
AOPT $p = 0.01$	2.64 MB	790.62 MB	31.71 MB
AOPT $p = 0.1$	2.94 MB	3796.25 MB	130.49 MB

4.4   Experimental Details

In all of our experiments, the local learning rate was 0.01, the batch size was 16, the communication rounds were 200. In AOPT-FL, the first 100 communication rounds were global aggregations and the last 100 communication rounds were in-cluster aggregations. The number of local epochs was 3 and the optimizer was SGD. The loss function was CrossEntropyLoss.

5 Conclusion

In the paper, we propose AOPT-FL, a communication-efficient Federated Learning method with clusterd and sparsification. Based on Newton's law of cooling, AOPT-FL dynamically adjusts the sparsity p and update the weight of the residual by depreciation rate λ. In addition, AOPT-FL uses the OPTICS algorithm to divide clients into different clusters, so that the clients in the cluster have similar data structures. At the same time, in order to avoid the overall dataset becoming smaller and affecting the training effect due to the division of clusters, we dynamically adjust the members of the cluster by monitoring the Silhouette coefficient parameters of the cluster to ensure knowledge sharing between each cluster.

Finally, we construct a Non-IID dataset and verify it with RNN, VGG9 and CNN networks, and the experimental results show that the AOPT-FL algorithm can improve the accuracy of Federated Learning on Non-IID data. The communication overhead of AOPT-FL is much smaller than that of the original algorithm. Compared to STC, AOPT-FL improves accuracy with similar communication overhead.

Research by Geming X. et al. [22] points to the growing attack on Federated Learning. In future work, we will explore the method of introducing differential privacy [23] into AOPT-FL algorithm, while better compressing the parameters of the model, improving communication efficiency [24] while improving the security of Federated Learning.

References

1. McMahan, B., Moore, E., Ramage, D., Hampson, S., Arcas, B.A.: Communication-efficient learning of deep networks from decentralized data. In: Artificial Intelligence and Statistics, pp. 1273–1282. PMLR (2017)
2. Sattler, F., Wiedemann, S., Müller, K.R., Samek, W.: Robust and communication-efficient Federated Learning from non-IID data. IEEE Trans. Neural Netw. Learn. Syst. **31**(9), 3400–3413 (2019)
3. Chaodong, Y., Jian, C., Geming, X.: Coordinated control of intelligent fuzzy traffic signal based on edge computing distribution. Sensors 5953 (2022)
4. Sattler, F., Wiedemann, S., Müller, K.R., Samek, W.: Sparse binary compression: towards distributed deep learning with minimal communication. In: 2019 International Joint Conference on Neural Networks (IJCNN), pp. 1–8. IEEE (2019)
5. Li,. X., Huang, K., Yang, W., Wang, S., Zhang, Z.: On the convergence of FedAvg on non-IID data. arXiv preprint arXiv:1907.02189 (2019)
6. Hsieh, K., Phanishayee, A., Mutlu, O., Gibbons, P.: The non-IID data quagmire of decentralized machine learning. In: International Conference on Machine Learning, pp. 4387–4398. PMLR (2020)

7. Agrawal, S., Sarkar, S., Alazab, M., Maddikunta, P.K.R., Gadekallu, T.R., Pham, Q.V., et al.: Genetic CFL: hyperparameter optimization in clustered federated learning. Comput. Intell. Neurosci. **2021** (2021)
8. Zhao, Y., Li, M., Lai, L., Suda, N., Civin, D., Chandra, V.: Federated learning with non-IID data (2018)
9. Ström, N.: Scalable distributed DNN training using commodity GPU cloud computing (2015)
10. Rice, R., Plaunt, J.: Adaptive variable-length coding for efficient compression of spacecraft television data. IEEE Trans. Commun. Technol. **19**(6), 889–897 (1971)
11. Aji, A.F., Heafield, K.: Sparse communication for distributed gradient descent. arXiv preprint arXiv:1704.05021 (2017)
12. Bernstein, J., Wang, Y.-X., Azizzadenesheli, K., Anandkumar, A.: signSGD: compressed optimisation for non-convex problems. In: International Conference on Machine Learning, pp. 560–569. PMLR (2018)
13. Ghosh, A., Hong, J., Yin, D., Ramchandran, K.: Robust federated learning in a heterogeneous environment (2019)
14. Kim, Y., Hakim, E.A., Haraldson, J., Eriksson, H., Silva, J., Fischione, C.: Dynamic clustering in federated learning (2021)
15. Cheng, X., Gang, L., Pramod, K., V.: Federated learning with soft clustering. IEEE Internet Things J. 7773–7782 (2022)
16. Khan, K., Rehman, S.U., Aziz, K., Fong, S., Sarasvady, S.: Dbscan: past, present and future. In: The Fifth International Conference on the Applications of Digital Information and Web Technologies (ICADIWT 2014), pp. 232–238. IEEE (2014)
17. Deutsch, P.: RFC 1951: Deflate compressed data format specification version 1.3(1996)
18. Ziv, J., Lempel, A.: A universal algorithm for sequential data compression. IEEE Trans. Inf. Theory **23**(3), 337–343 (1977)
19. Bryant, A., Cios, K.: RNN-DBSCAN: a density-based clustering algorithm using reverse nearest neighbor density estimates. IEEE Trans. Knowl. Data Eng. **30**(6), 1109–1121 (2017)
20. Ankerst, M., Breunig, M., Kriegel, H.P., Sander, J.: Optics: ordering points to identify the clustering structure. ACM SIGMOD Rec. 49–60 (1999)
21. Arisdakessian, S., Wahab, O.A., Mourad, A., Otrok, H.: Towards instant clustering approach for federated learning client selection. In: 2023 International Conference on Computing, Networking and Communications (ICNC), pp. 409–413. IEEE (2023)
22. Geming, X., Jian, C., Chaodong, Y., Jun, M.: Poisoning attacks in federated learning: a survey. IEEE Access 1 (2023)
23. Dwork, C.: Differential privacy: a survey of results. In: Agrawal, M., Du, D., Duan, Z., Li, A. (eds.) TAMC 2008. LNCS, vol. 4978, pp. 1–19. Springer, Heidelberg (2008). https://doi.org/10.1007/978-3-540-79228-4_1
24. Liu, R., Cao, Y., Yoshikawa, M., Chen, H.: FedSel: federated SGD under local differential privacy with top-k dimension selection. In: Nah, Y., Cui, B., Lee, S.W., Yu, J.X., Moon, Y.S., Whang, S.E. (eds.) DASFAA 2020. LNCS, vol. 12112, pp. 485–501. Springer, Cham (2020). https://doi.org/10.1007/978-3-030-59410-7_33

A Central Similarity Hashing Method via Weighted Partial-Softmax Loss

Mengling Li[1], Yunpeng Fu[1], Zhiyang Li[1,2]([✉]) [iD], Duo Zhang[1], and Zhaolin Wan[3]

[1] Dalian Maritime University, Dalian, China
lizy0205@dlmu.edu.cn
[2] Haikou University of Economics, Haikou, China
[3] Harbin Institute of Technology, Harbin, China

Abstract. Image hashing techniques that map images into a set of hash codes are widely used in many image-related tasks. A recent trend is the deep supervised hashing methods that leverage the annotated similarity of images measured point-wise, pairwise, triplet-wise, or list-wise. Among these methods, central similarity quantization (CSQ) introduces a state-of-the-art point-wise metric called global similarity, which encourages aggregation of similar data points to a common centroid and dissimilar ones to different centroids. However, it sometimes will fail and lead to several data points drifting away from their corresponding hash centers during training, especially for multi-labeled data. In this study, we propose a novel image hashing method incorporating pair-wise similarity into central similarity quantization, which enables it to capture the global similarity of image data and pay attention to drift points simultaneously. To this end, we present a novel learning objective based on the weighted partial-softmax loss and implement it with a deep hashing model. Extensive experiments are conducted on publicly available datasets, demonstrating that the proposed method has achieved performance gains over the competitors.

Keywords: Hash centers · Supervised deep hashing · Central similarity

1 Introduction

Nowadays, images on the internet have been growing at an explosive rate, which poses several challenges in people's daily lives, such as the problem of storing and retrieving these massive images efficiently. Inspired by the low storage space and high search speed of binary codes, the image hashing technique that transforms images into binary codes has provided a promising solution, which is widely used in many image-related areas, i.e., large-scale image retrieval [16], object detection [20], cross-modal remote sensing [3].

Supported by the Natural Science Foundation of China (Nos. 62102059, 61672379, 61370198), the National Key R&D Program of China (No. 2021YFF0900503), the High-Performance Computing Center of Dalian Maritime University.

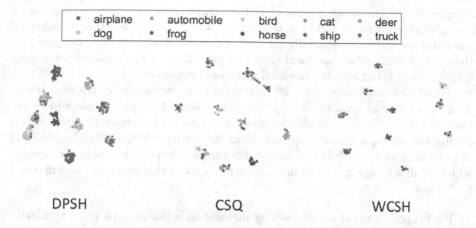

• airplane	• automobile	• bird	• cat	• deer
• dog	• frog	• horse	• ship	• truck

DPSH   CSQ   WCSH

Fig. 1. Two-dimensional t-SNE visualization for hash codes of 1,000 test images, generated by representative deep hashing methods and our method (WCSH) on CIFAR-10 database using 32 bits. DPSH [12] and CSQ [18] are two representative hashing methods based on pairwise similarity and point-wise similarity, respectively.

Generally, image hashing methods can be classified into two categories: data-independent hashing like locality-sensitive hashing (LSH) [10] and its variants [14], and data-dependent hashing like spectral hashing (SH) [17] and iterative quantization (ITQ) [6]. Among these data-dependent hashing methods, supervised hashing especially deep supervised hashing that incorporates the supervised information into the process of hash function learning, usually achieves promising retrieval performances even with small code-length of hash bits.

The common supervised information is the annotated labels of the images. The annotated similarity of images is usually measured pairwise, triplet-wise, or list-wise [8]. Although recent deep hashing methods using pairwise, and triplet-wise data similarity have shown a performance gain over traditional hashing methods. They often suffer from problems such as low-level similarity relations, insufficient data distribution coverage, and low effectiveness on unbalanced data.

One reason for the above problem is that the pairwise or triplet-wise data similarities are only the partial relationships between images. To this end, a global similarity metric is introduced by the central similarity quantization (CSQ) method [18]. CSQ measures the data similarity by images and their hash centers generated by the image labels beforehand, and encourages the aggregation of similar images to a common centroid and dissimilar ones to different centroids. Even when dealing with severe data imbalance, this point-wise similarity manner can still learn the hash function well from the global relationship while making the generated hash codes more distinguishable.

Since the hash code of an image is trained to converge to its hash center, the performance of the obtained hash codes largely depends on the quality of hash centers. Given the code length, CSQ tries to produce hash centers from the categories that are dispersed enough in the binary space. It is usually possible for a single-label database. However, it will become infeasible when the categories of the database increase, especially for multi-label databases. At this time, we find that many drift points are located at the periphery of the hash center during training, which has affected the final retrieval performance.

To address the above issue, we introduce pairwise similarity between image data which can be used to identify the drift points. The drift points will be set more weight in the loss calculation, and the weight will be dynamically adjusted during the multiple training epochs. Compared with CSQ, our method WCSH obtains hash codes with better inter-cluster separation and intra-cluster closure, which is shown in Fig. 1. The main contributions of this paper are summarized as follows.

- We propose a novel image hashing method incorporating pairwise similarity into central similarity quantization, which enables it to capture the global similarity of image data and pay attention to drifting points during the training process simultaneously.
- We propose a novel learning objective based on the weighted partial-softmax loss, and implement it by a deep learning model, so that the losses in the objective have jointly collaborated to optimize the hash codes.
- We conduct various quantitative experiments on three publicly available datasets, demonstrating that our method always performs best against the state-of-the-art.

2 Method

Generally, a labeled image dataset consists of n images $Y = \{y_i\}_{i=1}^n$ and a label matrix $L = \{l_i\}_{i=1}^n \in \{0,1\}^{c \times n}$, where $y_i \in R^d$ is the i-th image and its label vector $l_i = \{l_{i1}, l_{i2}, ..., l_{ic}\}^T \in \{0,1\}^c$ represents the category information to which the image belongs. Here, c is the total number of categories in the dataset, satisfying that $l_{ij} = 1$ if y_i belongs to the j-th category, and $l_{ij} = 0$ otherwise.

The goal of deep hashing is to find a function $f: R^d \to \{0,1\}^k$ using the deep network that maps each d-dimensional image to a k-bit binary code, satisfying that if two images y_i and y_j share the same label, their encoded binary codes $b_i = f(y_i)$ and $b_j = f(y_j)$ are also similar. By this way, the image set $Y = \{y_i\}_{i=1}^n \in R^{d \times n}$ is transformed to a binary set $B = \{b_i\}_{i=1}^n \in \{0,1\}^{k \times n}$.

2.1 Network Architecture Overview

It is unfeasible to learn the function f from the entire dataset Y when the size n is huge. Usually, a subset $X \subseteq Y$ is randomly chosen as the training set. For the given training set X and its label set L_X, the framework of our method is shown in Fig. 2. It consists of three key components: a hash center component, a feature learning component, and a loss function component.

The hash center component is to obtain the centers of the hash codes of images. An ideal case is that the hash centers are generated from the image categories and dispersed enough in binary space. Specifically, one category has its individual hash center, which will be taken as the ideal hash representation of the images belonging to this category.

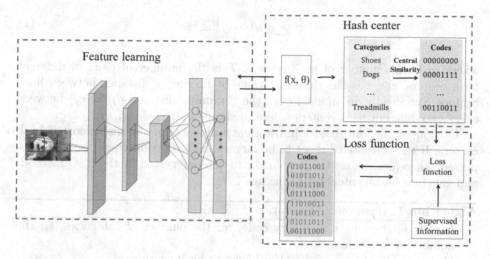

Fig. 2. The general framework of our model consists of three key components: the feature learning component, the hash center component, and the loss function component. The binary codes colored blue and green indicate that they belong to different hash centers, and the binary codes colored red indicate that they are drift points to their center during training. (Color figure online)

The feature learning component is to extract the features of images in Euclidean space. Besides the commonly chosen CNN networks such as AlexNet and ResNet, we also choose a transformer as the backbone network and employ a fully-connected layer as the output layer to project the features to the space with the same dimension as the hash codes.

The loss function component is to measure the difference between the output image feature and its hash center. To this end, given a set of training images, the parameters of the network can be learned by minimizing the loss above. The design of loss plays an important role in the learning process.

Noting that many drifting points are located at the periphery of the hash centers during training, illustrated as the hash codes colored red in Fig. 2, we propose a novel weighted partial-softmax loss that incorporates the pairwise similarity into central similarity quantization. The loss can take more consideration to the drift points, and achieve an obvious performance gain over the existing central similarity losses. In the following, we will discuss the implementation details of the three components.

2.2   Implementation Details

Hash Center Generation. The concept of hash centers is proposed in the central similarity quantization (CSQ) method, and later improved by several recent deep hashing methods. Motivated by their ideas, we define the hash center as a set of hash codes $C = \{c_i\}_{i=1}^m \subset \{0,1\}^k$ and satisfy the following constraints:

$$\frac{1}{T} \sum_{\substack{i \neq j}}^m d_H(c_i, c_j) \geq t \tag{1}$$

where m is the number of hash centers, T is the number of pairs of different hash centers, and t is the threshold value of Hamming distance between hash centers. Specifically, in an ideal case, the Hamming distance $d_H(c_i, c_j)$ between every two different hash centers c_i and c_j should satisfy $d_H(c_i, c_j) \geq t$.

Obviously, the threshold t has a great influence on the dispersion of hash centers. If the threshold t is set to be larger, the hash centers will be more dispersed. Therefore, we set the threshold t to $\frac{k}{2}$ as suggested by the CSQ method, and generate hash centers by Algorithm 1.

Algorithm 1. Hash center generation

   Input: k: the length of binary code, m: the number of categories, t_1: the number of iterations.
   Output: $C = \{c_i\}_{i=1}^m \subseteq \{0,1\}^k$: the set of hash centers.
1: **Initialization:** Construct $H_{2k} = [H_k, -H_k]^T$, where H_k is a $k \times k$ Hadamard matrix
2: **if** k is a power of 2 and $m \leq 2k$ **then**
3:    $C = H_{2k}(1:m,:)$
4: **else if** k is a power of 2 and $m > 2k$ **then**
5:    **for** each $i \in [1, t_1]$ **do**
6:       Compute L' with $m - 2k$ centers by random sampling.
7:       $C = \arg\max_{H_{2k} \cup L'} avg(\{d_H(c_i, c_j) | i > j\})$
8:    **end for**
9: **else**
10:    Compute C with m hash centers by random sampling.
11: **end if**

In Algorithm 1, lines 1 to 3 is to generate the hash centers by Hadamard matrix [9], which is known to be a binary matrix with orthogonal rows whose elements are -1 or 1. It is obvious that a subset of Hadamard matrix $H_{2k}(1:m,:)$ satisfies Eq. (1) above. Line 6 or 10 is to generate the hash centers by random sampling based on Bernoulli distribution to deal with the conditions that $m > 2k$ or k is not a power of 2. Specifically, for each bit x of the hash center c_i, a random real number $l \in (0,1]$ is generated. If $l \leq 0.5$, the current bit $x = -1$, otherwise, $x = 1$. The expectation of the distance between every two hash centers c_i and c_j satisfies $E(d_H(c_i, c_j)) = k/2$. That is to say, Eq. (1) is satisfied by expectation.

Suppose we have obtained the hash center set C for the categories by Algorithm 1. If the tag set $l_i = \{l_{i1}, l_{i2}, ..., l_{ic}\}^T$ of the i-th image contains a single category, we randomly assign its corresponding hash center to this image. However, if the tag set contains multiple tags, we will adopt a voting strategy. For example, the j-th image has three class tags, and their hash codes are $c_1 = [-1, 1, -1, 1]$, $c_2 = [1, -1, -1, 1]$ and $c_3 = [1, 1, 1, 1]$, and the hash center of the j-th image will be $c_{123} = [1, 1, -1, 1]$. If there is no dominant value for a particular bit, i.e., -1 and 1 have 50% of the votes each, we will set the value of that bit to 1 or -1 randomly.

Feature Learning. We employ multiple neural network models for feature learning. In the area of image hashing, CNN is the first choice to extract the features of images and is widely used in recent deep hash methods. We take two representative CNN networks AlexNet and ResNet as the backbone network. In addition, we replace the original fully-connected layer with a new hash layer that the image features can be converted to a k-dimensional continuous hash code.

Besides the above CNN networks, we also choose a transformer network [7] to learn the features more efficiently, inspired by the wide application and excellent robustness of the transformer in natural language processing. Our network involves the use of an encoder and a decoder with multiple transformer layers. Each transformer layer consists of a multi-head self-attention mechanism, a feedforward neural network, and layer normalization. We also add some convolutional and fully connected layers to extract more useful features from the original image.

Compared to CNNs like AlexNet and ResNet, the above transformer structure can better improve the performance of image hash coding. This is because a transformer, unlike CNN, can adaptively consider long-range dependencies when processing sequential data and can model the global information of the input data. Moreover, using a transformer, we do not need to consider the adjustment of parameters such as convolutional kernel size and step size, while we can control the complexity and training difficulty of the model more flexibly.

Considering the limitation of backpropagation in neural networks, we use the hyperbolic tangent function as the activation function of the last layer, to approximate the threshold process and compress the continuous codes into intervals, which reduces the gap between the continuous codes and the binary hash codes, thus improving the quality of the hash codes.

Loss Function Design. Upon obtaining the hash centers C and hash network $F(x; \theta)$ above, for a training images dataset $< X, L >= \{< x_i, l_i >\}_{i=1}^m$, the ideal output of the hash networks $F(x_i; \theta)$ should be $c(l_i) \in C$, where $c(l_i)$ represents the hash center of the tags set l_i. The hash representation $\{c(l_i)\}_{i=1}^m$ not only satisfies that images within the same class have the same hash code but also satisfies that the hash codes of images between different classes reach at least a Hamming distance of $\frac{k}{2}$. To achieve this, we designed a partial-softmax loss shown in Eq. (2).

.

$$L_C = \sum_i^m \sum_{j\in\Gamma_i} -log\frac{exp(\eta(cos(h_i^T,c_j)-\mu k))}{exp(\eta(cos(h_i^T,c_j)-\mu k)) + \sum_{q\in\psi_i} exp(\eta cos(h_i^T,c_q))} \quad (2)$$

where η is a hyperparameter, $\mu \in [0,1]$ is a predefined margin, Γ_i is the index set of the category to which x_i belongs, i.e., the index of element "1" in the label vector l_i, ψ_i is the index set of the category to which x_i does not belong, i.e., the index of element "0" in the label vector l_i. It can be found that Eq. (2) is composed of a series of softmax-like losses, and the denominator of each softmax-like loss contains part of the categories of the dataset, i.e., the categories to which x_i does not belong and the one category to which x_i does belong.

The softmax loss is widely used in the central similarity quantization (CSQ) method and its variants [15]. However, since multi-label data will generate quite a number of hash centers larger than $|C|$ by the mentioned voting strategy, the Hamming distance between the hash centers at this time will be far from the expected $d_H(c_i,c_j) = \frac{1}{2}(k - c_i^T c_j) = \frac{k}{2}$. In practice, it leads to many drifting points at the periphery of the hash center during training, which will seriously affect the final retrieval performance. To solve this problem, we introduce pairwise similarity of the dataset in the training process without losing the advantages of the central similarity.

$$L_C = \sum_i^m \sum_{j\in\Gamma_i} -w_i log\frac{exp(\eta(cos(h_i^T,c_j)-\mu k))}{exp(\eta(cos(h_i^T,c_j)-\mu k)) + \sum_{q\in\psi_i} exp(\eta cos(h_i^T,c_q))} \quad (3)$$

The improved loss function is given in Eq. (3). Specifically, during training, we use the loss function in the first epoch with a weight $w = 1$. In the subsequent training epochs, we calculate the loss weights for each data point based on the analysis of the previous epoch of training results, and introduce these weights into the softmax loss function to pay more attention to the drifting points.

$$w_i^k = \frac{\sum_{j\in I(i)} d_h(h_i^{k-1},h_j^{k-1})}{\sum_{i,j\in I(i)} d_h(h_i^{k-1},h_j^{k-1})} \quad (4)$$

The calculation of weights $w^k = \{w_i^k\}_{i=1}^n$ used in the k-th epoch is given in Eq. (4). Here, $I(i)$ represents the index of all the images sharing the same label with the image x_i, and $d_h(h_i^{k-1},h_j^{k-1})$ is the Hamming distance between h_i^{k-1} and h_j^{k-1}. If a data point is the drift point of a class, the sum of distances between it and other data points in this class will be larger, and therefore the weight in Eq. (4) and the corresponding loss in Eq. (3) will be larger. In this way, we can control the training process of each data point more precisely, improving the accuracy and retrieval performance of the hash code.

Since each hash center is binary, which creates inherent optimization difficulties, we introduce a quantization loss L_Q to refine the generated hash code

h_i. It is quantized by a bi-modal Laplace prior with a log cosh smoothing [18] in Eq. (5).

$$L_Q = \sum_i^m \sum_{k=1}^K (\log \cosh(|2h_{i,k} - 1| - 1))$$ (5)

Then the final loss function can be formulated as in Eq. (6). Here, λ is a hyperparameter to balance the two terms of hashing error L_C and quantization error L_Q.

$$\arg\min_\theta L = L_C + \lambda L_Q$$ (6)

3 Experiments

3.1 Datasets and Settings

Three datasets are used in this study to evaluate the image hashing models, ImageNet [5], MS-COCO [13] and NUS-WIDE [4]. ImageNet is a single-label dataset, and MS-COCO, NUS-WIDE are two representative multi-label datasets. The experimental settings for all datasets are summarized in Table 1.

Table 1. The datasets in experiments

Datasets	Size	Type	#Classes	#Train	#Test
ImageNet	128,495	single-label	10	10,000/all	5,000
MS-COCO	112,217	multi-labels	80	10,000/all	5,000
NUS-WIDE	149,685	multi-labels	21	10,500/all	2,040

For performance evaluation of hashing methods on single-labeled images, two images are considered ground-truth neighbors if they share the same label, while on multi-label image datasets, two images are considered ground-truth neighbors if they share at least one category label. Meanwhile, we adopt Hamming ranking as the search strategy and exploit the widely-used performance metric of mean average precision (mAP).

In our implementation, when training the image hash network, all parameters are initialized using the parameters of the pre-trained network, except for the last layer. The last layer of the image hash network is initialized with the Xavier initialization method to avoid the problem of gradient disappearance or explosion. To optimize the model performance, we use the small-batch stochastic gradient descent (SGD) algorithm and set the learning rate to $10e$-5. Also, we fix the batch size of training images to 128 and set the weight decay parameter to $10e$-5 to avoid the overfitting problem. In addition, we set the hyperparameters λ to 0.001, η and μ to 1 and 0.2 for ImageNet, and 0.2 and 0.2 for two multi-label datasets as default values, respectively, to further improve the performance of the model.

Table 2. The mAP of hashing methods on ImageNet.

Datasets	Methods	16-bit	32-bit	64-bit
ImageNet	LSH	0.059	0.128	0.173
	SH	0.171	0.300	0.456
	ITQ	0.266	0.436	0.576
	DPSH	0.285	0.405	0.452
	DHN	0.367	0.523	0.627
	HashNet	0.602	0.716	0.807
	DCH	0.652	0.737	0.758
	DSEH	0.735	0.780	0.801
	CSQ	0.717	0.763	0.804
	PSLDH	0.734	0.792	0.817
	WCSH	0.860	0.886	0.900

3.2 Experimental Results

For a comprehensive evaluation of our approach, this study is compared with ten classical or state-of-the-art methods. These methods include a data-dependent hashing method LSH [10], two unsupervised hashing methods SH [17], ITQ [6], and six deep supervised methods DPSH [12], DHN [19], HashNet [2], DCH [1], DSEH [11], CSQ [18], and PSLDH [15].

Table 2 shows the results of different methods with different code-lengths from {16, 32, 64} on ImageNet. ImageNet is a representative single-label database. For all the deep hashing methods, we choose ResNet as their backbone network. From the result, we can see that deep hashing methods outperform non-deep hashing ones like LSH, SH and ITQ. Among the deep hashing methods, the CSQ and its variant PSLDH always achieve a higher mAP value, and our method gets the highest mAP.

The experimental results on the multi-label datasets MS-COCO and NUS-WIDE are shown in Table 3. Compared to the single-label dataset, it is more challenging to retrieve images in these two datasets. Most hashing methods could not get satisfactory performance on them. To increase performance, we choose a transformer as the backbone network of deep hashing methods. Our method has achieved an obvious performance gain over other deep hashing methods. The reason may be that the drift points frequently emerge during the training of the multi-label database, and our proposed loss function is more efficient in this scenario than other losses like BCE loss and softmax loss.

3.3 Ablation Study

The learning objective of our method WCSH includes two terms: the hash error L_C and the regularization error L_Q. In L_C, a weight w is introduced to control the training process more precisely. In this part, we discuss the roles of L_Q and w. The results on MS-COCO and NUS-WIDE datasets are summarized in Table 4.

Table 3. The mAP of hashing methods on two multi-label datasets.

Datasets	Methods	16-bit	32-bit	64-bit
MS-COCO	LSH	0.350	0.419	0.498
	SH	0.492	0.534	0.549
	ITQ	0.566	0.562	0.502
	DPSH	0.634	0.676	0.726
	DHN	0.719	0.731	0.745
	HashNet	0.696	0.741	0.761
	DCH	0.700	0.691	0.680
	DSEH	0.735	0.773	0.781
	CSQ	0.742	0.806	0.829
	PSLDH	0.782	0.835	0.853
	WCSH	0.828	0.874	0.893
NUS-WIDE	LSH	0.388	0.406	0.183
	SH	0.444	0.534	0.502
	ITQ	0.656	0.713	0.660
	DPSH	0.812	0.821	0.831
	DHN	0.712	0.759	0.771
	HashNet	0.757	0.775	0.790
	DCH	0.773	0.795	0.818
	DSEH	0.812	0.827	0.825
	CSQ	0.801	0.818	0.835
	PSLDH	0.820	0.843	0.851
	WCSH	0.838	0.862	0.865

Our method WCSH using L_Q and w achieves the highest mAP values in most cases. When removing the weight w, i.e. $w = 1$, it can be seen that the mAP value will decrease significantly on various code-lengths in these two datasets. It shows the positive effect of the introduction of w. Moreover, when removing the quantization term L_Q, i.e., $\lambda = 0$, the mAP value will decrease slightly in the dataset of MS-COCO, and almost remain the same in the dataset of NUS-WIDE. The reason is that the proposed central loss L_C in Eq. (3) encourages the hash code of an image converging to its hash center which has played a similar role as L_Q.

Table 4. Ablation study

Datasets	w	λ	16-bit	32-bit	64-bit
MS-COCO	1	1e-3	0.782	0.835	0.853
	✓	0	0.793	0.867	0.887
	✓	1e-3	0.828	0.874	0.893
NUS-WIDE	1	1e-3	0.820	0.843	0.856
	✓	0	0.840	0.862	0.868
	✓	1e-3	0.838	0.862	0.865

4   Conclusion

In this study, a novel image hashing method is presented, which takes both the pairwise similarity and central similarity of images into account. The pairwise similarity can be used to identify the drift points, so as to control the training process more precisely and produce hash codes with better inter-cluster separation and intra-cluster closure. Extensive experiments on three benchmark datasets have demonstrated that the proposed method outperforms several state-of-the-art methods.

References

1. Cao, Y., Long, M., Liu, B., Wang, J.: Deep cauchy hashing for hamming space retrieval. In: Proceedings of the IEEE Conference on Computer Vision and Pattern Recognition, pp. 1229–1237 (2018)
2. Cao, Z., Long, M., Wang, J., Yu, P.S.: Hashnet: deep learning to hash by continuation. In: Proceedings of the IEEE International Conference on Computer Vision, pp. 5608–5617 (2017)
3. Chen, Y., Xiong, S., Mou, L., Zhu, X.X.: Deep quadruple-based hashing for remote sensing image-sound retrieval. IEEE Trans. Geosci. Remote Sens. **60**, 1–14 (2022)
4. Chua, T.S., Tang, J., Hong, R., Li, H., Luo, Z., Zheng, Y.: NUS-WIDE: a real-world web image database from national university of Singapore. In: Proceedings of the ACM International Conference on Image and Video Retrieval, pp. 1–9 (2009)
5. Deng, J., Dong, W., Socher, R., Li, L.J., Li, K., Li, F.F.: Imagenet: a large-scale hierarchical image database. In: IEEE Conference on Computer Vision & Pattern Recognition (2009)
6. Gong, Y., Lazebnik, S., Gordo, A., Perronnin, F.: Iterative quantization: a procrustean approach to learning binary codes for large-scale image retrieval. IEEE Trans. Pattern Anal. Mach. Intell. **35**(12), 2916–2929 (2013)
7. Han, K., Xiao, A., Wu, E., Guo, J., Xu, C., Wang, Y.: Transformer in transformer. Adv. Neural. Inf. Process. Syst. **34**, 15908–15919 (2021)
8. Hoe, J.T., Ng, K.W., Zhang, T., Chan, C.S., Song, Y.Z., Xiang, T.: One loss for all: deep hashing with a single cosine similarity based learning objective. Adv. Neural. Inf. Process. Syst. **34**, 24286–24298 (2021)
9. Horadam, K.J.: Hadamard matrices and their applications. In: Hadamard Matrices and Their Applications. Princeton University Press, Princeton (2012)

10. Indyk, P., Motwani, R.: Approximate nearest neighbors: towards removing the curse of dimensionality. In: Proceedings of the Thirtieth Annual ACM Symposium on Theory of Computing, STOC 1998, pp. 604–613. ACM, New York (1998)
11. Li, N., Li, C., Deng, C., Liu, X., Gao, X.: Deep joint semantic-embedding hashing. In: IJCAI, pp. 2397–2403 (2018)
12. Li, W.J., Wang, S., Kang, W.C.: Feature learning based deep supervised hashing with pairwise labels. In: Proceedings of the Twenty-Fifth International Joint Conference on Artificial Intelligence, pp. 1711–1717 (2016)
13. Lin, T.Y., et al.: Microsoft coco: common objects in context. In: Fleet, D., Pajdla, T., Schiele, B., Tuytelaars, T. (eds.) ECCV 2014. LNCS, vol. 8693, pp. 740–755. Springer, Cham (2014). https://doi.org/10.1007/978-3-319-10602-1_48
14. Liu, H., Zhou, W., Zhang, H., Li, G., Zhang, S., Li, X.: Bit reduction for locality-sensitive hashing. IEEE Trans. Neural Netw. Learn. Syst. 1–12 (2023)
15. Tu, R.C., Mao, X.L., Guo, J.N., Wei, W., Huang, H.: Partial-softmax loss based deep hashing. In: Proceedings of the Web Conference 2021, pp. 2869–2878 (2021)
16. Wang, L., Pan, Y., Lai, H., Yin, J.: Image retrieval with well-separated semantic hash centers. In: Proceedings of the Asian Conference on Computer Vision, pp. 978–994 (2022)
17. Weiss, Y., Torralba, A., Fergus, R.: Spectral hashing. In: Advances in Neural Information Processing Systems, vol. 21 (2008)
18. Yuan, L., et al.: Central similarity quantization for efficient image and video retrieval. In: 2020 IEEE/CVF Conference on Computer Vision and Pattern Recognition (CVPR), pp. 3080–3089 (2020)
19. Zhu, H., Long, M., Wang, J., Cao, Y.: Deep hashing network for efficient similarity retrieval. In: Proceedings of the AAAI Conference on Artificial Intelligence, vol. 30 (2016)
20. Zou, Z., Chen, K., Shi, Z., Guo, Y., Ye, J.: Object detection in 20 years: a survey. Proc. IEEE (2023)

AIFR: Face Recognition Research Based on Age Factor Characteristics

Biaokai Zhu[1] , Zhaojie Zhang[1], Yupeng Jia[1], Xinru Hu[1], Yurong Shen[1],
Manwen Bai[1], Jie Song[2(✉)], Ping Li[3], Sanman Liu[1(✉)] , Feng Li[4],
and Deng-ao Li[4]

[1] Shanxi Police College, No. 799 Qingdong Road, Taiyuan 030401, Shanxi, China
hongtaozhuty@gmail.com, lsm601719@126.com
[2] Intelligent Policing Key Laboratory of Sichuan Province, No. 186, Luzhou 646000,
Sichuan, China
scjcxysj@163.com
[3] Anhui University, Hefei 230601, Anhui, China
20145@ahu.edu.cn
[4] Taiyuan University of Technology, No. 79 West Street Yingze, Taiyuan 030024,
China
lidengao@tyut.edu.cn

Abstract. As we all know, the facial appearance change caused by age change leads to the low accuracy of face recognition, which is a significant difficulty in cross-age face recognition tasks. How to overcome the age problem, face feature extraction has become the key. This paper proposes a cross-age face recognition method based on deep learning. This method uses the Arcface loss function to realize cross-age face recognition by improving the residual neural network, combining it with the attention mechanism. Firstly, the face image is enhanced, and the Retinaface algorithm detects the face to complete the look preprocessing. Then the preprocessed face image is extracted by this model to achieve the purpose of cross-age face recognition. In addition, due to the lack of Asian face datasets in public data sets, this paper makes a self-use dataset based on the public data sets. It conducts experiments with FG-NET and CALFW datasets to confirm the universality of this method. The effect of the experimental training set reaches 92.67%, which makes other progress in cross-age face recognition.

Keywords: Deep learning · Retinaface · Arcface loss · AIFR
(Age-invariant face recognition)

1 Introduction

As an important research hotspot in computer vision, face recognition has a wide range of application scenarios in real life, such as face unlocking of smartphones. As a branch of face recognition, cross-age face recognition also plays an extremely important role in real life, such as finding lost children, identifying criminals who have absconded for many years, and optimizing face recognition systems.

Although face recognition technology has entered a mature period, it has endured many influences, such as illumination, posture, object occlusion, age and so on. At present, face recognition technology is not very suitable for face recognition with age change. With the increase of age, a person's appearance will change greatly, which will directly lead to the decline of recognition rate. Aiming at the problems of low recognition accuracy and poor robustness of AIFR, we modify it based on Resnet network model and extract multi-level features. Using attention mechanism to focus on the characteristics of important areas of the image and suppress irrelevant information can efficiently and quickly analyze the information in complex environment and enhance the weight of its identity characteristics. Due to the lack of data sets and the difficulty in obtaining them, most researchers use GAN to generate faces, so that the generated face images and the target face age group images can be matched in similarity. We also set up a unique data set of young Asia for this event. When classifying AIFR recognition faces, the dataset itself has a large age span, and the distance between classes will be greater than the distance between classes. In order to remove this situation, we use arcface loss to reduce this kind of influence.

The main contributions of this paper are as follows:

(1) Aiming at the problems of reduced model robustness and difficulty in feature extraction caused by uncertainties in cross-age recognition, the research group proposed a cross-age face recognition model combining deep learning and loss function on the basis of the proposed system model. The model can effectively acquire image face features and improve the performance and recognition rate of cross-age face recognition systems. In addition, the model is optimized on the basis of the Resnet network model, and multi-level features are extracted to make it more robust.
(2) In order to strengthen the weight of identity features, an attention mechanism is introduced, which focuses attention on the features of important areas of the image, suppresses irrelevant information, and can efficiently and quickly analyze information in complex environments.
(3) In order to solve the problem of the dataset, we introduce a self-made cross-age face dataset and use the public datasets FG-Net and CALFW for experimental training, which achieves the expected effect and proves the effectiveness and reliability of this experiment.

We also compare the model designed in this paper with existing methods to demonstrate the breadth and feasibility of the system.

2   Related Work

Looking back at past face recognition, early research [1] focuses on how traditional face recognition methods can be applied to cross-age face recognition, such as using handmade models to make features and methods based on traditional machine learning. Image processing technology is used to extract face features for matching, but these traditional methods not only have low accuracy in cross-age tasks, but also are limited by the design of artificial features, which has a series of problems. Based on the mainstream framework of traditional machine learning [2] methods to identify faces, extract features, perform face recognition, and finally compare similarities, although this mode plays a pivotal role in the field of cross-age face recognition, it is still susceptible to external factors such as lighting, posture, and occlusion, and cannot guarantee identity features. All in all, traditional face recognition methods are not suitable for recognizing faces with significant age changes.

Deep learning in the field of face recognition is essentially a method of learning features, compared with traditional artificial methods, deep learning through multi-layer neural self-learning of data to obtain the best features, no longer need to manually design face features. The following are two applications of deep learning in cross-age face recognition.

2.1 Cross-Age Face Recognition Based on Generative Adversarial Networks

Generative Adversarial networks (GAN) are an important generative model in the field of deep learning. By extracting the changing features of face age and generating face image through GAN network, similarity matching is carried out with target face age group image, which has an absolute contribution to cross-age face recognition. Chen et al. [3] used GAN to generate face age data of the same person at different ages for sample enhancement, and trained age-invariant face recognition models, which improved recognition accuracy to a certain extent. Wang et al. [4] proposed a future face prediction system for missing children, combining StyleGAN2 and FaceNet methods to quickly confirm the biological relationship between parents and any possible children at a low cost. The system injects the face image of the missing child and the face image of the related family member before the disappearance, uses StyleGAN2 mixed face image, and compares the similarity of the two face images by FaceNet to generate the current age appearance of the missing child. Zhao et al. [5] proposed DR-RGAN network, which uses non-entangled representation learning and residual universal adversarial network. Based on the analysis of facial features and age changes, the dual encoder structure and the unique training discriminator generate facial features to overcome the age interference and obtain the real ideal aging effect and high precision face verification results.

The age accuracy of the images generated by the Gan network is not ideal. There are issues such as lack of authenticity, preservation of generated image identity features, insufficient diversity, and fine-grained facial age images, which are currently heavily dependent on in most studies. If the output of a certain model is incorrect, it will affect the performance of the entire model. The unique dataset established has to some extent alleviated these issues.

2.2 Cross-Age Face Recognition Based on Convolutional Neural Networks

In recent years, Convolutional Neural Networks (CNN) are a common method for deep learning to analyze and process images, which have the ability of representation learning and can realize end-to-end learning. Methods for cross-age face recognition based on the convolutional neural networks model have been proposed successively.

In 2012, Hinton [6] and his students increased the depth of convolutional neural network on the famous ImageNet problem, improved the training mode of convolutional neural network, greatly reduced the error rate, achieved good results, and made great strides in image recognition. In 2014, FaceBook [7] proposed that DeepFace adopted the process of "face detection - face alignment - CNN feature extraction - classification" to achieve good results in face recognition, which can be regarded as the foundation work of CNN in face recognition. Li et al. [8] proposed JLA (joint learning approach) model, and used deep convolutional neural network to construct joint learning of features and matching, so as to learn the best measure function and improve the accuracy of cross-age recognition and matching.

In this paper, based on deep learning method, transfer learning model is introduced to integrate the residual neural network architecture Resnet34 and Arcface loss function in convolutional neural network. Convolutional neural network Resnet34 was used to extract features from input images, and Arcface algorithm was combined to conduct training and classification to realize face recognition across ages. The methodological content of this article will be detailed later.

3 Basic Principles

3.1 Convolutional Neural Network - Residual Neural Network (Resnet)

Convolutional neural network system is usually output by one or more fully connected layers after alternating convolutional layer and pooling layer. However, the early stacked network structure models, such as AlexNet [9], VGGNet [10], etc. these models have simple structures and are easy to deepen the depth of the model. When they finally reach a certain depth, there will be gradient disappearance or gradient explosion or even model degradation. In 2015, He Kaiming [11] proposed Resnet network model structure, which solved this degradation problem to a certain extent and became a sensation in the world's major classification tasks at that time.

The residual mechanism module introduced by the Resnet network uses the skipping mechanism to directly transmit the information to the output part to realize a deeper network structure. For the common stacked network structure, the feature g (x) will be learned when x is input, and it is difficult to directly learn g (x). Therefore, the residual network wants to learn the RESNET function: F (x). It is worth noting that the residual function F (x) can only be added when the dimensions of the jump connection X are the same. When the dimensions are different, the dimensions of X can be adjusted by 1*1 convolution operation to make the dimensions consistent.

3.2 Attention Mechanism

Attention mechanism has made rapid progress in deep learning research in recent years, and has been deeply applied in computer vision fields such as target classification and detection, semantic segmentation, etc. It has quickly become a research hotspot in the field of deep learning.

This mechanism has the characteristic of dynamic extraction. Attention mechanism selectively pays attention to and learns all the information obtained, ignores other information, and efficiently collects key information for relevant processing.

The existing attention mechanism can automatically learn, so that the network model can focus on the main features, eliminate noise and quickly obtain the most important information, thus improving the recognition accuracy of the model. The attention mechanism further strengthens the weight of identity features. We try to combine the advantages of the channel and spatial attention mechanisms, so that the network model knows what to pay attention to, while also enhancing the representation of specific regions, improving the correlation of each feature in the channel and space, and better extracting effective features. And because of its high degree of lightweight, it can bring stable performance improvement after embedding the model. Moreover, the mechanism can be better embedded in the CNN model due to its strong versatility and high portability to achieve end-to-end training. That is, in our research, we found that the combination of dual attention mechanisms is very effective.

3.3 Loss Function

In our investigation, we found that Additive Angular Margin Loss (Arcface) function was published in January 2018 by Deng Jian-kang et al. [12] which is an improvement of the traditional Softmax Loss function.

Traditional Softmax loss function is shown in Eq. (1):

$$L = -\frac{1}{N} \sum_{i=1}^{N} \log \frac{e^{W_{y_i}^T x_i + b_{y_i}}}{\sum_{j=1}^{n} e^{W_j^T x_i + b_j}} \tag{1}$$

CosFace [13] is to normalize the feature vector x and weight w on the basis of softmax loss, and then subtract a positive m from the cosine space, and multiply the feature scale S on this basis, and finally obtain $s(cos(\theta - m)$. As shown in Formula (2):

$$CL = \frac{1}{N} \sum_{i} -\log \frac{e^{m(cos(\theta_{j,i})-t)}}{e^{m(cos(\theta_{j,i})-t)} + \sum_{j \neq y_i} e^{m \cos(\theta_{j,i})}} \tag{2}$$

Arcface is similar to Cosface. The core idea of Arcface is to normalize the feature vector X and weight W first on the basis of softmax loss, so that the predicted value only depends on the angle between the feature and weight. The angle interval q is added to the Angle θ between the feature vector x and the true weight w, which will reflect the maximization of the decision boundary between classes in the Angle space. On this basis and multiplied by the characteristic scale S.

Finally, the original softmax function is substituted to more reasonably and effectively calculate the probability of each category and finally calculate the loss value. As shown in Formula (3):

$$AL = -\frac{1}{N} \sum_{i=1}^{N} \log \frac{e^{m(\cos(\theta_{y_i}+t))}}{e^{m(\cos(\theta_{y_i}+t))} + \sum_{j=1,j\neq y_i}^{n} e^{m\cos\theta_j}} \qquad (3)$$

Both are similar. The comparison between arcface and cosface shows that the arcface function trains the angle between the depth feature and its corresponding weight in an additive manner, and the angle interval is more direct than the cosine interval in terms of the impact on the angle, In addition, Arcface Loss is different from CosFace and SphereFace [14] in that the geodesic distance and Angle have accurate correspondence, so it has good geometric attributes and constant linear Angle. In addition, Arcface Loss is based on the segmentation of Angle space, so it has the minimum distance on the hypersphere instead of the minimum distance on the feature points, and the range of feature comparison is more comprehensive. Compared with other loss functions, Arcface has the advantages of high performance, easy programming, low complexity, and high training efficiency, which is an improvement of its functions. This improvement allows Arcface to not need joint supervision with other loss functions, the model will converge more easily on different datasets, and the training process will be more stable.

4 Data Sets and Experimental Design

4.1 Experimental Data Set

Cross-age facial recognition plays an important role in finding missing persons, tracking down criminals who have been on the run for years, and identifying people of different ages. However, in the process of deep learning research, the lack of multi-background and multi-posture database has been a difficult problem that cannot be ignored. So the experimental group collected Asian face data from a public database to build its own dataset. In this experiment, the cross-age face databases used for training and detection were FG-NET, CALFW and self-made data sets.

Self-made Dataset: Due to the influence of FG-NET data set and CALFW data set, face images related to Asian race could not meet the accuracy of experimental results. Therefore, in order to ensure the accuracy and universality of the experiment, the experimental group collected FG-NET data set and CALFW data set and made data set about Asian race. The homemade data set contains images of the same person at different ages, as well as a variety of scenes, poses and lighting, while ensuring image quality. The data set obtained 191 individuals with face age information, a total of more than 1300 cross-age face images, the age of the selected images ranged from 5 years old to 25 years old, an average of 7 images per person, the average age span is 3 years old, the maximum age span of a single individual face image is 5 years old. In the training set, some photos of people of different ages were selected, and the images of each age group were selected to form positive sample image pairs and negative sample image pairs.

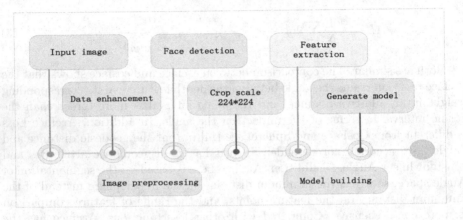

Fig. 1. Flow chart of cross age face recognition experiment

4.2   Experimental Design and Deployment

The deployment diagram of the cross-age face recognition based on deep learning designed in the experiment is shown in Fig. 1:

1) According to the ratio of 7:3, the used face database data is divided into training set and test set;
2) According to the data enhancement method described below, process the collected face image data in the database, then use the Retinaface algorithm to detect the face, and obtain the preprocessed face image;
3) The processed image is input into the improved Resnet network to obtain the image features of its training set. The detailed acquisition steps are shown below, and the training is conducted on this basis.
4) By introducing attention mechanism and Arcface loss function, the Resnet network model can improve the performance of the model and make the model more robust, so as to realize cross-age face recognition with high accuracy.
5) Finally, the results are compared with the existing AIFR algorithms.

Face Detection. Face detection is the first step to realize face recognition. We use a single-stage face detection algorithm called Retinaface [15]. This algorithm has better detection performance than cascade methods (such as MTCNN) and faster detection speed than two-stage face detection method, which is a method to balance detection speed and performance. Through the network backbone network, the image to be detected is extracted through three stages of feature extraction. Feature Pyramid (FPN) and SSH network are used to add features in the output layer of feature extraction, and then the prediction results are obtained from the features, and the prediction results are adjusted. Finally, redundant prediction frames are removed by maximum suppression (NMS).

After the process of face detection, we obtain the region image of the detected face, and then carry out face recognition with the help of face alignment.

4.3   Model Building

In the work of cross-age face recognition, the experiment put the preprocessed face images into convolution neural network for training, and optimize the model to get the corresponding results. Experimental results believe that the depth network can reasonably and effectively extract the age-invariant features of the input face images. However, when the network depth reaches a certain degree, the corresponding error will increase, and the over-fitting phenomenon will appear, and the model effect will become worse. Gradient disappearance and gradient explosion will increase sharply, which makes the network model unable to converge and leads to the failure of normal training of deep network model. In related research, we find that Resnet network model can solve this problem well. Resnet model uses residual network to transform the function G(x) learned by original convolution network into F(x)+x. This optimization proposed by Resnet greatly improves the training speed. At the same time, the attention mechanism is added to the Resnet network structure of the experimental group, which can save parameters and computation and significantly improve its performance. Classification is also an important research direction of cross-age face recognition. When solving the classification problem of cross-age face recognition, experiment designer decided to adopt the idea of Arcface Loss to solve the situation that the distance within classes is greater than the distance between classes when the age span is large. In addition, the experimental results significantly improve the intra-class similarity and inter-class differences of face recognition, and reduce the impact on the performance of face recognition system.

The experimental model adopts the Resnet-50 model in Resnet. The original Resnet 50 network model refers to the VGG19 network model and is modified on this basis. The 7*7 convolution kernel and convolution with step size of 2 are used for downsampling, and the average pool layer is replaced by the fully connected layer. Resnet also adds a short circuit mechanism between every three layers of the normal network, which leads to residual learning. The activation function is ReLu.

Summarize the following model architecture: First model input data for training a 112×112 face image, Through 3×3 layer convolution kernel (step size 1) and normalized PRuLe activation function, At the beginning and end of the network, channel attention and spatial attention mechanisms are connected by residual structure. Then, after four residual structure layers (3, 8, 11, 3), after a series of normalization and pre-activation functions, the matrix is processed in one dimension by flat layer, and then input into fully connected layers. After multiple classifications, output results. Finally, the R-Resnet-50 face recognition model is established.

5   Experimental Settings and Basic Parameters

5.1   Experimental Equipment

All the experimental operations are carried out and completed under the Windows 11 64-bit operating system and Pytorch environment, and the training model and test model use the unified GPU server environment. The main hardware and software parameters are shown in Table 1.

Table 1. Parameters of experimental equipment

Name of software and hardware	Equipment
CPU	Intel(R) Core(TM) i7-10700K CPU @ 3.80GHz 3.79 GHz
GPU	NVDIA GeForce RTX3080
Operating system	Windows11 21H2 X64
Machine with RAM	32.0G
Hard disk	Western Digital WDBEPK0020BBK-2TB
Experimental environment	Pytorch Pycharm

5.2   Experimental Parameters

For experiments, In the experiment, SGD is regarded as an optimizer, and dynamic deep learning is carried out under the condition that the learning rate range is preset as (1E-2, 1E-4). Optimizer internal parameters momentum 0.9 and weight_decay 5E-4. In order to ensure that the convergence loss value is the global optimal solution of Resnet neural network, COS principle is used as the periodic attenuation strategy of learning rate. The sample size of data captured in a training session (Batch_size) is set to 32. The batch size has an epoch value of 200.

6   Experimental Results and Evaluation

6.1   Experimental Comparison on Different Data Sets

This paper has made extensive attempts in the existing public data sets and self built data sets, and found that it performs well in each data set. The experimental comparison results are shown in Table 2.

In FG-Net database, the model method designed is also compared with other cross-age face recognition methods, including JLA, MHFA, PCA+SVM, GAN, HFA and other currently proposed methods. The comparison results are shown in the chart. As can be seen from the data in the table. The accuracy of the training set of the proposed method reaches 92.67%, which is better than most methods and can effectively realize cross-age face recognition, as shown in Table 3

Table 2. Accuracy of training sets of different data sets

Data	Accuracy of Rank-1
Self made data set	99.9%
Fg-Net	99.7%
Maegage	99.7%

Table 3. Cross age face recognition training set accuracy of different methods

Algorithm	Accuracy of Rank-1
JLA	93.65%
MHFA	87.94%
PCA+SVM	91.25%
GAN	86.5%
HFA	91.14%
R-Resnet	92.67%
LF-CNN	88.1%

6.2   Experimental Evaluation

After repeated experiments, research found that the accuracy rate of the training set reached 99.9%, the accuracy rate of the lfw test set reached 92.67%, and the difference between positive and negative values was between 90.56% and 95.83%. In this paper, the face threshold distance is used as an indicator for experimental research. Through Openface to calculate the distance of face threshold, find the main feature point of the face, through the calculation of the feature point of the real sample, the distance value from the database sample feature point, the smaller the face distance, the higher the face recognition rate, the successful threshold of this experiment recognition is 1.26. The loss value is as low as 0.02, and the Roc curve also indirectly proves the feasibility of the model. According to the above experimental results, the designed R-ResNet-50 model is compared with the traditional ResNet-50 model. Obviously, the improved model achieves better results. Therefore, it can be seen that the success of cross-age face recognition is inevitable.

Accuracy: Fig. 2 depicts the Accuracy results of the training set and validation set of R-Resnet-50 respectively.

Loss: The size of the loss value generated by the loss function is an indicator to measure the quality of the designed model. The smaller its value, the more successful the model is Fig. 3 depicts the loss value results for R-Resnet-50.

Confusion matrix: used to observe the performance of the designed model on each test category. Figure 4 depicts the confusion matrix results for some classes.

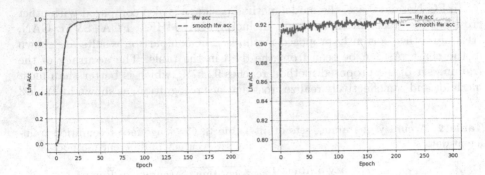

Fig. 2. Experimental accuracy curve

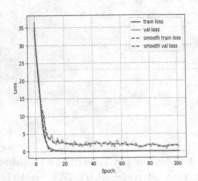

Fig. 3. Experimental loss value curve

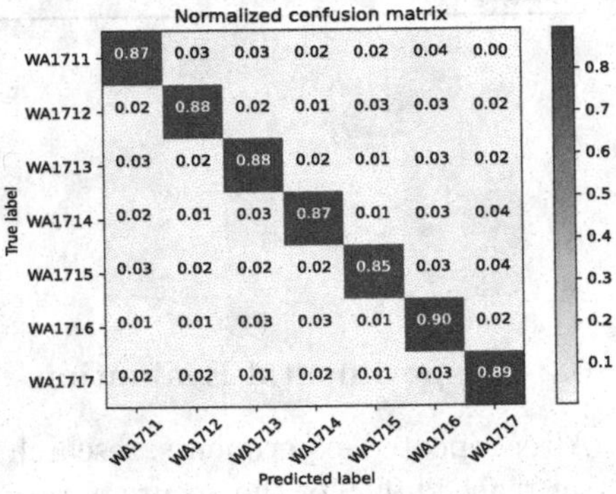

Fig. 4. Some types of cross-age face detection models and their detection accuracy

7   Conclusion

This paper first proposes a cross-age face recognition model combining deep learning and loss function, which improves the residual neural network. We incorporate the attention mechanism to extract features and better achieve stable performance effectively. Experiments on public FG-NET and CALFW datasets demonstrate the feasibility of the recognition model. At the same time, we also propose a self-made dataset containing Asian faces, which achieves the experimental group's expected effect and effectively improves the accuracy and universality of cross-age face recognition. The experimental results show that the proposed method can achieve the high recognition rate of the current cross-age face task and achieve relatively ideal results.

Limited by environmental conditions, this paper still needs to further improve and improve some contents, as follows:

(1) There are still some shortcomings in the self-made data set, among which there are few data that can meet the large age span, mainly concentrated in 5 to 25 years old. Most of the data are positive face effect, and do not involve multi-pose face image. Therefore, further collection of qualified diversity data will be more conducive to the final effect of the experiment.

(2) At present, the accuracy of the test set needs to be enhanced, and the extraction of relevant features of cross-age faces may need to be refined. In the next step, face features can be decomposed into identity features and age-invariant features.

Declarations.

– Funding

The research is supported by NSFC (No. 62306207), Intelligent Policing Key Laboratory of Sichuan Province (No. ZNJW2022KFZD004), Basic Research Plan of Shanxi Province (No. 202303021211339), Virtual Teaching and Research Office of Cyber Security (BJPC) of Ministry of Education (No. WAXVKF-2202), Anhui Natural Science Foundation (No. 2108085MF207), Shanxi Provincial Higher Education Teaching Reform and Innovation Project, Teaching Reform Project of Shanxi Police College.

– Competing interests

The authors declare that they have no competing interests.

– Availability of data and materials

Not applicable.

References

1. Ouarda, W., Trichili, H., Alimi, A.M., Solaiman, B.: Face recognition based on geometric features using support vector machines. In: 2014 6th International Conference of Soft Computing and Pattern Recognition (SoCPaR), pp. 89–95. IEEE (2014)

2. Filali, H., Riffi, J., Mahraz, A.M., Tairi, H.: Multiple face detection based on machine learning. In: 2018 International Conference on Intelligent Systems and Computer Vision (ISCV), pp. 1–8. IEEE (2018)

3. Chen, S., Zhang, D., Yang, L., Chen, P.: Age-invariant face recognition based on sample enhancement of generative adversarial networks. In: 2019 6th International Conference on Systems and Informatics (ICSAI), pp. 388–392. IEEE (2019)

4. Wang, D.-C., Tsai, Z.-J., Chen, C.-C., Horng, G.-J.: Development of a face prediction system for missing children in a smart city safety network. Electronics 11(9), 1440 (2022)

5. Zhao, S., Li, J., Wang, J.: Disentangled representation learning and residual GAN for age-invariant face verification. Pattern Recogn. 100, 107097 (2020)

6. Krizhevsky, A., Sutskever, I., Hinton, G.E.: Imagenet classification with deep convolutional neural networks. Commun. ACM 60(6), 84–90 (2017)

7. Taigman, Y., Yang, M., Ranzato, M.A., Wolf, L.: Deepface: closing the gap to human-level performance in face verification. In: Proceedings of the IEEE Conference on Computer Vision and Pattern Recognition, pp. 1701–1708 (2014)

8. Li, H., Hu, H., Yip, C.: Age-related factor guided joint task modeling convolutional neural network for cross-age face recognition. IEEE Trans. Inf. Forensics Secur. **13**(9), 2383–2392 (2018)
9. Simonyan, K., Zisserman, A.: Very deep convolutional networks for large-scale image recognition, arXiv preprint arXiv:1409.1556 (2014)
10. He, K., Zhang, X., Ren, S., Sun, J.: Deep residual learning for image recognition. In: Proceedings of the IEEE Conference on Computer Vision and Pattern Recognition, pp. 770–778 (2016)
11. Mnih, V., Heess, N., Graves, A., et al.: Recurrent models of visual attention. In: Advances in Neural Information Processing Systems, vol. 27 (2014)
12. Deng, J., Guo, J., Xue, N., Zafeiriou, S.: Arcface: additive angular margin loss for deep face recognition. In: Proceedings of the IEEE/CVF Conference on Computer Vision and Pattern Recognition, pp. 4690–4699 (2019)
13. Wang, H., et al.: Cosface: large margin cosine loss for deep face recognition. In: Proceedings of the IEEE Conference on Computer Vision and Pattern Recognition, pp. 5265–5274 (2018)
14. Liu, W., Wen, Y., Yu, Z., Li, M., Raj, B., Song, L.: Sphereface: deep hypersphere embedding for face recognition. In: Proceedings of the IEEE Conference on Computer Vision and Pattern Recognition, pp. 212–220 (2017)
15. Deng, J., Guo, J., Ververas, E., Kotsia, I., Zafeiriou, S.: Retinaface: single-shot multi-level face localisation in the wild. In: Proceedings of the IEEE/CVF Conference on Computer Vision and Pattern Recognition, pp. 5203–5212 (2020)

Author Index

Z. Tari et al. (Eds.): ICA3PP 2023, LNCS 14493, pp. 353–354, 2024.
https://doi.org/10.1007/978-981-97-0862-8